T0355512

Great Lakes Rocks

Great Lakes Rocks

4 Billion Years of Geologic History in the Great Lakes Region

STEPHEN E. KESLER

University of Michigan Press
Ann Arbor

Copyright © 2019 by Stephen E. Kesler
All rights reserved

This book may not be reproduced, in whole or in part, including illustrations, in any form (beyond that copying permitted by Sections 107 and 108 of the U.S. Copyright Law and except by reviewers for the public press), without written permission from the publisher.

Published in the United States of America by the
University of Michigan Press
Printed and bound by CPI Group (UK) Ltd, Croydon, CR0 4YY
Printed on acid-free paper
First published May 2019

A CIP catalog record for this book is available from the British Library.

Library of Congress Cataloging-in-Publication data has been applied for.

ISBN: 978-0-472-07380-1 (Hardcover : alk paper)
ISBN: 978-0-472-05380-3 (Paper : alk paper)
ISBN: 978-0-472-12374-2 (ebook)

CONTENTS

ACKNOWLEDGMENTS

This book is an expansion of material used in a class on the geology of the Great Lakes region at the University of Michigan. I am grateful to the generation of students who participated in that class, came along on field trips, and asked lots of questions. I am also indebted to the host of geologists who have studied the geology of this area and written about it, in both research journals and popular press, and to the many of them who have taken me and my students on field trips in the region. I have also made liberal use of geology guidebooks for Minnesota by Dick Ojakangas, Wisconsin by Robert Dott and John Attig, and Ontario by Nick Eyles, as well as the wonderful summary of geology around Lake Superior by Gene LaBerge and the treasure trove of abstracts and guidebooks archived from meetings of the Institute on Lake Superior Geology.

During the writing, lots of people helped by reading chapters or sections, contributing images, and discussing aspects of the geology, often in exhaustive detail. They include Robert Ayuso, Pat Bickford, Terry Boerboom, Ted Bornhorst, Bill Cannon, Dan Fisher, Jamie Gleason, Bill Harrison, Daniel Holm, Gordon Medaris, Grahame Larson, Kacey Lohmann, Jeff Mauk, Phil Meyers, Howard Mooers, Ted Moore, Dick Ojakangas, Klaus Schultz, John O'Shea, Henry Pollack, Lana Pollack, Howard Poulsen, Francois Robert, Larry Ruff, Anthony Runkel, Klaus Schulz, Fried Schwerdtner, Nathan Sheldon, Phil Thurston, John Valley, Ben van der Pluijm, Rob van der Voo, Ed van Hees, David Wacey, Tom Waggoner, Bruce Wilkinson, Laurel Woodruff and Grant Young. Errors of fact or interpretation that remain after all of this help are purely my responsibility. Thanks also go to Scott Ham with the University of Michigan Press, who provided editorial guidance and support throughout the project, and Dale Austin, who drew almost all of the maps and diagrams. Finally, I am grateful to Bill Kelly, who lured me to Michigan and Great Lakes geology, and to my wife, Judy, who has been to almost as many outcrops in the region as I have, and whose careful reading of the manuscript helped smooth over bumps that might have slowed the reader.

CHAPTER 1

Introduction

Geologic Time Travel in the Great Lakes Region

1.1. Great Lakes Rocks Tell the History of the North American Continent

Great Lakes rocks hold an amazing story that goes back to the dawn of time on Earth, a trail that we will follow in this book. As we work our way back in time, the trail will take us through the history of early human settlement of North America, through periods of mountain building, rifting continents and meteorite impact, to the appearance of life when the continent itself was just beginning to grow.

It might be surprising that the Great Lakes region has such a long and interesting story to tell. For most people today, mention of the Great Lakes region conjures up an image of lakes, shorelines, and little else. But that is not the way native peoples or early European explorers saw it.[1] To them the Great Lakes region was the gateway to central North America. The St. Lawrence River and the Great Lakes provided easy access to the surrounding continent. In this book, we take the same expansive view of the Great Lakes region. As you can see in figure 1.1, it covers the five Great Lakes and extends northward toward James Bay, the southern extension of Hudson Bay; southward toward the Ohio River; and westward to the start of the prairie in Minnesota and Manitoba.

This expanded Great Lakes region includes a wide array of geologic features that built the North American continent and, in fact, the world as we know it. The Great Lakes themselves are the crowning finale in a long series of geologic events that created mountain belts and deep basins, formed continents and ripped them apart, generated vast mineral wealth, and finally scraped it all clean with large glaciers. By reaching out to this enlarged Great

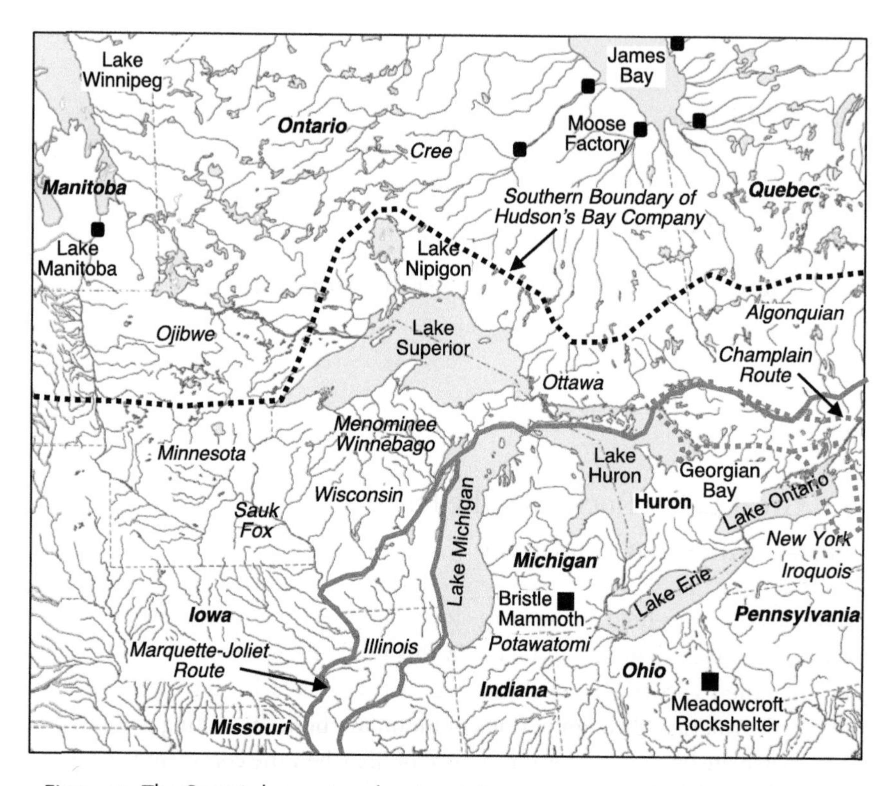

Figure 1.1. The Great Lakes region showing states, provinces, major lakes, and rivers, along with routes taken by early explorers, the southern boundary of Hudson's Bay Company land, the approximate location of Native American and First Nations tribes in about 1600, and the Meadowcroft Rockshelter and Bristle Mammoth sites of early human activity in the region. Squares show the locations of Hudson's Bay Company trading posts, known as factories. (Modified from May 1967.)

Lakes region, we can better understand how geologic processes played a role in drawing people into it.

1.2. Our Geological Story Follows the Early Explorers

In our tour through the Great Lakes region, we will be following in the footsteps of early explorers, many of whom left their names on lakes, rivers, and towns. The first explorers were the aboriginal people who came from the far eastern part of Asia during the last retreat of the glaciers (the earliest trace that we see of these immigrants is in the Meadowcroft Rockshelter and Bristle Mammoth site). They and their descendants, along with waves of other aboriginal immigrants, colonized the Great Lakes region, eventually producing the tribes, councils, and confederations that greeted the European explorers (fig. 1.1).[2] Their names get pride of place on four of the Great Lakes and Lake Nipigon, as well as many important rivers and towns, including Chicago, Milwaukee, Ottawa, and Toronto. My personal favorite is Wawa, Ontario, at the northeast corner of Lake Superior, named for the Canada goose (see fig. 1.2A).[3]

Among the first Europeans were the French missionaries and traders, who moved up the St. Lawrence into the Great Lakes region establishing missions and trading posts. Their language graces towns such as Detroit, Eau Claire, La Crosse, and Marquette, and they even captured two Great Lakes wannabes, St. Clair and Champlain. Among the earliest was Samuel de Champlain, known as the "Father of New France," who explored the eastern part of the Great Lakes system and founded the city of Quebec in 1608. The missionary Jacques Marquette went farther west, establishing missions at Sault Ste. Marie and St. Ignace in what is now Michigan and near Ashland, Wisconsin. In 1673 he and Louis Jolliet headed south into the continent (fig. 1.1). Following directions from Native Americans, they paddled up Green Bay and the Fox River to a two-mile portage that took them into the Wisconsin River and then the Mississippi River at today's Prairie du Chien. Their return trip, through what is now Chicago, explored one of today's important connections between the Great Lakes and the Gulf of Mexico.[4]

The English also came into the Great Lakes region, but from the north via Hudson Bay and its southern extension, James Bay. Although English money was involved, leadership was provided by Pierre-Esprit Radisson and Médard Chouart des Groseilliers, residents of New France, whose explorations had extended far north of Lake Superior. By 1670 their group, known then as the "Governor and Company of Adventurers of England Trading into Hudson's Bay" and today as "The Bay," had been incorporated in England with a monopoly on trade in an enormous region that covered most of central Canada and extended southward into Minnesota and North Dakota (fig. 1.1).[5] The company established trading posts, known as factories, throughout the vast region and constructed Prince of Wales fort to protect them (fig.1.2B).[6]

As more Europeans entered the region, attention shifted to its agricultural potential and mineral resources. Nick Eyles, Peter Newman, and others have pointed out that the rocky, glaciated land north of Lake Superior impeded migration of farmers from the south thus preserving what was to become central Canada. Other migrants were attracted by the possibility of mineral wealth; first by tales of the mysterious Ontonagon cop-

Figure 1.2. (**A**) Giant statue of a Canada goose at Wawa, Ontario; (**B**) Prince of Wales Fort near Churchill, Manitoba (photo by Peter Fitzgerald, 2012); (**C**) boulder of pure copper found along the Ontonagon River in Michigan (SIA ACC. 11–009, Smithsonian Institution); (**D**) statue of a giant 10 meter high iron miner, Mesabi Iron Range, Chisholm, Minnesota.

per boulder (fig. 1.2C) and later by the huge Mesabi and other iron ranges (fig. 1.2D).

The geologic history that controlled European settlement of the Great Lakes region stretches back in time for almost 4 billion years. Two processes that dominate this geologic history, both past and future, are global climate change and plate tectonics. So, before starting our journey through Great Lakes geologic time, we need to review just how much time we are dealing with and the role of these two processes in geologic history.

1.3. Great Lakes Rocks Span Most of Geologic Time

Although the Great Lakes formed only a few thousand years ago, they are the result of much older processes. In fact the geologic history of the Great Lakes region covers almost the entire 4540-million-year (Ma) span of time since the Earth formed (box 1.1).[7] Some of the oldest rocks in the world, aged 3800 to 3400 Ma, are found in the Minnesota River valley and in the Watersmeet and Carney Lake areas in the Upper Peninsula of Michigan, and even older rocks are found along the shore of Hudson Bay in Quebec. At the other end of the scale, most of the glacial ridges, valleys, and lake beds that form our present landscape are only a few thousand years old, and the Great Lakes shorelines are evolving today.

BOX 1.1. HOW OLD IS EARTH?

Just when did Earth form? Was it when planetisimal debris first coalesced into a protoplanet Earth or when the protoplanet separated into a core and mantle? Or should we say that Earth really formed only after the giant collision that formed the Moon? How long did a magma ocean last before a solid crust formed and was that really the birth of our planet? Finally, even if we can agree on which of these events represents the beginning of Earth, how can we measure its age? One way would be to measure parent-daughter relations in some isotope system, but what isotope system should we use and what material should we analyze? Another way would be to construct and test theoretical models based on the physics of coalescing planetisimals, but what size should they be and over what period of time did they collect? Although progress is being made on many fronts, these questions continue to fascinate cosmologists, physicists, astronomers, and geologists. At this point, the accumulation of planetisimals to form Earth is thought to have started at 4568 Ma, based on Pb isotope measurements on meteorites, and to have taken only a few million years to form planets. The next big event, the impact that formed the Moon, took place sometime between 4530 and 4520 Ma, based on the Hf-W isotope system. A more widely reported age of 4540 Ma, which averages the ages obtained by several isotope systems, is commonly cited as the age of Earth (https://pubs.usgs.gov/gip/geotime/age.html). As we will see in chapter 8, the formation of Earth's earliest crust happened sometime between 4440 and 4430 Ma.[8]

This enormous span of time can be put into perspective with the geologic time scale (fig. 1.3), which is divided into four large periods known as eons—the Hadean, Archean, Proterozoic, and Phanerozoic. Where rocks fit in the time scale is determined by fossil evidence and geologic relations, including the laws of superposition and cross-cutting relations (fig. 1.4A,B,C in color section).[9] These give us relative ages, such as the obvious fact that beach deposits along the Great Lakes must be younger than the ancient hard rock on which they rest, and that basalts[10] of the Midcontinent Rift are younger than the granites that they cut across. We can also use isotopic analyses,[11] which give us absolute ages based on the decay of radioactive elements such as uranium (fig. 1.4D in color section). It is isotopic age measurements that provide the 3800 to 3400 Ma ages of the Minnesota River valley and Carney Lake rocks.

Many geologic stories begin with old rocks like those in the Minnesota River valley. In this book, however, we start at the present and work our way backward. This allows our time traveler to start with features, like waterfalls, lakes, caves, deltas, and glaciers, that are relatively familiar and to use insights about these familiar features to understand how sedimentary basins, lava-filled rift valleys, and colliding continents behaved farther back in time.

Figure 1.3 shows how the chapters of this book work their way back through time, starting with chapter 2, which concerns current geologic features that formed largely during the Holocene epoch after the last glaciers retreated from the Great Lakes region about 10,000 years ago. Chapter 3 moves back to Pleistocene time, about 10,000 to 2.5 million years (Myr) ago, when glaciers ground down much of North America and left the southern part with thick deposits of sand and gravel. Then, in chapter 4, we jump back to Mesozoic and Paleozoic time, about 145 to 542 Myr ago, when oceans flooded the continents, covering them with sedimentary basins from which we get oil, natural gas, salt, gypsum, and limestone.

The last part of the book deals with Proterozoic and Archean time, which go by the informal designation of Precambrian. Chapter 5 reviews a time, about a thousand million years ago (1000 Ma), when the North American continent began to split apart, forming the Midcontinent Rift and its large copper deposits. Chapter 6 moves farther back in time to the Paleoproterozoic Era, between about 1600 and 2500 Ma, when the edge of the North American continent was located in the Great Lakes region. This was a busy time for Earth. We gained an oxygen-rich atmosphere, vast iron and uranium deposits, and the Penokean mountain range and were struck by the enormous Sudbury meteorite. Chapter 7 goes all the way back to Archean time, before 2500 Ma, and the story of how early continental or craton fragments

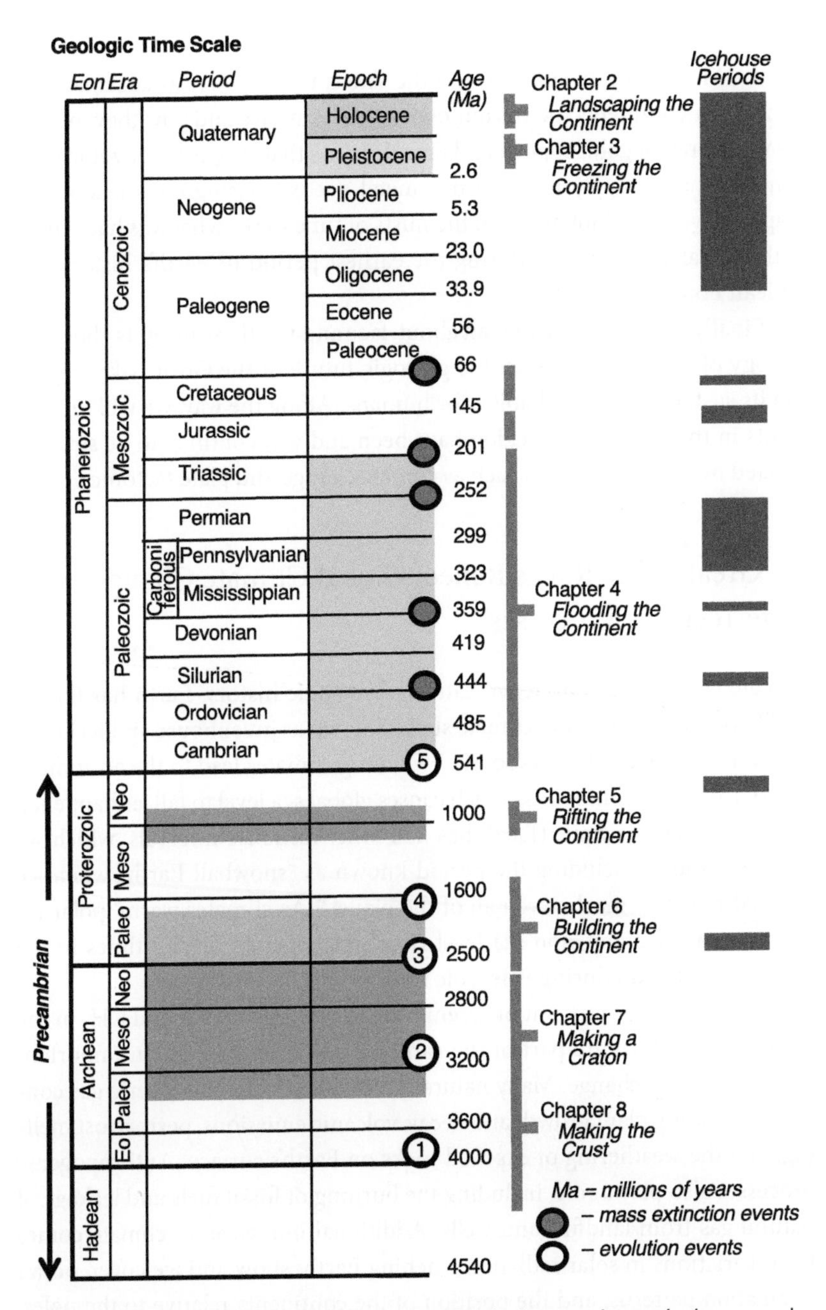

Figure 1.3. Geologic time scale showing time periods discussed in each chapter and their relation to global icehouse periods, mass extinctions, and evolutionary events discussed in the book: (1) earliest life, (2) photosynthesis and cyanobacteria, (3) onset of the Great Oxidation Event, (4) eukaryotes, (5) Cambrian explosion. (Modified from Dyson 1999; Nesbitt et al. 2003; Dalrymple 2004; Cockell 2007; Committee on the Importance of Deep-Time Geologic Records 2011; Lyle 2016.)

like the Minnesota River valley and the Wawa terrane were amalgamated to form larger continents such as our own North America and how they formed the wild array of gold, copper, and zinc deposits that support many northern economies and supply many of our metal needs. Closing out the story, in chapter 8 we try to look through the mists of time to see what was happening in the Great Lakes region during the earliest period of Earth's history, the Hadean Eon.

Finally, in chapter 9 we do an about-face and use these insights about the geology of the area to predict the geologic future of the Great Lakes region and its sustainability as a habitat for humans. Along the way, we will see that events in the Great Lakes region have been and will continue to be strongly affected by global processes such as climate change and plate tectonics.

1.4. Great Lakes Rocks Reflect Global Climate Change and Plate Tectonic Processes

Climate change is a long-term feature of geologic history. Earth has fluctuated between hot and cold climate states, known as greenhouse and icehouse respectively, since at least 2500 Ma.[12] Cooling climates lead to the accumulation of ice on the continents, which causes global sea level to fall, as happened during Pleistocene time. (Earth has had other icehouse intervals, which we will visit briefly, including the period known as "snowball Earth" when ice covered much or maybe even all of the planet.) As climates warm, polar ice melts, oceans expand, and sea level rises, as happened several times in the Great Lakes region during Paleozoic time.

Changes in the amount of greenhouse gases such as CO_2 and CH_4 in the atmosphere, which are part of the carbon cycle (box 1.2), are an important cause of climate change. Many natural processes affect atmospheric gas contents, including plant growth and decay, volcanic emissions, permafrost melting, and the weathering of exposed rocks on Earth's surface. Anthropogenic processes also play a role, including the burning of fossil fuels and leakage of natural gas from landfills and wells. Additional influence on climate comes from variations in solar radiation reaching Earth, snow and ice cover, ocean circulation patterns, and the position of the continents relative to the poles, which leads us back to plate tectonics and continental drift.

Changes in global climates also reflect changes in the deeper Earth. You can see in figure 1.5 that the continents consist of rock up to 4 billion years old whereas the oldest ocean rocks are less than 300 million years old. This huge

BOX 1.2. THE CARBON CYCLE

The carbon cycle is one of the most effective ways that Earth has to control global climate and sea level. It involves the flow of carbon among the planet's four major reservoirs: the atmosphere, hydrosphere, lithosphere, and biosphere. In the atmosphere, most carbon is present as CO_2, an important greenhouse gas. (CH_4, also a greenhouse gas, was part of the early Earth atmosphere.) CO_2 is taken out of the atmosphere by plants, which send the carbon back to the atmosphere when they decay or burn or to the lithosphere when they are buried in sediment. Carbon in the atmosphere can also be dissolved in water to form carbonic acid, which is the main agent of rock weathering. Dissolved carbon is taken out of the ocean by plants and animals that form limestone, which accounts for about 80 percent of the carbon in the lithosphere; the other 20 percent is in buried plants and other organic matter. Carbon in the lithosphere returns to the atmosphere and hydrosphere when rocks are metamorphosed or melted to form magmas that release CO_2.

The flow of carbon among these reservoirs is closely linked to global climate. High flows of CO_2 or CH_4 into the atmosphere lead to global warming, which melts glaciers, causing sea level to rise. Volcanoes are an important source of carbon for the atmosphere, especially large-scale basalt eruptions. Conversely, increased weathering or plant growth consume atmospheric carbon, causing cooling that might lead to glaciation and a lower sea level.

These processes are also linked to the supercontinent cycle, which is discussed below. When continents collide to form supercontinents, the resulting orogenies and mountain belts lead to increased weathering, which draws CO_2 out of the atmosphere. When seafloor spreading breaks up the supercontinents, this can be reversed by the increase in volcanism.

Processes and places that produce carbon are called sources, and the carbon is consumed at sinks. When carbon sources and sinks are in balance, planet Earth cycles gently between cool and warm periods. If the balance is disturbed, climates and sea level can take extreme excursions, possibly including icehouse events and mass extinctions. So far these cycles have been controlled by natural processes, although we are now undertaking an experiment involving anthropogenic transfer of huge amounts of carbon from the lithosphere (fossil fuels) to the atmosphere.[13]

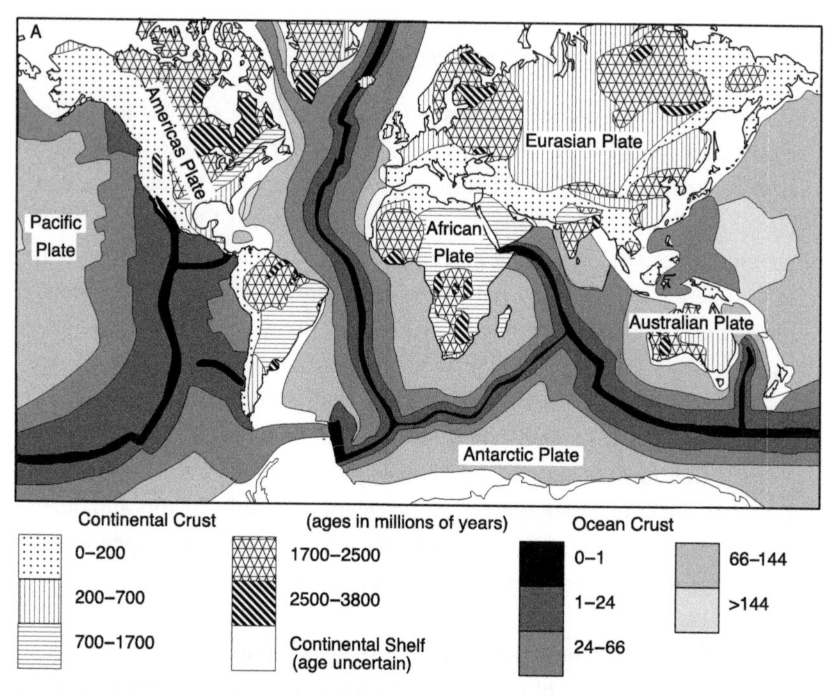

Figure 1.5. Present configuration of the tectonic plates showing zones of seafloor spreading (represented by black strips) in the oceans. Note that most of the ocean crust is less than about 200 million years old, whereas some parts of the continents are up to 4300 million years (Myr) old.

difference exists because new ocean crust is being formed continually at the mid-ocean ridges (the thin black zones in fig. 1.5) and old ocean crust is sinking back into the deeper Earth (the mantle) at deep ocean trenches known as subduction zones. This process, which we call plate tectonics,[14] has operated throughout much of Earth's history, moving continents across the globe.

The present arrangement of continents and tectonic plates at Earth's surface is just the latest in a sequence of arrangements that Earth has gone through for billions of years. Occasionally this has brought the continents together into clusters called supercontinents. The assembly of supercontinents takes tens of millions of years and involves continent-scale collisions (orogenies) in which rocks are pushed up to form mountains, which erode and shed sediments into the ocean. The accumulation of heat beneath the supercontinents eventually causes them to break apart through rifting, forming new ocean basins that grow by seafloor spreading at mid-ocean ridges. The

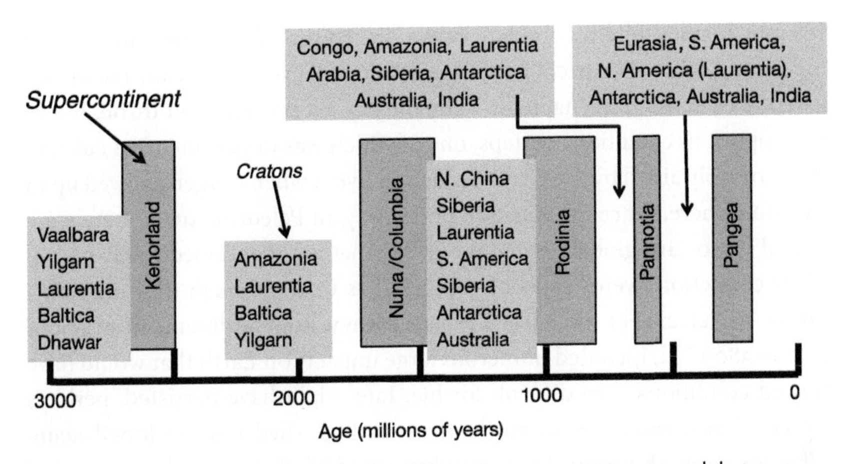

Figure 1.6. Supercontinent cycle showing the major supercontinents and, between them, names of the continental fragments (cratons) that assembled to form them. (Modified from Nield 2007; Murphy and Nance 2013; Li et al. 2016).

increased number of spreading ridges during the breakup of supercontinents leaves less room in the ocean for seawater, causing it to flood the continents. If the supercontinents are close to the poles, ice might accumulate, leading to global glaciation and a decrease in sea level. If the continental fragments are close to the equator, shallow seas might evaporate to form evaporite deposits consisting of salt and other minerals (discussed further in chapter 4).

The repeated joining and breaking up of supercontinents, known as the supercontinent cycle (fig. 1.6), has had a strong impact on the geologic history of the Great Lakes region. This cycle is so repetitive that we can see ancient processes in today's events. For instance, Africa and India are pushing against Eurasia from the south today, trying to create a new supercontinent and forming high mountain ranges like the Alps and Himalayas. In chapters 5 and 6, we will see that similar processes during the Rodinia-Pannotia supercontinent cycle formed the Midcontinent Rift and the Grenville Mountains in the Great Lakes region.

1.5. Great Lakes Fossils and Rocks Record the Evolution of Life

While the continents were forming and jostling for position, life began to evolve, and the Great Lakes region preserves some of the fossils that record this long history. Life required an ocean, which probably formed during the Great Rain when Earth cooled enough to condense water from the atmosphere, as discussed in chapter 8. Early life, once it started, went through several important evolutionary steps, one of which was development of eukaryotes, probably aided by the Great Oxidation Event when oxygen showed up in the atmosphere. Once life was well under way, in Paleozoic time, there were several important global extinction events that are mentioned in chapter 4. Older extinction events probably occurred as well, although they are much harder to detect. For instance, the Late Heavy Bombardment, from about 4100 to 3800 Ma, included numerous large impacts on Earth that would have created conditions very difficult for life. Life might have persisted, perhaps protected by ocean water, or maybe it was extinguished and developed again. What we know about this long-term history of life is discussed at the end of each chapter.

Now that we have outlined the "big picture" it is time to see how Great Lakes rocks have helped us understand the story. Keep in mind as we go along that everything we discuss, every rock and mineral and fossil, had to be found by someone walking along the ground. And then their significance to the larger story had to be outlined and interpreted. Each new rock exposure adds some information and helps clarify the story. So, after reading this, step outside and see what you find.

Notes

1. The term *native peoples* is used here to refer to Native American, First Nations, and other indigenous peoples. The Great Lakes region does not extend far enough north to include Inuit peoples.

2. Some of the tribes shown in figure 1.1 moved into the Great Lakes region from the east, first driven by internal efforts to expand, then by conflict with other tribes, and finally by pressure from European immigrants (Clifton and Porter 1987; Warren 2009; Schmitt 2016).

3. Although *wawa* is widely said to be the Ojibwe name for the Canada goose, *nika* is the name given in the *Ojibwe People's Dictionary*.

4. See Chmielewski (2017), Harkins (2009), and Skinner (2008) for further information on French activities in the Great Lakes region.

5. According to Peter Newman (2005), the Hudson's Bay Company is the oldest continuous commercial enterprise in the world. A few other organizations, including the Storra Koppaberg copper mine in Sweden (1288), the Löwenbrau brewery in Munich (1383), and the Banco di Napoli (1539), were formed earlier, but they have not operated continuously or in the same business. The English royal charter that established the Hudson's Bay Company did not consider either the aboriginal or other European claims to the land. French attempts to enter the area, which stimulated construction of Prince of Wales fort, were unsuccessful and ceded at the Treaty of Utrecht in 1713. In 1870, the

Hudson's Bay Company ceded its land control to Canada but continued its fur and trading activities.

6. See Elle (2009) and Newman (2005) for histories of the Hudson's Bay Company, and Maurice (2004) for an entertaining account of one of the last company traders.

7. The abbreviation Ma (mega-annum) is used throughout the book to indicate ages in units of millions of years.

8. If you want to read more about the age of Earth, try the summary by Rubie et al. (2015) and chapter 10 in Condie (2016).

9. The two main geologic laws governing relative geologic ages are the laws of superposition and cross-cutting relations. The law of superposition, which applies to layered sedimentary and volcanic deposits, indicates that the layer on top is youngest. The law of crosscutting relations, which applies to intrusive igneous rocks, indicates that older rocks are crosscut by younger ones.

10. Basalt is a volcanic (extrusive) igneous rock with a mafic composition (enriched in iron and magnesium). It makes up most of the ocean crust. Continental crust consists largely of granite, a felsic intrusive igneous rock (enriched in sodium and potassium) and andesite, a volcanic rock with a composition intermediate between basalt and granite (see also chapter 5 note 4).

11. In simple terms, isotopic ages (also known somewhat inaccurately as radiometric ages) are determined by measuring the abundance of a radioactive parent isotope and a daughter isotope in a sample. If all of the daughter isotope was derived from decay of the parent, as in the case of the decay of K^{40} to Ar^{40}, the age can be determined by the parent-daughter ratio and the half-life of the parent isotope. If some of the daughter isotope in the sample was not derived from decay of the parent (i.e., it was already present when the rock formed), it is necessary to analyze several samples and determine the age by means of the isochron method. For more insight into these and other isotopic age methods, see (https://pubs.usgs.gov/gip/geotime/radiometric.html).

12. See Nesbitt et al. (2003) and National Academies Press (2011) for more information on the global climate story.

13. For more information on the carbon cycle, see the summary on the National Atmospheric and Space Administration (NASA) web page (https://earthobservatory.nasa.gov/Features/CarbonCycle/) and Committee on the Importance of Deep-Time Geologic Records (2011).

14. Plate tectonics (fig. 1.7) involves 15 to 200 km thick, relatively rigid plates made up of crust and uppermost mantle, which slide over the underlying deeper mantle. (Two types of crust [ocean and continental, see note 10] overlie the mantle, which consists of ultramafic rocks, and the central core, which consists largely of iron.) Convective heat loss and resulting flow in the mantle drives plate tectonics. At the mid-ocean ridges, release of pressure on upward flowing mantle rocks causes partial melting to form basalt magma, creating new ocean crust. The new ocean crust and uppermost mantle migrate away from the mid-ocean ridges, and eventually sink into the mantle at subduction zones. During subduction, partial melting in the presence of water forms intermediate and felsic magmas that make up volcanic arcs consisting of andesite and granite. Volcanic arcs and sediments shed from them collect together to form continents. Because the continents are buoyant, they move about Earth's surface, colliding and rifting in the supercontinent cycle (fig. 1.6).

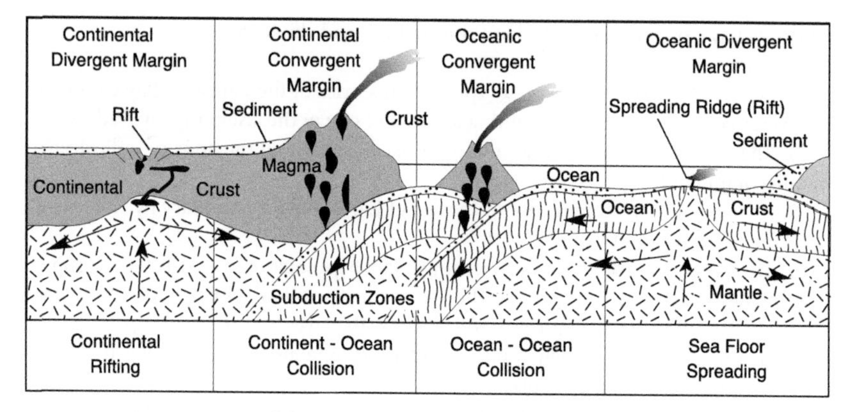

Figure 1.7. Cross section of the crust and upper mantle showing plate tectonic processes (rifting and collision) at continental and oceanic plate margins. At divergent margins (mid-ocean ridges), rifting pulls plates apart, breaking up continents and forming new ocean, and at convergent margins collision pushes them together, forming volcanic arcs and orogenic (mountain building) zones that coalesce to form continents.

CHAPTER 2

Landscaping the Continent

The Holocene and Anthropocene

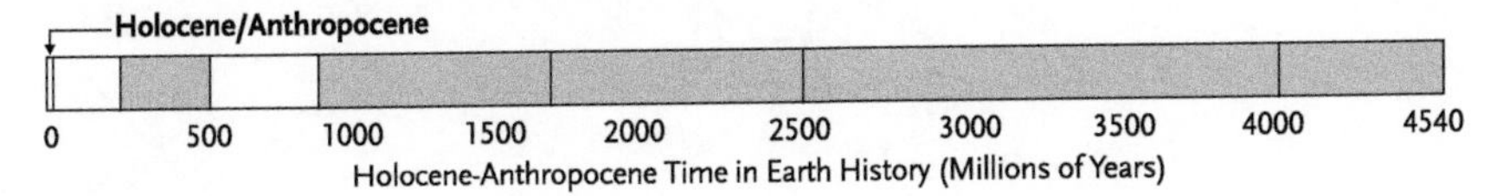

2.1. Surface Features Reflect the Underlying Geology

Mountains and valleys, waterfalls and lakes, and even sinkholes and caves are a direct reflection of the rocks that underlie the surface. These are the first things you see when you enter a new region, and they are definitely the best place to start if you want to understand its geologic history. This is certainly true for a time traveler, who would use them to identify a suitable landing spot.

Most of the surface features discussed in this chapter belong to the Holocene epoch, the most recent division of geologic time (fig. 1.3). The Holocene began about 10,000 years ago when the last glaciers retreated northward, leaving a new land surface behind. Most of the features on this recent land surface were formed by natural geologic processes, but some formed or were modified by human activity. In a 2005 study, Bruce Wilkinson showed that global human activity is eroding the average land surface about ten times faster than normal geologic processes. The fact that humans have become such an important agent of change on the planet has led some people to advocate changing the name of the Holocene to Anthropocene, a term coined by Eugene Stoermer, a professor at the University of Michigan. Others say that Anthropocene should refer to an epoch that started at the beginning of the twentieth century when human impacts burgeoned.[1] In this chapter, we will pay most attention to the natural features and processes but will note where and when they have been modified or influenced by anthropogenic processes.

Many large-scale surface features in the Great Lakes region are shown in the regional relief map in figure 2.1.[2] You can see that the region can be divided into three areas that turn out to have different geologic histories. In the far north the James Bay lowlands are flat and smooth. They are underlain by

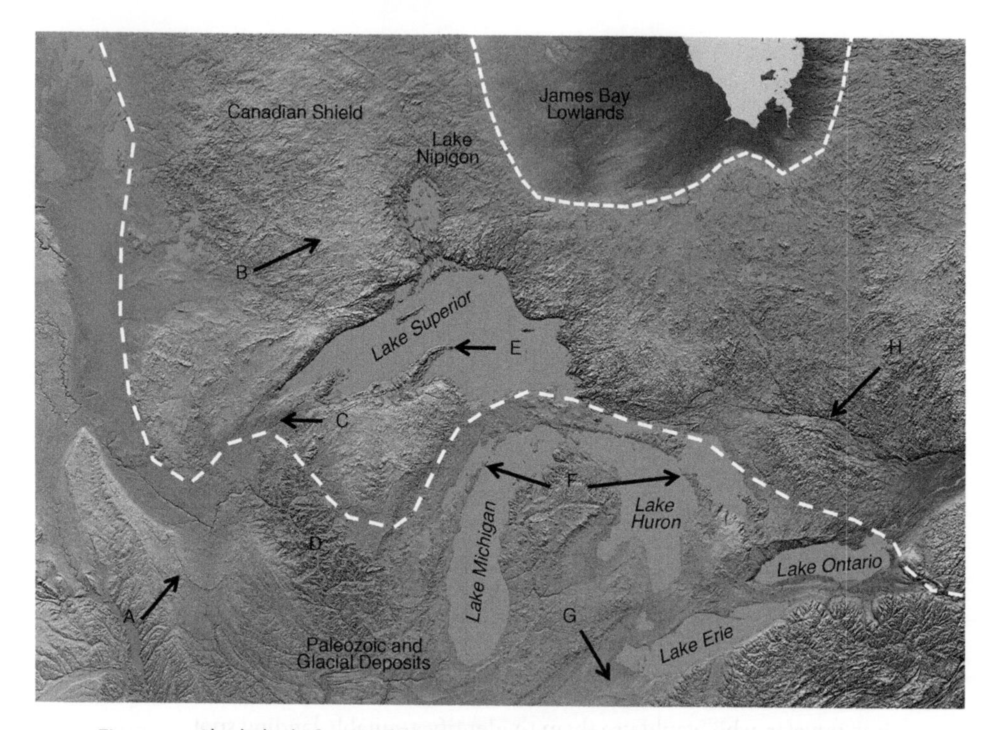

Figure 2.1. Shaded relief map of the greater Great Lakes region showing the three main areas, James Bay lowlands, Canadian Shield and the complex terrain to the south that is labeled here Paleozoic and Glacial Deposits. Letters and arrows are discussed in the text.

marine (ocean) sediments that accumulated in Hudson Bay after the glaciers melted. Beneath them are Paleozoic and Mesozoic sediments,[3] which we will learn more about in chapter 4. In 1611 the James Bay lowlands lured Henry Hudson southward from the bay that took his name. Hudson was searching for the Northwest Passage, which it was believed would take traders around North America to the Orient. Instead, he dead-ended in James Bay where his crew mutinied and left him and his son adrift in a small boat.

South of the James Bay lowlands, is an area with a very rough surface and relatively high altitude, especially around Lake Superior. It consists of Precambrian igneous and metamorphic rocks of the Canadian Shield that were scraped clean of soil by the glaciers. Canadian Shield rocks have a wide range of Precambrian ages, but most of them have a similar surface expression. The main exception is the Midcontinent Rift, which forms the flat, smooth zone around arrow C at the southwest end of Lake Superior and continues through the distinctive curve of the Keweenaw Peninsula (arrow E) to the southeast end of the lake at Sault Ste. Marie.

South of the Canadian Shield, the surface is more complex with a mixture of patterned hills and valleys that are underlain by Paleozoic and Mesozoic sedimentary rocks. The originally flat sedimentary layers have been tilted, and resistant layers stand out as ridges along the surface. The most distinctive of these is the almost circular ridge of the Niagara Escarpment, which runs through the Door, Garden, and Bruce peninsulas surrounding the Lower Peninsula of Michigan (arrow F).

All three of these areas were affected by glaciation during Pleistocene time, which began in the far north about 4 million years ago, reached as far south as Missouri about 2.5 million years ago, and retreated and readvanced numerous times after that. The most obvious features formed by the glaciers are the Great Lakes themselves, which were scraped out (scoured) by the glaciers. The glaciers also left deposits that form narrow ridges called moraines, which you can see north of Lake Superior (arrow B), at the end of the glacially scoured valley near Des Moines (arrow A), and at the west end of Lake Erie (arrow G). We will discuss these moraines and their role in the glacial history of the area in the next chapter.

One clear indication that glaciation affected the region is the lack of through going rivers, particularly in the hard Canadian Shield rocks where running water has not had enough time to erode river channels over the irregular surface. One of the few rivers that stands out on the map is the Ottawa River (arrow H), which early French explorers used to bypass the lower Great Lakes on their way to the upper lakes. Among them was Jean Nicollet, who in 1634 traveled up the Ottawa River, through Lake Nipissing and northern Lake Huron, and into Lake Michigan, where he met with Winnebago (Ho-Chunk) Indians around Green Bay. He, too, was searching for the Northwest Passage and had brought along silk garments to use in his first meeting with the Chinese. He returned by the same route with the sad news that the Winnebago had no knowledge of a route to China. Along the way, Nicollet would have noticed that the Ottawa River was too small to have formed such a large valley. In fact, the Ottawa occupies a down-dropped block of crust known as a graben, which we will hear more about in chapter 9, and it was created by a much larger river that drained the upper Great Lakes during late stages of Pleistocene glaciation.

The area denoted by arrow D on the relief map differs from all other parts of the region in having a relatively low elevation and strongly dissected surface with prominent stream valleys. This is the driftless area that the glaciers never entered. You can see on the map how the mountains in northern Wisconsin protected it from the south-flowing ice.

Although the relief map helps put things into regional perspective, we need to look at things in more detail to see the many processes that are sculpting the present land surface.

2.2. Lakes and Wetlands Are Our Most Obvious Surface Features

At the risk of stating the obvious, lakes are standing bodies of water surrounded by land. They can range in size from small ponds all the way up to Lake Superior, and they are amazingly abundant in the Great Lakes region (box 2.1). As the depth of water decreases, they merge into wetlands, although the boundary between lakes and wetlands is fluid.

BOX 2.1—WHO HAS THE MOST LAKES, AND WHY?

Ontario wins the prize for number of lakes with about 250,000, followed by Manitoba with 110,000. The United States is a poor second: Wisconsin claims 15,074 lakes, both Michigan and Minnesota have about 11,000, and the other states have much fewer. (The most common names are Fish Lake, Clear Lake, and even Mud Lake, which is number one in Minnesota and Wisconsin and a close third in Michigan.) If we adjust the count of lakes for the area of the states and provinces, Canada still wins. For every 100 km^2, Ontario has 23 lakes; Manitoba has 15; and Wisconsin, Minnesota, and Michigan have only 8.8, 4.9, and 4.4, respectively. This is an interesting statistic because the surface in Ontario and Manitoba was dominated by glacial erosion (scour), whereas Michigan, Minnesota, and Wisconsin were dominated by deposition of glacial sediment. So glacial scour is a much better way to make lakes than is glacial deposition.

Wetlands are commonly defined as areas that are saturated with water, either year-round or during wet periods. Year-round wetlands are called bogs if the water is acid and fens if the water is alkaline. Seasonal or intermittent wetlands are swamps if they are covered by forests and marshes if they are covered by herbaceous plants such as reeds and rushes. Wetlands commonly form along coasts or rivers in either salt or fresh water, although some form as the last stage in the filling of lakes. Most wetlands result from normal lake and stream processes, but some are helped along by beavers and other aquatic mammals that build dams.

The distribution of wetlands in the Great Lakes region is very much controlled by regional geology. In Canada wetlands make up more than 25 percent of the glacially scoured area of the Canadian Shield and at least 75 percent of the James Bay lowlands. In the United States, wetlands are most abundant along the margins of glacially deposited sediments, along the Mississippi River, and at the west end of Lake Erie (fig. 2.2).

2.2.1. *Lakes and Wetlands Form When Stream Gradients Are Blocked*

Although it doesn't seem that way in the Great Lakes region, natural lakes are actually very rare. Nature doesn't like them because they interrupt the smooth slope or gradient of a stream valley. Layers that are difficult to erode might produce some irregularity in the gradient of a stream, but these usually make waterfalls not lakes. In order to form a lake, the gradient of the stream has to be completely blocked, so that water must collect in a pond before it can continue downward. Many natural processes, including landslides, volcanoes, and shoreline processes, can form natural dams, but they are all uncommon.[4]

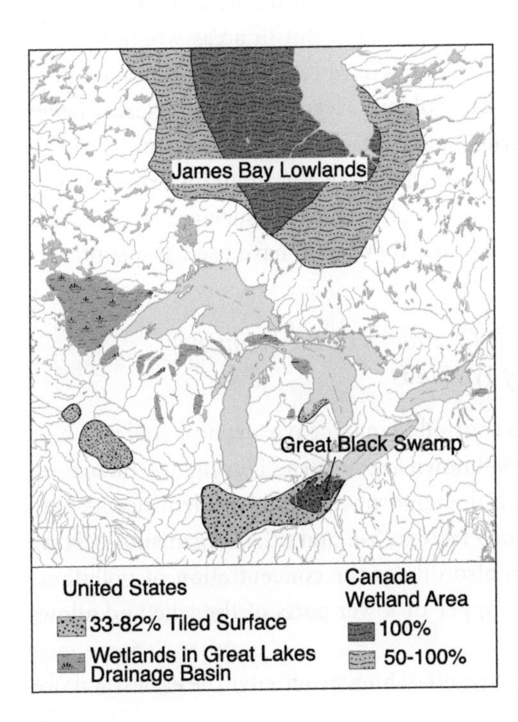

Figure 2.2. Distribution of wetlands in the Great Lakes region. Tiled areas in the United States are wetlands that have been drained and converted to agricultural use. (Based on data in Katz 1955; Fretwell et al. 1996; Kennedy and Mayer 2002; Sugg 2007.)

This makes it all the more curious that there are so many lakes in the southern United States, where these processes are largely absent. Here you are seeing engineered lakes that were made by damming streams for hydroelectric power generation, flood control, navigation, irrigation, and fish farming. Engineered lakes are so common that they have changed the rate of sediment flow to the ocean. There are about 45,000 dams more than 15 m high in the world, and the lakes behind them hold about 15 percent of Earth's river water. The dams collect sediment that would otherwise reach the ocean. In some rivers, such as the Nile and Ebro, more than 90 percent of the sediment never reaches the ocean. Worldwide, dams catch about 10 percent of the 15.5 billion tons of sediment that is transported in rivers.

Glaciation is really the only natural process that forms lakes over a large region. In areas of active mountain glaciers, it is common for valleys to be blocked by ice or moraines. But continental ice sheets are the really important lake formers. Glacial scouring and deposition leave an extremely irregular surface that has no continuous gradients. Water that collects on this surface, whether from precipitation or a spring, has no simple path to flow outward and must collect in ponds until it finds an outlet leading to the next low spot, where ponding happens again. In areas that are underlain by glacial deposits consisting of gravels and sands (as discussed in the next chapter), rivers can begin to erode more continuous channels. But in areas where the glaciers left only hard, scoured rock, channels have yet to form. This distinction can be seen in the greater abundance of long rivers south of the Great Lakes compared to that to the north (fig. 1.1). The lack of major rivers in the north impeded early travel. In fact some members of the Franklin expedition of 1845 to the Canadian Arctic islands might well have survived if there had been a major river in the far north that they could have used as a southerly escape route.

2.2.2. *Water Quality in Lakes Is Controlled by the Frequency of Overturn and the Health of Wetlands*

The quality of water in a lake is controlled in large part by how well and how often its water is mixed, a process known as overturn.[5] During overturn, deep water flows up into contact with the atmosphere, which replenishes its oxygen content, making it more hospitable to fish and other organisms. Mixing lake water from top to bottom also dilutes the concentration of pollutants that might be collecting in the upper or lower parts of the lake and allows them to be flushed out.[6]

Failure of overturn is often a result of human activity. The easiest way for humans to cause failure is to increase the density of the lake water, making it harder to circulate to the surface. Road salt and fertilizer are the two most common additives that increase the density of lake water, and their use has grown enormously during the last century. In 2008 Eric Novotny and others showed that there was an almost perfect correlation between the salinity of lakes in the Minneapolis–St. Paul area and the amount of salt that was used for road deicing, and that salinity had increased by more than 100 percent just since 1980. Nick Eyles and his associates have monitored Frenchman's Bay in Pickering, Ontario, just outside Toronto, and in 2010 they reported that about 3800 tons of salt flowed into this small body of water each year, largely from the main highway crossing the area. As long ago as 1971 Robert Bulbeck and others reported that road salt in runoff actually stabilized the lower water in Irondequoit Bay on the south side of Lake Ontario, preventing overturn. Although environmentally more desirable substitutes for salt are available, they are much more expensive and not widely used, suggesting that this effect will grow with time.[7]

Wetlands, through which much water enters lakes, rivers, and the ocean, are so important that they are sometimes referred to as nature's kidneys. They contribute to water quality by forming a sink for contaminants and pollutants, as well as environmentally important elements such as carbon, nitrogen, and sulfur.[8] Wetlands also store water that helps with flood control and groundwater recharge and provide coastal protection and zones in which to trap sediment before it enters lakes. Finally, they contribute to biodiversity in their role as a refuge for wildlife. At the geologic scale, accumulation of plant material in wetlands forms peat deposits, which can be transformed into coal with sufficient burial.

Wetlands are particularly important to the health of the Great Lakes. Destruction of wetlands has removed the buffer that purified runoff before it entered the lakes. This impact has been most severe around the Great Black Swamp, which originally occupied an area about 150 km long and 40 km wide just southwest of Lake Erie (fig. 2.2). Travel across the swamp was almost impossible in the early days of European settlement, and by 1850 it was being drained by digging ditches and laying drainage tiles.[9] The area is now heavily farmed, and fertilizer-rich runoff from it is a source of nutrients to the west end of Lake Erie. This has caused eutrophication,[10] resulting in the massive toxic algae blooms (fig. 2.3) that have caused several municipalities, including the city of Toledo, to temporarily close their water systems. The unusually shallow depths of Lake Erie exacerbate the problem.

Figure 2.3. Satellite image of the 2011 algae bloom, which impacted more than half the Lake Erie shoreline. (MERIS/ESA, processed by NOAA/NOS/NCCOS—http://www.noaanews.noaa.gov/stories2014/20140710_erie_hab.html.)

2.3. The Great Lakes Are the Largest Lakes in the Region

2.3.1. The Great Lakes Chain Is More Than 2000 km Long

The Great Lakes contain 22,250 km^3 of water, 20 percent of the world's surface freshwater. They are a major transport artery into the central part of North America, as well as a major recreation and fishing resource. The Great Lakes system begins at the west end of Lake Superior and flows more than 2000 km to lower and lower lake levels until it exits the east end of Lake Ontario. From there water travels another 870 km through the St. Lawrence River to the Atlantic Ocean. The famous children's book *Paddle to the Sea* by Holling C. Holling follows a toy canoe along this path from the headwaters of Lake Nipigon (fig. 2.1), which drains into Lake Superior, through the lower lakes, and finally into the ocean. Because the toy canoe had no paddler, it was moved along its journey by long-shore drift currents, which are a major agent of erosion and sediment transport, as we will see later in this chapter.

With the exception of Erie, which is very shallow, the Great Lakes extend to depths considerably below sea level (fig. 2.4). Their great depths reflect the fact that the glaciers were able to scour deeply into Paleozoic rocks that underlie the lake basins. For lakes Michigan, Huron, and Ontario, the glaciers cut into the soft shales, carbonate rocks, and evaporites (including the Mackinac breccia) that are discussed in chapter 4. Lake Superior, which is not surrounded by Paleozoic rocks, owes its depth to glacial scour along the soft Nonesuch Shale, a part of the Midcontinent Rift that is discussed in chapter 5. Depths in the lakes are not continuous, however, and are commonly interrupted by glacial deposits that formed when the lakes were much shallower during glacial retreats. For instance, in Lake Michigan, the Southern Basin, and the Algoma Basin in the north are separated by submerged moraines that formed during the late Wisconsin glacial retreat.

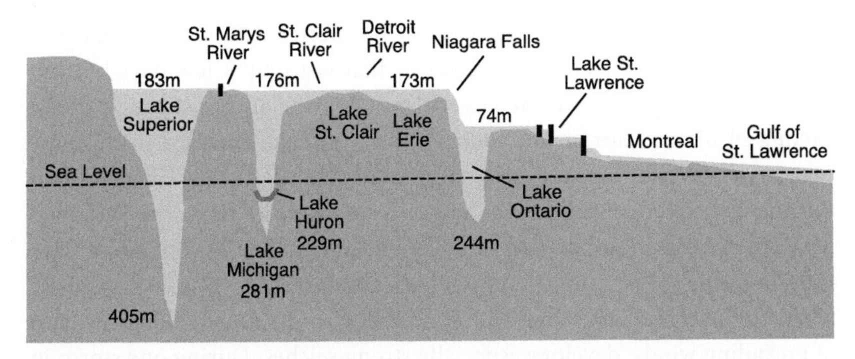

Figure 2.4. Schematic cross section of the Great Lakes drainage system showing the depth of individual lakes and the steplike decrease in their water levels toward the St. Lawrence River. (Modified from NOAA–Great Lakes Environmental Research Laboratory and U.S. Army Corps of Engineers data.)

Islands and promontories in the Great Lakes are underlain by rocks that were resistant to erosion. Most obvious are the Silurian-age dolomites discussed in chapter 4, which make up the Door and Garden peninsulas on the west side of Lake Michigan and the Bruce Peninsula on the northeast side of Lake Huron. These rocks continue eastward to make the Niagara Escarpment in Ontario and New York, as well as the resistant layer that makes Niagara Falls.

2.3.2. *Lake Levels Vary in Response to Natural and Human Factors*

Water levels in the Great Lakes vary seasonally by a meter or two, with the highest water levels commonly found during summer when runoff is greatest. Although this level of variation might seem small, it is significant to docks, homes, and other facilities along the shoreline. Great Lakes scientists and administrators are also concerned about whether the lakes will embark on a new path with greater changes in lake levels as the climate warms.

The variation of Great Lakes water levels at human time scales depends on wind, precipitation, evaporation, streambed erosion, and glacial rebound. Human features also play a role, particularly dams on the St. Marys River between Superior and Huron and on the St. Lawrence River below Lake Ontario. In the summer of 2017, when Lake Ontario levels were unusually high, the St. Lawrence dam was the focus of debate between upstream interests that wanted water released to alleviate flooding and downstream interests that feared the released water would flood them and by shipping interests that wanted to retain high water in the river for navigation. Three diversions, which are mentioned in the next section, also have the potential to affect lake levels but are not as important as the dams and the natural factors.

Among natural factors, wind varies in effectiveness over the shortest time frame. Storms generate large waves that last only a few days but can be very large, especially in the largest lakes. During an investigation of the sinking of the ore carrier *Edmund Fitzgerald* in Lake Superior in 1975, the captain of one "salty" (oceangoing ship) in the same area at the time reported that the waves were the highest he had seen anywhere in the world. Sustained winds in one direction can push water downwind, generating a standing wave called a seiche in which water sloshes back and forth from one end or side of a lake to the other. Lake Erie, which is shallow and oriented along the path of prevailing winds, develops especially strong seiches. During one storm in November 2015, the water level at Buffalo rose 2 m within a few hours and at the same time the water level at Toledo dropped almost 2 m.[11]

Meteotsunamis are also related to wind but in a different way. They form when a storm leads to rapid changes in atmospheric pressure that cause a corresponding abrupt change in local water levels. Waves formed in this way sweep across the water just like a tsunami in the ocean; according to a study by Adam Bechle and his coworkers, this happens most commonly in April through June. Most meteotsunamis generate waves only about 1 m above normal, although much larger ones have been observed. On July 4, 1929, a 6 m wave hit the pier at Grand Haven, Michigan, drowning ten people, and in 1954 a 3 m wave swept the Chicago shoreline, drowning seven.

Precipitation causes lake level changes over longer periods of at least a year or more. In the simplest sense, levels should go up when there is a lot of precipitation and down when there is less. As climates have warmed over the region, however, evaporation has increased in importance. Evaporation is relatively continuous during the warm months, of course, but it also happens during winter months in years with low ice coverage, and this is becoming more common. The overall trend of maximum ice coverage for the Great Lakes has decreased from about 60 percent in 1974 to about 45 percent in 2017.[12] Jay Austin and others, working on Lake Superior, have shown that the lakes are actually warming faster than local average temperatures, possibly because the darker water (darker than ice) absorbs more heat and the warming causes more evaporation, loss of water, and lower lake levels. Increased variability in ice cover from year to year will probably lead to larger fluctuations in future water levels and increased problems with shoreline erosion.

There appear to be additional lake level changes under way beyond the annual time frame. For instance, many people are convinced that since about 1930 there has been a gradual decline in the average water level of Lakes Michigan and Huron (which are connected and act as a single body of water) and that there has been a corresponding increase in the level of Lake Erie over the same time frame. One possible cause of this change, if it is indeed true, is erosion of the St. Clair riverbed at the outlet of Lake Huron. The erosion is also thought, at least by some, to have been caused by the mining of sand and gravel from the riverbed as reviewed by Frank Quinn. A more recent study by Xiaofeng Liu and others that attempted to evaluate the erosion issue was inconclusive. They found sedimentary bed forms in the river that could have resulted from erosion but concluded that present flow rates in the river were not sufficient to account for them.

Isostatic rebound, which is discussed in more detail in chapter 3, refers to the rise of land from which a heavy weight has been removed. In our case, the heavy weight was that of the glaciers, and Earth's crust is still recovering from this load. This process is slow and operates barely within human time scales. As we will see in chapter 3, it has accounted for much of the long-term history of the Great Lakes, especially in northern regions where the ice was thickest and longest lived. Since the ice left the Great Lakes about 10,000 years ago, Thunder Bay, Ontario, has risen about 200 m whereas Toledo, Ohio, has risen about 14 m. Although the rate of uplift has slowed, it continues today and is highest in the north. In Lake Superior, Michipicoten Island on the north side is rising about 35 cm per hundred years faster than Marquette, Michigan, on the south side. A similar difference in rates is seen between the north and south ends of Lakes Michigan and Huron, which, as noted earlier, operate as a single unit. This tilting of the lakes to the south causes water to leave northern regions such as Georgian Bay and pile up along the southern shores. The result is receding shorelines for property owners in northern areas and increasing storm damage for those in the south (fig. 2.5A in color section). It can even move water from one lake basin to another. Studies of modern tilting indicate that the Michigan-Huron system is decanting into Lakes Erie and Ontario, which are gaining water.[13]

2.3.3. *Ownership and Management of the Lakes Is a Joint Effort of the United States and Canada*

Because of their location along the boundary between Canada and the United States, the Great Lakes are subject to international treaties and agreements. These are based on the Public Trust Doctrine, which holds that some natural resources including navigable waters and shorelines, and the land beneath them, should be held in trust by the government for public use. Thus, any use or sale of such land must be in the public interest.

Several important binational agreements related to the Great Lakes

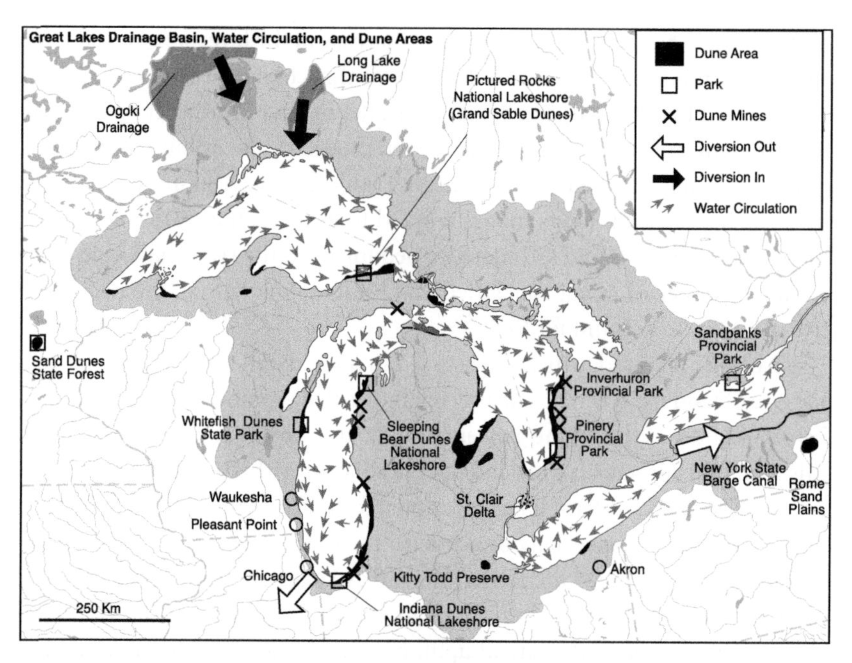

Figure 2.6. Map of the Great Lakes region showing the location of important deltas, dunes, and dune mining areas (including parks with these features), along with the Great Lakes drainage basin, water circulation patterns, and water diversions. (Modified from Lenders 2001; Schulte 2011; Lovis et al. 2012; Environmental Protection Agency 2017.)

have been negotiated between Canada and the United States, including the Boundary Waters Treaty of 1909, the Great Lakes–Saint Lawrence River Basin Sustainable Water Resources Agreement of 2005, the Great Lakes-St. Lawrence River Basin Water Resources Compact of 2008 and the Great Lakes Water Quality Agreement, which was most recently revised in 2012. Some of these agreements are binding and others offer good faith agreements among the two federal governments and in some cases the states and provinces as well. These agreements address issues as varied as water quality, out of basin diversion of Great Lakes water, shipping, hydropower and control of invasive species. Although Lake Michigan is entirely within the United States (fig. 2.6), it is included in international agreements because its waters are part of the larger lake system that is shared by both countries.

One of the most important Great Lakes ownership issues involves withdrawal of water from the lake (box 2.2). These can be returned after use or transferred out of the drainage basin. Withdrawals that are returned are by far the most important and include, in declining order of volume, electric power generation, public water supplies, industrial and domestic supplies, and irrigation. Most withdrawals take place within the drainage basin. Municipalities at Akron, Ohio, and Pleasant Prairie and Waukesha, Wisconsin, which are partly or totally outside the actual drainage basin, remove water but are required to return an equivalent amount as treated wastewater.

Transfers of water out of or into the Great Lakes drainage basin are strictly regulated. By far the largest transfer out of the drainage basin happens at Chicago, where withdrawals supply municipal water systems, as well as dilution of sewage treatment and flow to support navigation in the Chicago canal system, all of which flows southward into the Mississippi River (fig. 2.6). Almost all this water, totaling about 93 m^3/second, is diverted southward into the Mississippi River system. Other withdrawals from the Great Lakes systems, including the New York State Barge Canal, amount to less than 3 m^3/second. On the north side of Lake Superior, two systems actually transfer water into the drainage basin. These are at Long Lac and Ogoki (fig. 2.6), where rivers flowing toward Hudson Bay have been dammed for hydroelectric power generation. The flow from these systems, which is about 150 percent of the outflow at Chicago, goes into Lake Superior and accounts for about 6 percent of its inflow.

BOX 2.2. WATER WITHDRAWALS: THE GREAT LAKES AND ARAL SEA

Arid areas of North America look to the Great Lakes as a potential source of water, leading some to envision *Great Lakes Water Wars* as reviewed by Peter Annin. We already transfer oil and gas all the way across the continent, of course. So there are no real engineering limitations on moving water. But who owns the water? Does water belong to the area from which it evaporated or into which it is precipitated? Most of the water that falls on the Great Lakes region evaporated from states to the west or near the Gulf of Mexico. Do they have an ownership interest in this water? Setting aside this issue, there are more immediate concerns; namely, can water transfer be done without damaging the Great Lakes? Experience in the Aral Sea has poisoned the pond for many people. The 68,000 km^2 Aral Sea, located between Kazakhstan and Uzbekistan, was once the fourth-largest lake in the world in terms of surface area, larger than any of the Great Lakes except Superior. The diversion of rivers that originally flowed into the Aral Sea to irrigation projects starting in the 1960s caused the water level to decline. According to estimates by Peter Micklin, 90 percent of the lake area had become desert by 2014. The Aral Sea differs from the Great Lakes, however, in that it is located in a much drier climate and has less inflowing water, as summarized by Ian Boomer. Would it be possible to measure water flows in the Great Lakes well enough to remove just the right amount? The answer appears to be no. In a study of the balance of inflows and outflows in the Great Lakes, Brian Neff and James Nicholas of the US Geological Survey concluded that the uncertainties in estimates ranged from 1.5 to 45 percent. Without much more precise and accurate estimates, realistic evaluation of large-scale water transfers from the Great Lakes is highly unlikely. Karen Bakker has discussed issues related to larger-scale water transfers, in her appropriately titled book *Eau Canada*.

2.3.4. *Pollution Is a Growing Problem in the Great Lakes*

The drainage basin that supplies water to the Great Lakes is surprisingly small (fig. 2.6), measuring 800,100 km^2 (308,900 mi^2), only two times larger than the 244,100 km^2 (94,250 mi^2) area of the lakes themselves. Water in the lakes comes from a combination of precipitation on the lakes, underground (groundwater) flow, and surface runoff, in that order of declining importance, making precipitation over the lakes a very important part of their water budget. The continuous inflow pushes water down the lake system, causing water in each lake to be replenished. The time needed to replace all the water in each lake, known as retention time, varies with the size of the lakes from 191 years for Lake Superior and 99 for Lake Michigan to 22 for Lake Huron, 6 for Lake Ontario, and 2.6 for Lake Erie.

Replenishment of water in the lakes helps minimize pollution, although the degree of pollution still depends on the quality of water that is entering each lake from its surrounding drainage basin, and this is heavily dependent on the population density around the lake.

Some point sources of pollutants are important to the Great Lakes, but more widely dispersed road salt and fertilizers are major pollutants. In a recent study, Steven Chapra and others showed that pollution in the lower Great Lakes has increased enormously whereas pollution levels in Lake Superior have remained almost nil (fig. 2.7). The greater degree of pollution in Lakes Erie and Ontario reflects the significantly larger proportion of water that flows into these lakes from surface runoff. Efforts to minimize polluting runoff during the 1970s and 1980s helped reverse the increase in Lakes Erie and Ontario, although the trends have begun to increase again, as you can see in figure 2.7.

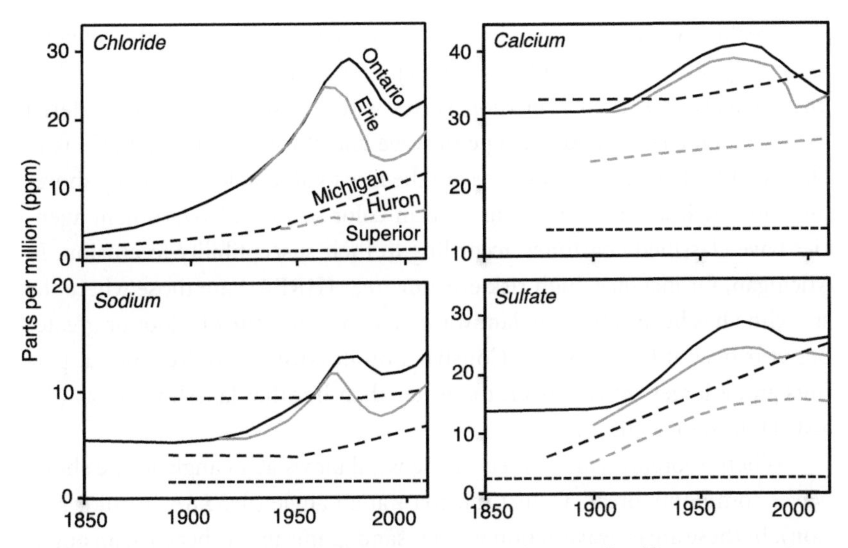

Figure 2.7. Long-term trends in chloride, calcium, sodium, and sulfate concentrations in the Great Lakes show that the lower lakes have been strongly polluted whereas Lake Superior has been largely unaffected. (Modified from Chapra et al. 2012.)

2.4. Shorelines, Beaches, and Deltas Surround the Lakes

Shorelines are places where water meets land and people. From the standpoint of recreation, this is the most important part of any lake. Shorelines range from cliffs to wetlands along the Great Lakes and other lakes in the region, and beaches range from mud through fine sand to large, rounded cobbles and even to hard rock. The character of beaches depends in large part on the material that makes up the shore. In the Great Lakes, wide sandy beaches are best developed along the east side of Lake Michigan where the shore is made up of dunes and other glacial sand. In northern Lake Michigan, as well as parts of Huron, the rocky shoreline consists of limestone and dolomite, which do not yield enough sand to make a beach.

Whether the shoreline develops a wide sandy beach also depends on the type of wave action it experiences.[14] If the coast is relatively steep, breaking waves will cut a shelf or terrace at the water level. Where the coast is flatter, terraces are less obvious. Terraces of all types are commonly coated with variable amounts of sand, gravel, or other material eroded from the shore. During periods of high wave action, especially in the winter, this cover is moved offshore where it accumulates in bars. During the summer, when wave action is gentler, the beaches are replenished.

Breaking waves have an enormous amount of energy, which they expend in eroding the shoreline, and this is a major concern around the Great Lakes. The intensity of shoreline erosion varies with the strength of the material making up the shore and the type of wave and storm activity that effects it. Dunes and other sandy shorelines are the most vulnerable, and rocky shorelines are the least vulnerable (fig. 2.5B in color section). Government agencies have classified coastlines according to their susceptibility to erosion. In Michigan, for instance, high risk erosion areas (HREAs) are those where the shoreline has been retreating landward at an average rate of 1 foot or greater per year over at least 15 years. Construction in HREAs requires special permits and a larger setback from the water than in other lakeshore areas (fig. 2.5C,D in color section).

Where shorelines are curved or the wind blows at an angle to the shore, waves commonly break against the coast at an angle (fig. 2.8A in color section). In these areas, wave action washes sand grains up the beach at an angle, whereas gravity washes them straight back down the beach when the wave retreats. Thus, a grain of sand will migrate along the beach in the direction that the waves are flowing onto shore, a process called longshore drift. The direction of waves is controlled by local wind, as well as the circulation of water in the lake, which is very prominent in the Great Lakes (fig. 2.6).

If the supply of sand and the strength of the current vary along the beach, longshore drift can deplete some areas of sand and enrich others. Many efforts to minimize erosion and longshore drift along coasts involve construction of barriers of some sort. One of the most common barriers is a jetty that is built out into the water to catch sand brought along by longshore drift (fig. 2.8B in color section). Unfortunately, these barriers deprive down-flow beaches of sand needed to replenish the sand that continues to be removed.

Natural barriers can also affect longshore drift. The St. Clair River at the bottom of Lake Huron is one such barrier. Much of the sand that is moved southward by longshore drift down the west coast of Lake Huron is diverted into the St. Clair River at the south end of the lake instead of continuing eastward up its eastern shore (fig. 2.6). This sand moves down the St. Clair River, where it is dumped into the north end of Lake St. Clair, creating the St. Clair Delta (fig. 2.8C in color section), the second-largest freshwater delta in the world. There is no way that this huge delta could have been formed by sand eroded from the St. Clair River valley, which is far too short and flat.

2.5. Coastal Dunes Are Major Features around Lakes Michigan and Huron

Sand dunes are widespread in the Great Lakes region. Most of the dunes are at or near the present shoreline of the Great Lakes, although there are also some inland dune fields (fig. 2.6). The largest dunes are found along the south and east side of Lake Michigan and the Lake Huron shore. Several important national, state, and local parks are centered on dunes, including the Sleeping Bear and Indiana Dunes National Lakeshores along Lake Michigan, Sandbanks Provincial Park on Lake Ontario, Inverhuron and Pinery Provincial Parks on Lake Huron, and the Grand Sable dunes in the Pictured Rocks National Lakeshore on Lake Superior.

There are two types of Great Lakes dunes. The simpler type, foredunes, form ridges up to 5 m high paralleling the shore and on the lake side of the beach. These form during low stands of the lake that allow wind to sweep sand shoreward. Because they are directly on the beach, these dunes are very easily removed or modified by storms. The more complex type of dune, known as transgressive dunes, are much larger, reaching heights of 60 m, and are found farther onshore from the beach. Some transgressive dunes are actively moving, although most have been stabilized by vegetation (fig. 2.9A in color section). Some dunes reach additional elevations because they formed on top of sandy glacial deposits, where prevailing winds picked up sand to make what are called perched dunes. Both the Sleeping Bear and Grand Sable dunes in Michigan (fig. 2.9B in color section) are of this type. Some transgressive dunes, such as those south of Marquette, Michigan, are not found along coasts and appear to have formed when wind concentrated sand on plains of outwash that formed in front of the retreating glaciers.

These dunes are the product of glaciers, lakes, and wind. The glaciers ground the rock into sand grains, wave action along glacial and modern lakeshores concentrated the sand, and the wind piled it up into dunes. This process probably continued intermittently throughout the more than 2-million-year-long Pleistocene glacial epoch, which is discussed in chapter 3, although the dunes we see today formed during and after the last glacial retreat starting about 10,000 years ago, and they have provided an interesting window into the most recent postglacial history of the region. In a continuing program of study, Alan Arbogast and his associates have used isotopic and other methods to measure the age of sand in the dunes, as well as the age of ancient soils (paleosols) buried beneath the dunes. Using this information, they have shown that the first stage of dune formation coincided with the Nipissing high stand of the Great Lakes about 5000 years ago, which is discussed in the next chapter. At this time, Lakes Superior, Huron, and Michigan were at the same level, about 10 to 15 feet above the present level. Another phase of dune formation took place between about 3500 and 2000 years ago, and this was followed by a final phase of dune growth between about 1000 and 500 years ago.

The latest of these dune growth phases coincided with the Medieval Warm Period,[15] when Great Lakes climates were drier, and a growing body of work indicates a more regional control on the dunes. For instance, Walter Loope and others reported that dune growth happened when a drought that affected the entire Great Lakes Basin caused a loss of forest cover. Dune sand can also be distinguished from beach sand, and concentrations of these two types of sand in deposits in small lakes along the Lake Michigan shoreline have been used to determine the history of lake levels and periods of high wind activity. In a recent study, Timothy Fisher and others showed that periods of high dune sand in the lakes could be correlated with solar cycles, suggesting that dune mobility is controlled by climate. Because dunes change so rapidly, they are relatively unstable, and many dunes along shorelines are subject to landslides (box 2.3).

BOX 2.3. LANDSLIDE IN SLEEPING BEAR DUNES

Steep cliffs of dune sand along the Lake Michigan shoreline at the Sleeping Bear Dunes National Lakeshore collapsed and formed landslides in 1914, 1971, and 1995. During the last event, an estimated 35 million cubic feet (991,000 m^3) of sand slid into the lake, forming a sheet of sand, tree trunks, and other debris that extends more than 3 km offshore. At first it was suspected that the landslide occurred because sand had been removed from the base of the dune by longshore currents. But surveys of the lake bottom showed that the dune front had not been significantly disturbed. Instead it looks like the landslide was a result of excess fluid pressure in the sand. The landslide happened on an unusually warm day in February when water from melting snow at the top of the dune percolated downward into the sand. The water increased the weight of the sand dune, and it could not flow out at the bottom because it was sealed by a wall of ice-saturated sand along the face of the dune. The accumulating water actually began to inflate the dune slightly, removing some of the pressure between sand grains, making it easier for them to slide against one another.

2.6. Caves and Sinkholes Formed on Paleozoic-age Carbonate Rocks

Caves and sinkholes form where carbonate rocks (limestone and dolomite) are at or near the surface and can be dissolved by rainwater and soil water made acid by small amounts of dissolved carbon dioxide (CO_2).[16] Although this dissolution is slow, over thousands of years it can create a very irregular land surface with ridges, towers, fissures, and sinkholes that is known as karst (fig. 2.10). Below the karst, groundwater has usually dissolved more carbonate rock to make caves, which often form at one or more levels in the subsurface that reflect earlier levels of the groundwater table. Most caves contain speleothems, which are mineral deposits such as stalagtites (up) and stalagmites (down), which form by the precipitation of calcium carbonate from water that flows or drips through the cave. Even in areas where there are no obvious caves, karst terrain is easy to recognize because it is full of sinkholes formed by collapsed caves and "disappearing streams" that flow underground into the cave system. If the water table is near the surface, sinkholes fill with water and become lakes.

Large parts of the Great Lakes region are underlain by carbonate rocks (fig. 2.11) on which karst topography has developed to varying degrees. Karst features are most obvious at the surface where the cover of glacial deposits is thin. The Door Peninsula of Wisconsin and the Bruce Peninsula of Ontario, which lie along the Niagara Escarpment, have abundant karst features many of which are aligned along faults that control topography locally. Where glacial deposits are thick, underlying karst-cave terrain can be unrecognizable. In some such areas, however, sinkholes have migrated upward through the glacial deposits to the surface. Sinkholes of this type are common in the northeastern part of the Lower Peninsula of Michigan and include Shoepac, Rainy, Sunken, Devil's, and Long Lakes (fig. 2.11). Karst and caves are also found beneath present-day lake levels, particularly at the Thunder Bay National Marine Sanctuary near Alpena, Michigan, where an onshore karst terrain continues into Lake Huron.

Caves provide important information about the Pleistocene history of the region and even about the Cenozoic history, for which we have a very sparse rock record in the Great Lakes region. Some caves and sinkholes contain soils, sediments, and plant and animal remains that are useful in efforts to reconstruct the ancient climates and hydrology. Ages of events and fossil material can be determined by isotopic measurements on speleothem over-

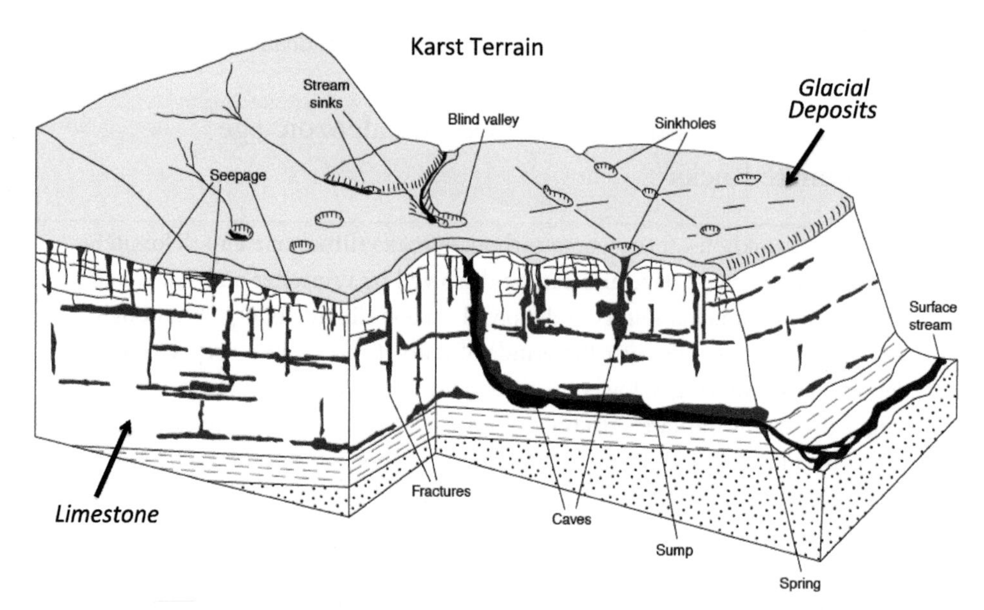

Figure 2.10. Karst terrain showing a limestone layer with caves, sinkholes, and springs, many of which are located along fractures that guided the groundwater flow that dissolved the limestone. The limestone layer is overlain by glacial deposits, which develop sinkholes where the underlying karst caves collapse, and underlain by shale (*dashed*) and sandstone (*stippled*). (From Alexander and Lively 1995.)

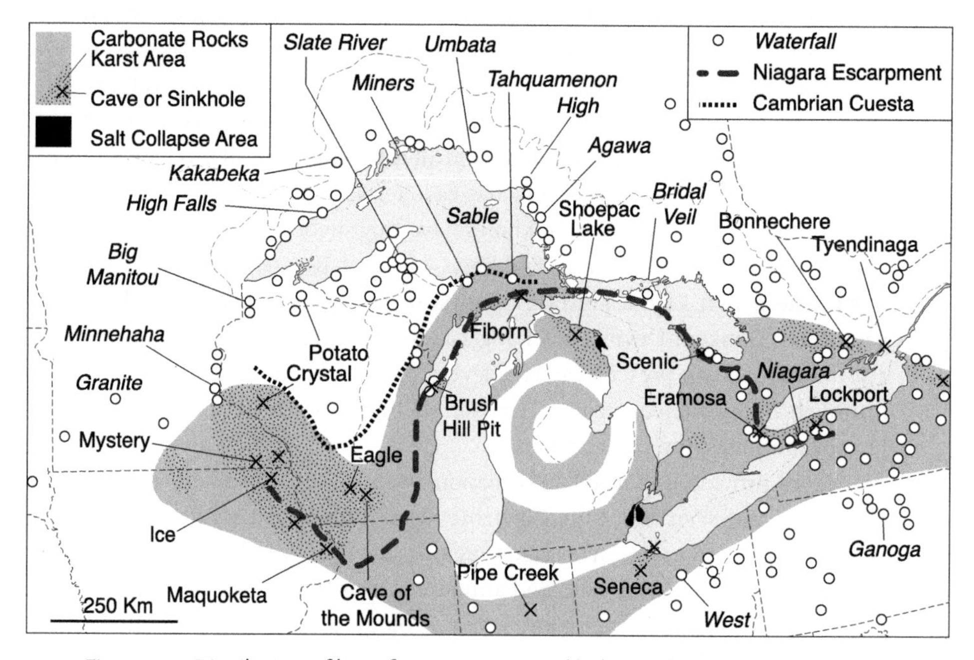

Figure 2.11. Distribution of karst features, caves, sinkholes, and waterfalls in the Great Lakes region. (Modified from Ege 1984; Weary and Doctor 2014.)

growths. In work at New Hope Cave in Wisconsin, for instance, John Luczaj and Ronald Stieglitz have found leaves and wood, as well as bones and teeth from animals ranging from shrews and bats to deer and bears. These remains have an age of about 1000 years, providing insight into climates that existed shortly before the advent of Europeans in North America. In a series of studies at the Pipe Creek Junior Sinkhole in Grant County, Indiana, James Farlow and his coworkers have found animal and plant remains, with an age of about 4 to 5 million years, that provide a much deeper look at past climates.[17] Cave deposits can even record directions of ancient groundwater movement, as shown by Calvin Alexander's work on 300,000-year-old speleothems at Mystery Cave in Fillmore County, Minnesota.[18]

Sinkholes and large-scale surface collapse can also form by dissolution of salt and other evaporite deposits that formed by evaporation of ancient seawater (fig. 2.11). Evaporite dissolution is also thought to account for at least some of the sinkholes and collapse features in northeastern part of the Lower Peninsula of Michigan, where limestone of the Devonian Traverse Group is underlain by salt and gypsum deposits of the Detroit River Group that are discussed in chapter 4. In southeastern Michigan, these same Detroit River Group evaporites come near the surface along the Detroit River between Michigan and Ontario and have caused numerous collapse features. As we will see in the next chapter, collapse features associated with these evaporites and overlying limestones formed weak rocks that were scoured to create the basins of Lakes Michigan and Huron.

2.7. Waterfalls and Potholes Recall Our Glacial Heritage

Waterfalls are very abundant in the Great Lakes region, and two processes account for many of them (fig. 2.11). One of these processes is related to changes in the water levels of the Great Lakes. We will see in the next chapter that the ancestral Great Lakes went through a long and complex history as the glaciers receded, with many deep lakes dammed against the retreating ice. Water levels in the lakes dropped when the last glaciers receded from the area, turning streams that had flowed directly into the lakes into waterfalls flowing down steep slopes to reach the lower lake level.

The other process is differential erosion. Where a stream crosses rocks with very different degrees of hardness, it will cut more deeply into the softer rocks, leaving the harder rocks as resistant features that become waterfalls.[19] In the southern Great Lakes region, differential erosion was most common along two Paleozoic rock units in the Michigan Basin. The most famous of these is the Niagara Escarpment, which extends along the south side of Lake Ontario in New York, through Ontario and the Upper Peninsula of Michigan, then into Wisconsin, and finally into Illinois where it is largely obscured by glacial deposits (fig. 2.11). The escarpment has a circular form because it consists of a bowl-shaped layer of dolomite, known as the Lockport Dolomite in New York and Ontario, which was deposited during Silurian time and downwarped into the Michigan Basin (as discussed further in chapter 4). Throughout the entire region, the dolomite is tilted downward toward the center of the Michigan Basin, and it stands out as a long sloped ridge or cuesta. The steep side of the cuesta faces north and east in New York, Ontario, and Michigan, and west in Wisconsin, and it forms waterfalls wherever rivers cross it. The largest of the Niagara Escarpment waterfalls is Niagara Falls.

The other important pair of hard and soft layers is in the Upper Peninsula of Michigan where sandstones of the Cambrian-age Munising Formation are overlain by more resistant limestones and dolomites of the Ordovician-age Au Train Formation. The ridge formed by these rocks is known as the Cambrian cuesta (fig. 2.11), and it forms waterfalls at Au Sable, Miners, Munising, Laughing Whitefish, and Au Train. The more famous Tahquamenon Falls (fig. 2.12A in color section) is at two levels in the Cambrian Munising Formation, where there are variations in the hardness of the layers.

Farther to the west, in Minnesota, numerous waterfalls along the valley of the Mississippi River formed where hard carbonate rocks of the Platteville Formation overlie soft sandstone of the St. Peter Formation (fig. 2.12B in color section). The best known of these are Minnehaha Falls, which occupies a tributary of the Mississippi River, and St. Anthony Falls, which is in the Mississippi River itself.

In the older, Precambrian rocks to the north, many waterfalls are also related to major differences in the hardness of layers. The highest waterfall in Minnesota, High Falls of the Pigeon River, cascades over a layer of diabase that overlies shales of the much softer Rove Formation,[20] discussed in chapter 6. Kakabeka Falls (fig. 2.12C in color section) near Thunder Bay, Ontario, is held up by a chert-carbonate layer that overlies softer shales in the Gunflint Formation, one of the iron-formation sedimentary units discussed in chapter 6.

Waterfalls are not static features; they migrate upstream as erosion breaks down the hard rock layers (box 2.4). In some streams, the abrupt change in slope disappears; in others it remains as a feature known as a nick point. Niagara Falls is a great example of a migrating nick point. After the last recession of the glaciers, this cataract began at the Niagara Escarpment that

Figure 2.13. (*A*) Relief map of Niagara Gorge and Falls. The waterfall originated at the Niagara Escarpment after the last glacial retreat and has since migrated up the river toward Lake Erie. (*B*) View up the gorge of the Niagara River from a point below the falls (buildings show size of the gorge).

bordered ancestral Lake Ontario (Lake Iroquois). Since then the falls have migrated upstream as falling water eroded the softer underlying Rochester Shale, causing the Lockport cap rock to collapse (fig. 2.13A). During the last 12,000 years or so, the falls have migrated 11.4 km up the river toward Lake Erie, forming a gorge (fig. 2.13B). Eventually the gorge will reach Lake Erie, raising the possibility that it will be drained, as discussed in chapter 9.

BOX 2.4. ST. ANTHONY FALLS AND GEOLOGIC TIME

St. Anthony Falls is now in the Mississippi River in downtown Minneapolis, but it originated about 10 km to the south at the junction of the Mississippi and Minnesota rivers. Herbert Wright, in a very readable summary of Minnesota rivers, has explained how St. Anthony Falls migrated northward up the Mississippi to its current position since the last ice retreat.[21] The position of St. Anthony Falls in the Mississippi River was first mapped by Father Louis Hennepin in 1680 and later by other explorers. In 1888 Horace Winchell, the first state geologist of Minnesota, used these observations to calculate the rate of northward migration of the falls over the preceding 200 years. Using that rate, he estimated that the falls migrated from the Mississippi-Minnesota River junction to its present position in about 8000 years, a number close to modern estimates based on isotopic age measurements. This was one of the first quantitative estimates of the magnitude of geologic time, a subject that was just beginning to find its way into philosophical and theological discussions.

Giant potholes are very unusual river features that are found in only a few places in the world, and the Great Lakes region is almost the world capital. These features are cylindrical holes that have been drilled into the bed of a river by stones (cobbles or boulders) that were caught in eddies and spun around like drills by fast-flowing water.[22] Giant potholes are much larger than those found in most modern rivers. At Taylors Falls Interstate Park along the Wisconsin-Minnesota border and at the Grand River Conservation Authority in Ontario, some potholes measure about 30 m deep (fig. 2.12D in color section). Smaller but still anomalous potholes are found at Devils Lake, Wisconsin; Temperance River State Park, Banning State Park, and Big Stone National Wildlife Refuge in Minnesota; Archbald Pothole State Park in Pennsylvania; and Hilton Falls Conservation Area and Potholes Provincial Park in Ontario. In addition to their size, these potholes stand out because they are far above present river levels, and the rivers have relatively low flow rates. The much larger river flows needed to account for the high elevations and large sizes of the potholes, likely came when water flowed out of lakes along the margins of the retreating Wisconsin glaciers. It is even possible that some of these potholes formed during immense catastrophic flow events caused by the collapse of ice dams.

2.8. Soils and Dune Sand Are Important Surface Mineral Resources

2.8.1. Soils Form Where Rocks Meet the Atmosphere and Hydrosphere (Water)

Most of our mineral deposits in the Great Lakes region were formed by older processes, but one in particular is definitely related to the present-day surface. That is soil. Estelle Dominati and her colleagues have argued that soil is the most important source of environmental services on the planet and thus outranks all other mineral resources in importance. In addition to its obvious role as a medium for plant growth, soil provides a reaction zone for the purification of water and air and a habitat for organisms. From an engineering perspective, soil is just unconsolidated material at the surface that does not make a good foundation. From a simple geologic perspective, however, soil is the zone of weathered material where the atmosphere and hydrosphere come into equilibrium with rocks. The type of soil that forms depends on the type of rock that undergoes weathering but also on the amount of water available to facilitate weathering reactions, topography that controls how well the water can percolate through the rocks, and how long these reactions have been going on.[23]

The important point, in terms of the geologic history of the Great Lakes region, is that most of our soils are very young. They are very poorly developed in most of the northern part of the region. Driving north from Sault Ste. Marie, Ontario, you find that much of the surface consists of hard igneous and metamorphic rocks with essentially no real soil. Even in forested areas, hard rock is present just below a carapace of dead vegetation. In this area, soil that can support agriculture has developed only on fine-grained glacial sands and gravels, as discussed in the next chapter. In the southern Great Lakes region, soil development is further advanced but is still largely on glacial deposits rather than hard rock. Soils that developed on hard rocks are present only in the southernmost parts of our region that the glaciers did not reach and to a lesser extent in the driftless area (fig. 2.1). We learn from this that it takes Earth thousands of years to develop soils in cool, temperate climates. Erosion takes much less time to remove soils. So we must take good care of the soils we have because we cannot wait for new soils to form.

2.8.2. Dunes Provide Sand for Industrial and Energy Markets

Although the dunes owe their origin to the glacial period thousands of years ago, as discussed in the next chapter, they have continued to move. In fact some dunes are moving today. In 2017 the *Muskegon News* reported that a house near the community of Silver Lake along the east shore of Lake Michigan was buried by a migrating sand dune. The dunes and the sand resources that they contain are definitely part of the modern surface features of the area.

Dune sands are an important source of industrial sand, which is used to make molds for casting metal parts (foundry sand), as a raw material for making glass, and as an abrasive material for sandblasting. Great Lakes dune sand has two important favorable qualities: it consists almost entirely of grains of quartz, with very few grains of other minerals; and the grains are nearly the same size. Also, it is unconsolidated, which makes it easier to mine than the sandstone layers discussed in later chapters. The main areas of dune mining are along the east coasts of Lakes Michigan and Huron (fig. 2.6). Their location near parks and other dune-related areas has led to public pressure to preserve the dunes. Christy Fox has reviewed the history of the Sand Dune Protection Act in Michigan, and Alan Arbogast and Brad Gammon, in a recent report, reviewed the new approaches that are being taken in Michigan to regulate and manage all types of activity on the dunes, from industrial to recreational.

2.9. Surface Life and the Threat of a Sixth Extinction

2.9.1. Lakes and Wetlands Provide a Record of Early Humans

Humans have used lakes and wetlands as sources of water and food since they came into the region during the late Pleistocene. Early aboriginals even used lakes to preserve food. Numerous modern excavations, usually for farm ponds or drainage systems, have unearthed remains of animals in partly filled ancient lakes and wetlands. Dan Fisher has described sites where remains of mammoths and mastodons show good evidence for human caching, including weights used to hold the bones (and attached meat) down in the water and upright logs that may have been used to mark their position for later recovery.[24] Some of the finds also show marks and cuts on the bones, which are interpreted to be the result of butchering. Fisher followed up these discoveries with an imaginative experiment in which he submerged horsemeat in a lake and then ate it at two-week intervals for several months through the winter. These observations have added to the controversy about whether early humans caused the extinction of North American large animals, or megafauna, as they are known in the scientific world. This question is part of a much larger controversy about the interaction of early humans with megafauna throughout the world, which is discussed further in the next chapter. Resolution of the debate depends on the discovery and study of new fossil remains, which happens frequently (box 2.5).

BOX 2.5. THE BRISTLE MAMMOTH

In 2015 two men excavating a pit for installation of a drainage pipe at a farm in Southeast Michigan found bones. Fortunately, they brought in University of Michigan paleontologist Dan Fisher, who has worked on mammoth and mastodon remains all over the world. Working with the excavators, Fisher and his team recovered almost half of the bones of an enormous mammoth that appears to have been dismembered and butchered and then stored at the bottom of a lake. As the lake filled and became dry land, the mammoth bones were covered and preserved at a depth of several meters. As can be seen in figure 2.14 (in color section), recovery required a large machine to lift the head and tusks, which were still connected. This find turned out to be very important for two reasons. First, it contained further evidence that use of these large animals sometimes contributed to human subsistence. Second, early work on the mammoth suggests that it lived more than 15,000 years ago, considerably earlier than the time frame usually thought to represent extensive human activity in North America.[25]

2.9.2. *The Study of Mammoth and Mastodon Remains Helps Us Understand Episodic Global Extinctions*

In the following chapters, we will review briefly the major developments in the origin and evolution of life for each time period. One of the important issues about life throughout geologic time is extinction. Extinction is a basic "fact of life" in the fossil record, of course, as we know from the fate of the dinosaurs. But it's an entirely different matter to suggest that extinction happened to so many species that there was a major hiatus in life on the planet, and that it happened episodically. As more and more studies accumulated, however, this idea gained support, and it was put on a strong footing by a detailed survey of the available data by David Raup and Jack Sepkoski. They identified five major extinction events in which a large fraction of biodiversity alive at the time was lost to extinction. According to a recent summary by Norman MacLeod, these events happened at about 440 Ma (End Ordovician), 365 Ma (Late Devonian), 252 Ma (End Permian), 201 Ma (End Triassic), and 66 Ma (End Cretaceous) time.

In *The Sixth Extinction*, Elizabeth Kolbert has summarized the chorus of evidence that human impact on the planet is leading toward another period of widespread loss of species. Proponents of this view point to the extinctions of numerous megafauna species and to the widespread bleaching and dying of coral reefs over much of the world. Others ask how we would recognize an extinction if it were actually happening. Anthony Barnosky and others attempted to evaluate the sixth extinction quantitatively and concluded that it might occur within a few centuries if we lost all currently endangered and vulnerable species. But Pincelli Hull and co-workers have noted that the real key to estimating mass extinction is not species loss but mass loss—the decline in the number of animals and plants—and they note that we have lost much more mass than species during the Anthropocene. And there is the question of timing: how much time does it take to achieve a massive extinction? Douglas Erwin has pointed out that the duration of individual extinction events has been shown to be shorter and shorter as isotopic age measurements improve and that this suggests that extinction events might happen too fast for us to see the foreshadowing.

We will return to this debate in the next chapter when we look at the controversy about the disappearance of large mammals during Pleistocene time. For the moment, more recent events in the Great Lakes region provide a microcosm of the extinction issue. In *The Death and Life of the Great Lakes*, Dan Egan reviewed enormous recent changes in Great Lakes fish and other fauna and the role that humans played in these changes. Almost all the changes are due in one way or another to the opening of the Great Lakes to oceangoing shipping through the St. Lawrence Seaway. Before that time, the Great Lakes had been isolated from the ocean for thousands of years. The invasion or introduction of foreign aquatic organisms, including the sea lamprey, alewife, Coho and Chinook salmon, and zebra and quagga mussels, led to booms and busts of population that completely changed life in the Great Lakes.

In a sense, we are doing the same thing on a global scale, although that project is so much larger that it is taking more time for our efforts to become obvious. Human involvement in global extinction has several possible outcomes. We might become victims and extinguish ourselves. If so, we can look to Alan Weisman's *The World Without Us* for a hint of the future. More likely, we will stick around and cause a change in global climate that might even start another ice age. Changes in global ocean circulation patterns caused by the melting of polar ice caps, as reiterated recently by L. G. Henry and others, could well lead to an ice advance. Some of these potential future scenarios will be reviewed in the final chapter of this book. In the meantime, if Earth is to return to, or possible avoid, another ice age, we should look to the last one for some guidance on what happened—and when and why. And that leads us to the next chapter, on the Pleistocene and Earth's last major ice age.

Notes

1. For discussions of the Anthropocene and human impacts, see Walling (2006); Steffen et al. (2011); and Waters et al. (2016).
2. *Relief* refers to the difference in elevation between the highest and lowest parts of a region.
3. For a review of geologic ages, see figure 1.3 and summary diagrams at the start of each chapter.
4. There are many ways to form natural lakes, and most depend on blocking a stream valley. For instance, Tangjianshan Lake in China formed as the result of a landslide caused by an earthquake in 2008. Lava flows or even growing volcanoes can also block valleys, as happened at Lake Nicaragua. Streams that flow into a lake or the ocean can build up a sand bar that isolates a lake, such as happened at Lake Ponchartrain in Louisiana. Lakes can also form because of subsidence of the land surface. In areas with abundant caves, some can reach the surface as sinkholes that form lakes, like many of those in central Florida. In volcanic areas, collapse of the summit can form a caldera lake like Crater Lake in Oregon. Finally, glaciers can form lakes by ponding water against ice as the glacier moves or by ponding water in low areas left by the glacier (including against moraines). Costa and Schuster (1987) have reviewed many of these lake-forming processes and their failure rates. The only natural process that forms widespread lakes is glaciation, which means that natural lakes are rare features of the landscape and are common only in glaciated areas.
5. Lake overturn usually happens because the density of water in the lake is uniform

from top to bottom. This happens when water has its maximum density at a temperature of 4°C. Water cooler or warmer than this is less dense and will therefore seek the top of the lake, causing overturn. In temperate zones, the water temperature passes through 4°C during cooling in the fall and warming in the spring. If the lake is deep enough, water at the bottom maintains a constant temperature of about 4°C throughout the year. Thus, twice a year the surface water will pass through a temperature (and density) similar to the bottom water, leading to overturn. Wind can aid this overturn by mixing surface waters. Overturn might stop if the density of the water is increased by addition of salt or other dissolved material.

6. Lakes in the Great Lakes region differ in their degree of overturn. Most smaller lakes undergo mixing twice a year and are known as *dimictic lakes. Monomictic lakes* mix only once a year, with mixing continuing through the fall and winter into the next spring. The Great Lakes are sometimes classified as monomictic because well-defined stratification takes place only during the summer, when cold bottom water remains at depth (Boyce et al. 1988). The interface between cold and warm water in the Great Lakes actually begins as a near vertical boundary that separates warm, shallow, near-shore water from deeper, offshore water. The boundary, which can be seen in satellite images, and sometimes from high shorelines, is known as a *thermal bar* by limnologists. Fishermen call it a scum line because debris and insects tend to concentrate along the boundary. Lakes that do not mix completely are known as *meromictic*, and many are a result of human activity. The most common meromictic lakes in the Great Lakes region have a high ratio of depth to area, and are so deep that overturn reaches only part of the way down. Flooded open pit mines commonly are of this type, especially where the pits were cut into strong rock that allowed for steep pit walls, such as Miners Pit Lake at Ely, Minnesota, and Steep Rock Lake near Atikokan, Ontario.

7. Road salt is halite (NaCl), most of which comes from evaporite deposits, which are discussed in chapter 4. Substitutes for road salt include calcium chloride, which causes the same environmental problems as salt and costs much more; and organic compounds such as calcium-magnesium acetate, which also has a high cost but is less environmentally damaging.

8. See Kesler and Simon (2015) for a discussion of global cycles of elements such as sulfur.

9. Similar drainage projects converted most wetlands into usable agricultural land in the temperate part of North America, although their original distribution can be estimated from the location of areas where a significant proportion of the surface has been drained (usually by burying tiles or pipes).

10. Eutrophication is the process by which nutrients such as phosphorus and nitrogen accumulate in water, causing an increase in fertility and excessive plant growth such as algae blooms. These block light from reaching into the lake water and absorb oxygen that is needed by aquatic life.

11. For a graph showing the change in water levels during this event, see https://www.mlive.com/news/grand-rapids/index.ssf/2015/11/lake_erie_phenomenon_seiche_ca.html referenced August 12, 2018.

12. For data on Great Lakes annual maximum ice coverage, see https://www.glerl.noaa.gov/data/ice/#historical referenced December 3, 2018.

13. This summary is based largely on data summarized by Mainville and Cramer (2005) and Bruxer and Southam (2010). Additional information on long-term variations in lake levels and their relation to isostatic uplift can be found in Baedke and Thompson (2000); Lenders (2001); Lewis et al. (2005); Breckenridge (2013).

14. Waves are formed largely by wind, which pushes the water along in a circular motion that gets smaller with the depth of the water. When this circular motion runs into a shallow coastline, water in the bottom of the circle is stopped and the upper part of the circular motion forms a wave that breaks.

15. The Medieval Warm Period, which lasted from about 900 to 1300, was a period when climates were one to two degrees centigrade warmer in the North Atlantic region and possibly also in China (Diaz and Hughes 1994).

16. The rock known as limestone consists largely of the mineral calcite ($CaCO_3$) and the rock known as dolomite (or dolostone) consists of the mineral dolomite ($CaMg[CO_3]_2$).

17. See also Rogowski and Farlow (2012); Ochoa et al. (2016).

18. As will be discussed in chapter 4, some karst topography formed on these carbonate rocks during the formation of unconformities in Paleozoic time, although the speleothems formed more recently.

19. Waterfalls can also form where faults cut across a stream, raising the upstream side, but this is not common in the Great Lakes region.

20. Diabase is a mafic intrusive rock with a composition similar to basalt. Diabase in the Pigeon River area was emplaced as part of the magmatic activity associated with the Midcontinent Rift (discussed in chapter 5).

21. At the time St. Anthony Falls formed, the Minnesota River was the site of Glacial River Warren, which drained Lake Agassiz, the largest of the lakes that formed along the margins of the Laurentide Ice Sheet as it retreated into Canada. When Lake Agassiz found other outlets, flow (and the water level) in Glacial River Warren dropped and it became the Minnesota River. The Minnesota River could not fill the old Glacial River Warren valley, and this left tributary streams like the Mississippi River flowing down a steep gradient into the new, diminished river. Where these gradients were capped by hard rocks, they became waterfalls.

22. There has not always been complete agreement on the origin of giant potholes. For an interesting and entertaining discourse on the history of thinking about them, see the web paper by Alan Morgan (https://uwaterloo.ca/wat-on-earth/news/glacial-potholes) referenced December 3, 2018.

23. The classification and genesis of soils is discussed in Buol et al. (2011).

24. According to Fisher, fresh carcass parts are actually heavier than water and should sink, but bacteria produce CO2 that accumulates in the meat, making it float. This requires that the meat be anchored in some way to stay below water level out of reach of predators.

25. Although the Bristle Mammoth provides good evidence for unusually early human occupation of North America, recent discovery of butchered mastodon remains in California raise the possibility of human occupation as early as 130,000 years ago, far earlier than commonly thought (Holen et al., 2018).

CHAPTER 3

Freezing the Continent

Pleistocene Glaciation of the Great Lakes Region

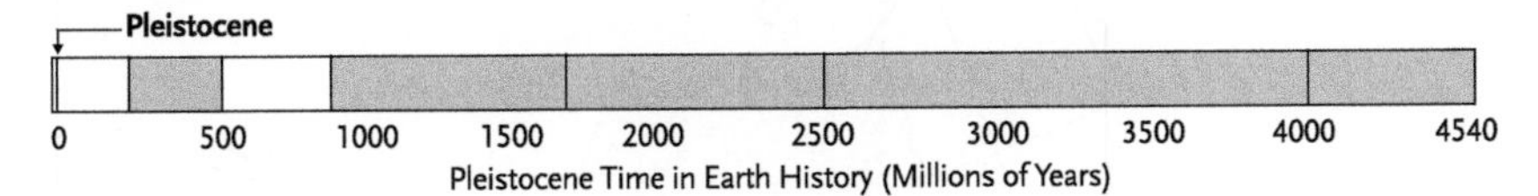

As we move further back into geologic history, the first things we encounter are glaciers, which advanced and retreated across the Great Lakes region as part of an icy onslaught on North America (fig. 3.1). It would have been an unwelcoming time for time travelers visiting our planet, much of which was covered by ice and snow.

There's no question that we would consider it a catastrophe if the Great Lakes region was covered again with a thick sheet of ice. But pause for a moment and think about what a contribution the glaciers have made to those of us who showed up after their last retreat. The glaciers formed the lakes that are the basis of much of our tourist industry, that support our commercial fishing industry, and that hold our huge supply of fresh water. Groundwater, which supplies homes and communities throughout the region, comes mostly from wells in the glacial sediments. Where glaciers scraped down to bare rock, we find and mine metal deposits to support our standard of living. Where they melted and left glacial deposits, we mine them to produce sand and gravel for highways and buildings, and we ship sand, gravel, and cement on barges all over the upper Midwest. And most of our good farmland and even our airports are located on the flat sediments that were deposited by the glaciers and the lakes that bordered them. Clearly, life would be very different for us if the Great Lakes region had not been covered with a thick sheet of ice.

But, do you really believe it? Did an immensely thick ice sheet cover the Great Lakes region? Is there any proof? For that, you have only to look at local boulder collections. In most towns, boulders as large as cars are a central attraction, and they are frequently coated with paint to celebrate sports victories and other events (figs. 3.2A, C in color section). The paint makes it hard to see that these are actually rocks. But scraping through the paint usually

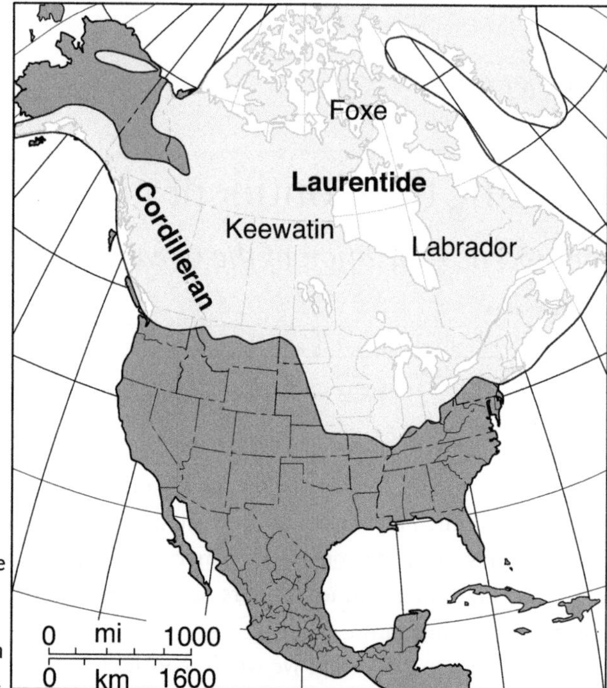

Figure 3.1. Laurentide and Cordilleran Ice Sheets showing the three main centers in the Laurentide sheet.

reveals strange rocks unlike anything in the immediate area. In many cases, we can show that the rocks came from far away. For instance, some boulders in Illinois consist of native copper that came from the Keweenaw Peninsula of Michigan, the only place in North America with rocks of this type. These rocks are called erratics because they are geologic errors that seem out of place. And it's not just the boulders that came from the far north. The 16.25-carat Eagle diamond, which was recovered from glacial gravels in Waukesha County, Wisconsin, probably came from diamond deposits near James Bay, which are discussed in chapter 7.

Glacial erratics are too numerous and widespread to have been pushed by a flood and far too heavy to have been blown by wind. The only way to transport them is by embedding them in ice. This seems farfetched even now, but imagine having the idea 200 years ago and suggesting that the Great Lakes region was invaded from the north by a layer of ice thick enough to carry boulders. This is exactly the problem faced by early Great Lakes geologists. In grappling with this conundrum, they were aided by European geologists who had studied glaciers and glacial deposits in the Alps. The most vociferous of these was Louis Agassiz, a professor of natural history in Switzerland, who became convinced that glaciers had covered much of northern Europe.

When he presented his conclusions to the Swiss Society of Natural Sciences in 1837, there was almost universal condemnation of the idea. But Agassiz persisted, and by the time he moved to the United States in 1847 to take up a post at Harvard, there was widespread acceptance of the revolutionary idea that glaciers had covered much of the continents in the northern hemisphere at some point in the recent past.

The Great Lakes region posed an additional complication for this line of thinking, however. There were no high mountains from which the glaciers might have advanced onto the plains. In Europe at least some of the erratic-strewn plains were in sight of the Alps and other high mountains that still contained glaciers. Even given the widespread erratics, more evidence was needed to convince skeptics that glaciers had covered the Great Lakes region, and it turned out to be in plain sight.

3.1. Evidence for Glaciers in the Great Lakes Region Comes from Landforms, Erosional Features, and Glacial Drift

The land surface alone tells us that something big happened. In the north, it consists largely of bare rock with a few patches of sediment consisting of silt, sand, gravel, cobbles, and even boulders. As you move southward, bare rock is harder to find, and the sediment layer becomes much thicker. The simplest explanation for this large-scale pattern is that something scraped off a lot of rock in the north and dumped it on the area to the south.

This was a continent-scale process, and satellite images give a good idea of its enormous scale. The land surface around Lake Nipigon north of Lake Superior (fig. 3.3A) is very rough and irregular, just like it would be if it had been scraped clean of soil. In contrast, land farther south has a smoother surface because it consists mostly of sediment released and then sculpted by the glaciers. The most amazing examples of this are the two long, smooth valleys that extend southward from Manitoba all the way to Des Moines and Sioux City, Iowa (fig. 3.3B). Take a good look at these valleys; they are exceptional features and will be mentioned again as we go along. The valley on the right, the Des Moines lobe, was formed by an enormous lobe or tongue of ice that extended all the way from Winnipeg to Des Moines as the ice was finally receding from the region. It was, in a way, a last gasp by the main ice sheet, which was by that time centered north of Winnipeg. A similar valley at the southwest the end of Lake Erie (fig. 3.3C) looks almost like it was scraped by a smooth-bottomed shovel, and it was the ice that did it. Notice also that the

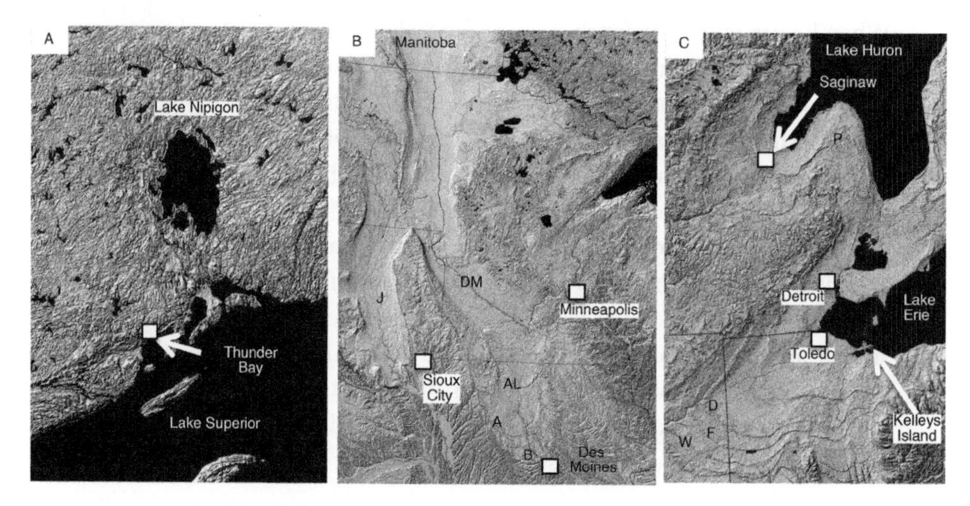

Figure 3.3. Shaded relief maps showing (**A**) the rough, rocky land surface around Lake Nipigon, where bare rock was left by glacial erosion; (**B**) the smooth land surface west of Lake Superior where the Des Moines (DM) and James (J) lobes of glacial ice deposited drift (terminal moraines in the Des Moines lobe include the Algona [AL], Altamont [A], and Bemis [B]); (**C**) glacial deposits in the Toledo and Saginaw valleys, including the Port Huron (P), Defiance (D), Fort Wayne (F), and Wabash (W) moraines. (Images from http://photojournal.jpl.nasa.gov/catalog/PIA03377.)

Des Moines and Lake Erie valleys have arcuate ridges at their ends. These are end moraines, which formed when the ice paused during its retreat; we will have more to say about them shortly.

By now you should be convinced that our land surface owes much to the glaciers. But earlier geologists did not have this advantage and had to look for evidence of glaciation as they walked around on the surface. In fact they found two lines of evidence for the passage of glaciers. One line relates to erosion of the land that the ice passed over. Erosion shows up as glacial striations and plucking. Striations form when gravel and cobbles embedded in the ice scrape the bedrock over which the glacier advanced. Very large striations are known as glacial grooves, and some of the world's largest are exposed on Kelleys Island in western Lake Erie (fig. 3.2B in color section). On these scraped surfaces, you can also see evidence that the ice plucked fragments of rock and included them in the moving ice. Some cobbles and pebbles that were embedded in the bottom of the moving ice sheet have smooth, flat (faceted) sides that were ground off as the ice moved over the underlying rock (fig. 3.2D in color section). Through grooving and plucking of this type, the ice sheet eroded the rock it passed over and transported it southward.

The other line of evidence for past glaciation comes from sediment deposited by the glaciers. All such glacial deposits, whether created by ice or the water released when ice melts, are known as drift, and they cover much of the Great Lakes region. Drift deposited directly by ice is known as till;[1] drift deposited by water is referred to as outwash. Figure 3.4 (in color section) shows what these deposits look like. Till deposits are unsorted and lack layering, whereas outwash deposits are sorted and layered.[2] Figures 3.4A and 3.4B (in color section) shows various types of till deposits, and figure 3.4D (in color section) shows a layered, sorted outwash deposit. The two types of sediment are transitional, of course, because melting ice forms water, which can redistribute some till deposits. Figure 3.4C (in color section) shows till that has experienced this sort of reworking by meltwater.

3.1.1. Till Deposits Create Many Land Forms, Including Moraines and Drumlins

Most of the common glacial landforms are summarized in the two diagrams in Figure 3.5, which were originally drawn by Bill Farrand. Moraines, which are the best-known till deposits, form because the movement of ice in a glacier (always flowing toward the front of the glacier) is not the same as the movement of the front of the glacier, which depends on a balance between the rate of ablation[3] at the front of a glacier and the rate at which the ice flows forward. If the flow of ice matches the rate at which the ice front melts, the front of the glacier will remain static and dump a pile of till, which forms an end or terminal moraine.[4]

Moraines stand out as prominent ridges in the satellite images in figure 3.3B,C. The sinuous ridges at the south end of the Des Moines lobe in figure 3.3B are moraines that formed when the flow of ice down the lobe from the ice sheet to the north was not strong enough to match the ablation at the south end of the lobe. As the long tongue of ice retreated toward Manitoba, it paused long enough to form the Bemis moraine, and then farther up the valley it paused again to form the Altamont moraine, and finally the Algona moraine formed when the ice paused near the Minnesota border. The lack of moraines north of the Algona indicates that the ice tongue did not pause again as it retreated northward. The Wabash, Fort Wayne, and Defiance moraines at the west end of Lake Erie (fig. 3.3C) are also terminal moraines that formed as the Erie ice lobe retreated toward the northeast and Lake Ontario.

We have spent so much time on moraines that you might think that there are no other important types of till deposits. In fact there are many.

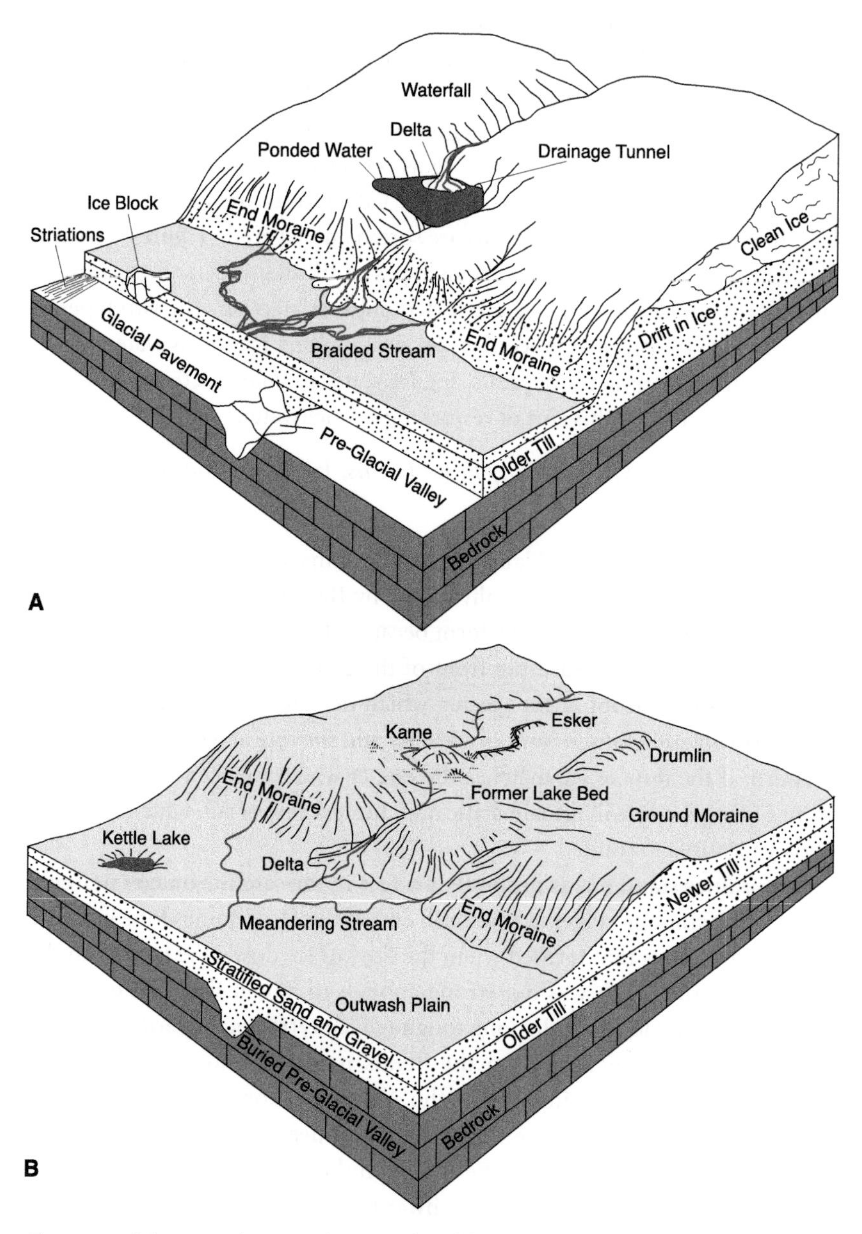

Figure 3.5. Schematic diagram showing glacial features (**A**) during advance and (**B**) after retreat of a continental glacier. (Modified from Farrand 1988.)

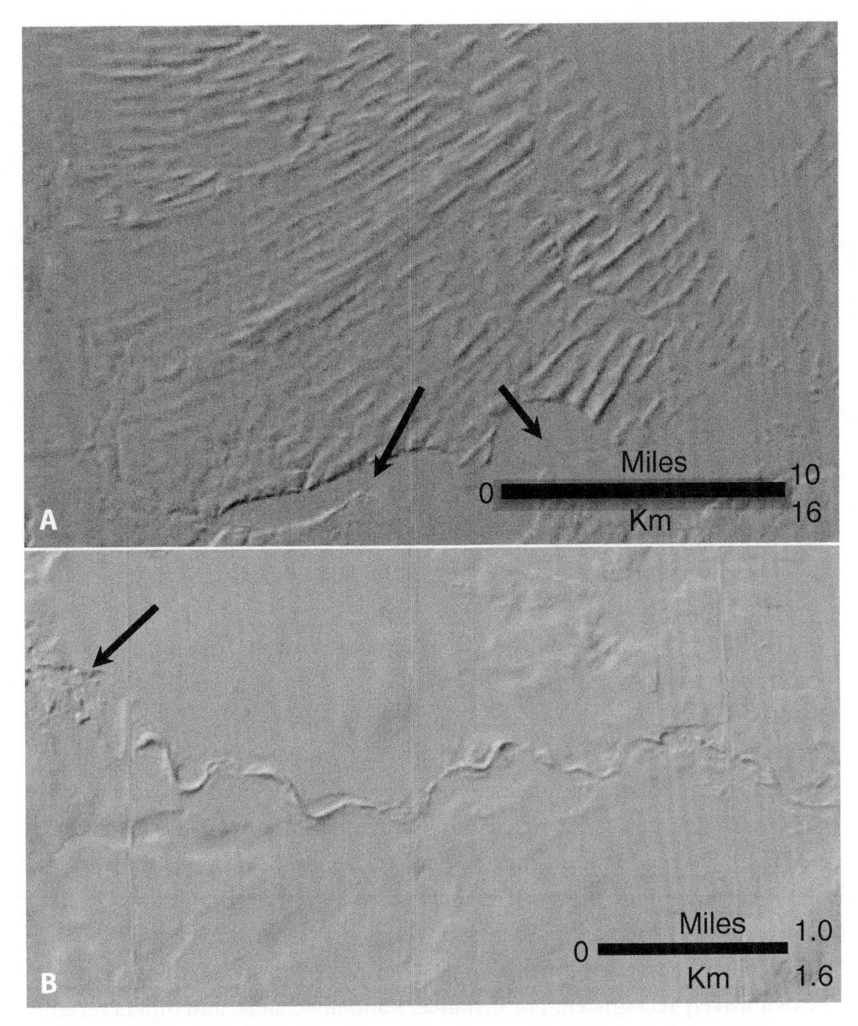

Figure 3.6. Digital topographic images of (A) part of the Wadena drumlin field between Sebeka and Wadena, Minnesota (the sinuous, flat area [*arrows*] is a river valley; (B) the Ripley esker near Fort Ripley, Minnesota (the disturbed area at the arrow is a mine that extracts gravel from the esker).

Till can also be deposited beneath the ice, forming thinner deposits known as till plains.[5] The movement of ice over this till can form unusual streamlined hills known as drumlins (fig. 3.6A), which are aligned in the direction of the flow of ice. One of the largest drumlin fields in the world is located around and even under the water in Lake Ontario, as described by Michael Kerr and Nick Eyles.

3.1.2. Outwash Deposits Include Outwash Plains, Kames, Eskers, and Kettles

Melting ice forms lots of water, of course, and it reworks the till to form outwash deposits. Reworking by flowing water separates grains of different sizes and densities, producing sediment that is sorted and stratified (layered) (fig. 3.4D in color section). The most common sediments of this type are found in outwash plains and lake deposits that form at the front of the glacier (large lakes, known as proglacial lakes, are common in front of glaciers because the weight of the glacier depresses the land surface, a point we will return to later). In some places, such as the Port Huron moraine in the Saginaw Bay area of southeastern Michigan, water-deposited sediment actually piled up against the front and side of the stationary glacial margin, forming ridges that look like moraines from the air but consist largely of sorted, stratified sediment. These water-lain moraines are known as heads of outwash, and their process of formation is explained by William Blewett and others in their summary of the glacial geology of Michigan.

Meltwater streams are also present beneath, within, and on top of some glaciers, as you can see in the diagram in figure 3.5. Sediment left by these streams forms eskers, which remain as long sinuous ridges when the ice melts. Figure 3.6B shows a digital image of an esker in northern Minnesota, which looks just like a meandering stream except that it is a ridge rather than a valley. Eskers are surprisingly common in the southern Great Lakes region, especially lower Michigan, but they are even more widespread in northern Ontario, where they are used in mineral exploration.[6] Kames are isolated hills, many of which are remnants of deltas that formed in proglacial lakes. Because of their relatively small size, they are often the last land forms to be definitively recognized. For instance, Randall Schaetzl and others recently reported newly discovered kames in central Michigan that mark the location of a previously unknown glacial lake. Kettle lakes are a special type of lake that forms in drift when blocks of ice melt, leaving a hole that is often filled later with water. Herbert Wright has suggested that the many lakes in the Minneapolis–St. Paul area formed by a similar process from ice that filled rivers draining south from the Lake Superior glaciers. Although most of the ice contained large amounts of sediment, parts with little or no sediment melted to form depressions that became the lakes.

3.1.3. *Drift Deposits Show the Paths Taken by Ice during the Final Retreat from the Great Lakes Region*

Now, take a look at figure 3.7, which shows the distribution of these various types of drift deposits in the Great Lakes region. There are two parts to the figure. In figure 3.7A, you see the distribution of major terminal moraines that formed during the last retreat of the glaciers. Moraines to the south, like the Shelbyville moraine, are the oldest, and those to the far north, like Lac Seul and Sachigo, are the youngest. Some of the moraines form tongues or lobes that protrude down the Great Lakes basins, suggesting that they had something to do with the valleys the lakes occupy today. Figure 3.7B shows the location of eskers and drumlin fields; you can see that eskers, including the Munro esker, are much more abundant in the far north, whereas drumlins, including the field around Lake Ontario, are more common in the south where there was more sediment to be pushed around and streamlined by the ice. If you want to see some of these features firsthand, you are in luck; drift deposits are the focus of many of our parks and recreation areas (box 3.1).

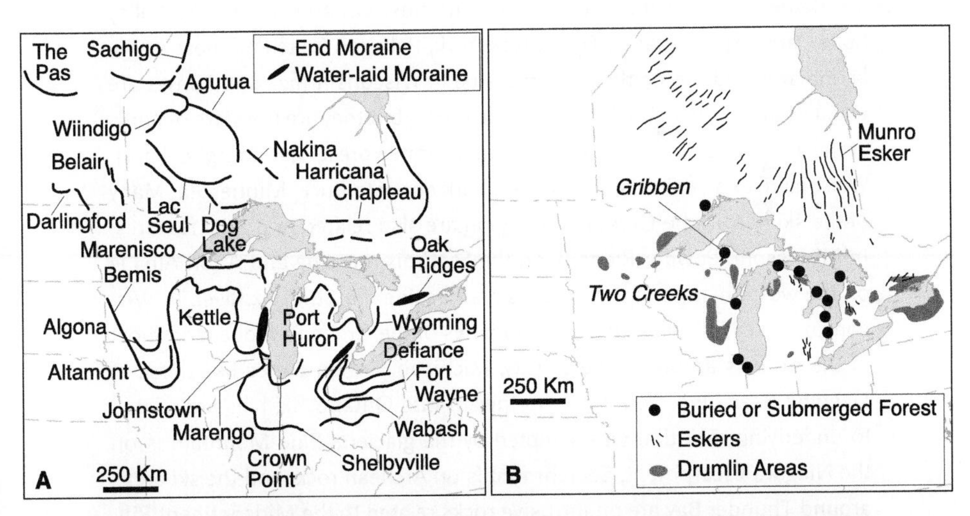

Figure 3.7. Important Pleistocene glacial features in the Great Lakes region include (**A**) moraines; and (**B**) buried and drowned forests, drumlins, and eskers. (Compiled from Zoltai 1965; Prest et al. 1968; Sado and Carswell 1987; Barnett 1992; Pregitzer et al. 2000; Larson and Schaetzl 2001; Soller 2001; Fulletron et al. 2003; Mickelson and Colgan 2003; Dyke 2004; Hunter et al. 2006; Blewett et al. 2009; Curry et al. 2011; Larson, 2011.)

BOX 3.1. PARKS AND GLACIAL GEOLOGY

National and regional parks in the Great Lakes region provide good opportunities to see glacial deposits and processes firsthand. For a long walk with lots of glacial features, try the Ice Age National Scenic Trail, which wanders through much of Wisconsin. A small part of that trail overlaps the Glacial Drumlin State Trail between Cottage Grove and Waukesha, which is a great place to see drumlins, as is the Drumlin Trail in Ferris Provincial Park in Ontario. Moraines can also be seen at the Glacier Ridge Metro Park in Columbus, Ohio, and the Moraine Nature Preserve near Valparaiso, Indiana, where the large Valparaiso moraine loops around the south side of Lake Michigan. In and around Camden State Park in southwestern Minnesota, you can see topography of this type on the Bemis and Altamont moraines, which were formed by the Des Moines lobe. Another park with varied glacial geology is Sleeping Bear Dunes National Lakeshore near Empire, Michigan, where you can see older glacial deposits, including moraines, as well as the perched sand dunes mentioned in chapter 2. Good places for kettle lakes are Kettle Lake Provincial Park near Timmins, Ontario; Chenango Valley State Park near Chenango Forks, New York; Pokagon State Park near Angola, Indiana; and the Kettle Lakes moraine in Wisconsin (fig. 3.7). Kames are usually hard to distinguish as special features, but they are the stars of parks at the ambiguously named Kettle Moraine State Forest near Campbellsport, Wisconsin, and at Glacial Lakes State Park near Starbuck, Minnesota. Many of the ski hills in the Great Lakes region are also related to glacial features. In Minnesota, Powder Ridge is on the St. Croix moraine and Afton Alps is on the western edge of the driftless area beside the St. Croix River. In Wisconsin, Cascade Mountain is on moraines of the Green Bay lobe, and most of the ski hills around Traverse City, Michigan, are on dunes like those at Sleeping Bear. Once you move northward into Ontario, ski hills are related to underlying rock that was sculpted by the glaciers. Blue Mountain is on the Niagara Escarpment, Searchmont is on Archean rock, and the ski areas around Thunder Bay are on intrusive rocks related to the Midcontinent Rift.

By mapping the distribution of drift deposits, we can reconstruct the glacial history of the Great Lakes region. The drift deposits are clearly telling us that the global climate was very cold in the past. But how old are they and when was it so cold? This is our first real encounter with geologic time. Are these deposits millions of years old or thousands or even hundreds? It matters. If we are to understand global climates, we have to reconstruct the glacial deposits in both time and space. That brings up the question of just what we can learn from the Great Lakes region and how it relates to the rest of the world.

3.2. Pleistocene Glaciation Was Part of Global Climate Cycles That Are Seen in the Marine Isotope Stages (MISs) of Ocean Sediments

Recall from chapter 1 that Earth's climate has fluctuated between icehouse and greenhouse conditions and that we are currently in an icehouse period. Icehouse periods are much less common than greenhouse periods, and they do not have a constant cold condition.[7] The global glacial record shows that ice advanced and retreated repeatedly during our current icehouse period, and we are all waiting to see whether the present interglacial period will give way to another advance of the ice or to a gradual return to greenhouse conditions. To deal with either outcome, and to understand Earth's climate better, we need to learn as much as we can about past glacial history, and that is where the record of ice deposits in the Great Lakes becomes important to all of us.

Glaciers that advanced over the Great Lakes region were part of the large ice sheets that covered much of North America, Europe, and Asia. Because of their global importance, this glacial period is given its own name, the Pleistocene epoch. According to recent age measurements (discussed a little later), it began at about 2.588 Ma and ended at about 0.0117 Ma (11,700) years ago, at the start of the Holocene epoch, the interglacial period we are experiencing today.[8] As discussed in chapter 2, some people use the term Anthropocene for all or part of the Holocene to emphasize the importance of human activity during this time.

Ice sheets advanced and retreated many times during the Pleistocene epoch, and we would like to know the entire record in order to reconstruct the history of our current icehouse period. It is hard to find the entire record on land, however, because younger glaciers overrode the older glacial deposits, destroying them. We can get around this by looking at sediment in the ocean and layers of ice in the polar ice caps. Both of these layers are relatively undisturbed and show the sequence from the earliest ice ages to the most recent ones. We can even see how temperatures fluctuated from layer to layer by analyzing the relative abundance of two isotopes of oxygen that are preserved in the ice and the shells of small marine animals that grew in the ocean.[9]

Detailed oxygen isotope measurements in successive layers of ocean sediment, fossils, and polar ice that were deposited over the last few million years show that global ice volumes have varied cyclically from cool to warm many times during this period. They also show that average global ice volume increased and associated temperatures decreased during this period. Finally, the amplitude or range of the cycles from cool to warm periods also increased, especially beginning about 2.5 million years ago. This increased amplitude has been interpreted to reflect greater degrees of ice accumulation and melting related to the growth and destruction of continental glaciers.

The ocean and the ice cap record of temperature changes over the last 2.5 million years has been divided into a series of cycles known as Marine Isotope Stages (MISs), and these stages have been linked to the episodic advances and retreats of the glaciers on land. The cyclic nature of these changes supports the early suggestion of the Serbian geophysicist Milutin Milanković that Earth's orbital and rotational patterns, known now as Milanković cycles, exert some control over global climate.[10] Thus, one of the goals of the study of glacial geology is to clarify the correlation between MISs and the advances and retreats of the continental ice sheets and to use this to reconstruct the history of global climate. To do this in the Great Lakes region, we need information about the location and age of continental glacial deposits that can be compared to the marine isotope record.

3.3. Drift Deposits Tell the Glacial History of the Great Lakes Region

3.3.1. Great Lakes Drift Deposits Can Be Compared to MISs

Our first test of the correlation between MISs and continental glaciation is simply whether Great Lakes glaciation began about 2.5 million years ago, as suggested by the marine isotope record. The answer is a qualified yes. The oldest known Pleistocene drift deposits in midcontinent North America, which were reported from Missouri by Greg Balco and Charles Rovey, consist of till that is about 2.4 million years old. Older deposits are known elsewhere, however. Debris that looks like it was transported by ice rafts (which probably requires a glacier to make the raft) has been collected in 3- to 4-million-year-old sediment in the ocean, and 3.5-million-year-old glacial deposits on land were reported by Cunhai Gao and others from just east of James Bay. So 2.5 Ma might mark the time when the ice sheet was large enough to cover significant parts of North America.

The oldest glacial deposits in North America are widespread in Iowa, as well as Nebraska and Kansas, and they peek out along the Ohio River from beneath the youngest deposits (fig. 3.8A). Originally, it was thought that these old deposits in the Great Lakes region could be divided into two old glacial ice advances (episodes), known as the Kansan and Nebraskan, and that they were followed by more recent Illinoian and, finally, Wisconsin ice advances, in other words that there were four major ice advances. However, detailed study has shown that the older ice advance record is much more complex. Recent age measurements on the old deposits range between about 2 and 0.3 Ma (300,000 years) (table 3.1). As a result, the terms Kansan and Nebraskan are not in wide use, and the older deposits are referred to simply as pre-Illinoian in age.

Glacial deposits that have formed since about 300,000 years ago are sufficiently widespread and continuous in the Great Lakes region to outline a clearer history. These deposits are especially well exposed in Illinois and Wisconsin, where they have been divided into the Illinoian glacial episode, followed by the Sangamon interglacial and finally the most recent Wisconsin glacial episode, which retreated from the Great Lakes region at the beginning of the Holocene about 11,700 years ago. Even during these episodes, the climate varied greatly. For instance, during the Wisconsin glacial episode large parts of the southern Great Lakes were free of ice between about 64,000 and 25,000 years ago. The final retreat of the ice began about 25,000 years ago and consisted of a sequence of episodic retreats that are recorded by the terminal moraines.

To test whether these advances and retreats correlate with the MISs (and to try to relate Great Lakes glaciation to global glaciation), we have to measure the age of the glacial deposits. Relative ages of buried deposits can be determined by simply looking at the sequence of layers in a pit that cuts downward through a sequence of glacial layers; the youngest ones are on top. But absolute ages are needed. This is usually done by means of isotopic analysis of buried plant material or quartz, as explained in box 3.2. Not just anything can be analyzed; the sample must have a clear relation to the local glacial history and a composition that is suitable for age measurements. A log buried in thinly layered sediment would be a good candidate for age analysis. By doing some geologic mapping around the sample, it might be shown that the layered sediment was deposited in a kettle or glacial lake that formed in front of a newly formed moraine. Because moraines mark the still-stand that occurred during the retreat of a glacier, this age would represent a maximum age for one stage in the retreat (the log might be older than the sediments, but it cannot be younger).

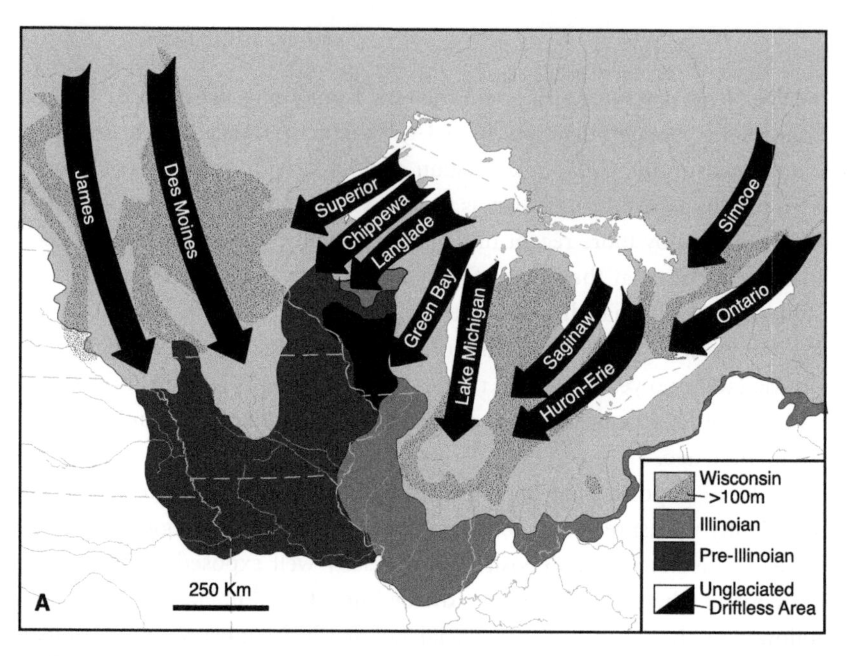

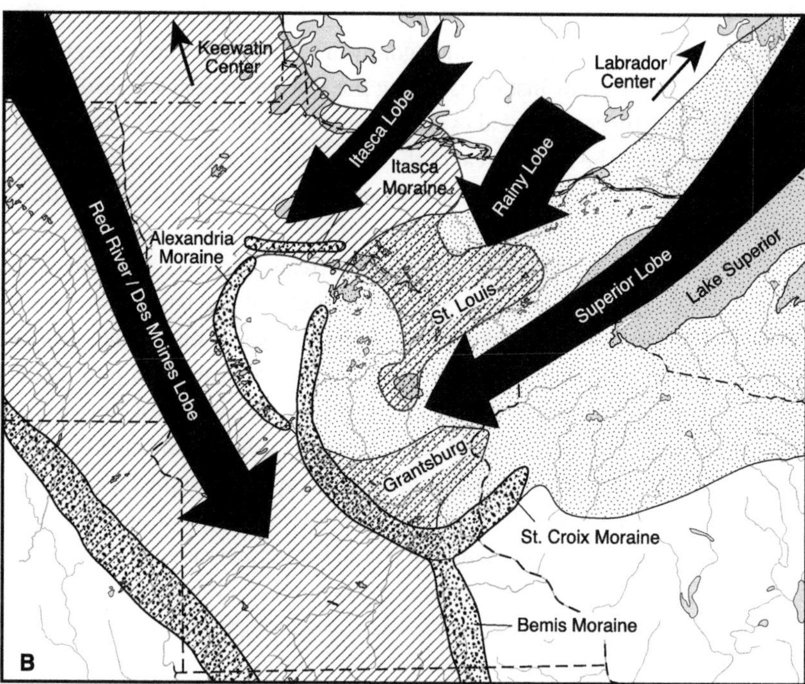

Figure 3.8. (**A**) Ages and thickness of glacial deposits in the Great Lakes region (black arrows show the location of Wisconsin-age ice streams); (**B**) multiple Wisconsin-age ice lobes in Minnesota. (Compiled from Fullerton et al. 2003; Mooers et al. 2005; Knaeble 2006; Jennings and Johnson 2011; Quaternary Geologic Map of Minnesota.)

Table 3.1. Maximum Age Ranges of Late Pleistocene Glacial and Interglacial Stages and Ages of Pre-Illinoian Deposits in the Great Lakes Region

Glacial Period	Age Range (Ka)	Comments
Wisconsin Glacial Stage	79 to 11	
Sangamon Interglacial Stage	132 to 79	Ended at 115 Ka in Ontario
Illinoian Glacial Stage	302 to 132	

Source: Based largely on Cambridge Global Stratigraphical Correlation Table.

Note: In Illinois and some other areas, the warm period that preceded the Illinoian ice advance is known as the Yarmouthian interglacial. Pre-Illinoian deposit ages are~300, ~450, ~560, ~650, ~830, ~900, ~1200, ~1700, ~1800, ~1900, and ~2500 Ka.

Numerous measurements of the age of older drift deposits in the Great Lakes region have shown that glacial advances and retreats correlate relatively well with the MIS record.[11] For a more detailed understanding of the history of a glacial episode, we need deposits that have not been disturbed by later advances. For that we have to turn to the last part of the most recent episode, the Wisconsin.

BOX 3.2. MEASURING THE AGE OF GLACIAL DEPOSITS

Carbon-14 Age Measurements

The most common method used to determine the age of glacial deposits is based on the decay of carbon-14 (^{14}C), which is produced in the atmosphere by cosmic ray bombardment of ^{14}N. Carbon-14 can combine with oxygen to produce CO_2, which is taken up by plants. This ^{14}C is radioactive with a half-life of about 5,730 years. Once a plant dies, it ceases to take in CO_2 and its ^{14}C begins to decay. The age of a sample can be determined by comparing its concentration of ^{14}C to that of living matter. The ^{14}C method can be applied to plant or animal remains as old as 50,000 to 60,000 years, which makes it a perfect tool for the investigation of late Wisconsin glacial history. All you need is a fragment of organic matter, whether tree or mammoth, that was living at the time and was then buried in a glacial deposit. Be careful in using these ages, however, because they can be quoted in either calendar or ^{14}C years, which differ considerably for reasons we cannot delve into here. Programs to convert between the two ages are available on the internet. All ages quoted here are calendar years.

Aluminum-26 and Beryllium-10 Age Measurements[12]

Illinoian and earlier glacial deposits are too old for the ^{14}C method. Fortunately, another method that also depends on cosmic ray bombardment has been developed. In this case, cosmic rays passing through the atmosphere create neutrons that bombard rocks and minerals, creating a wide range of nuclides, some of which are radioactive. Two of the most useful nuclides are ^{10}Be, with a half-life of 1.387 million years, and ^{26}Al, with a half-life of 0.717 million years, which are produced by bombardment of oxygen and silicon, respectively. Quartz, one of the most common minerals in the crust, consists of silicon and oxygen, making it a perfect medium for these measurements. Measurements can be made in two ways. The simplest way is to derive an age from the concentration of ^{10}Be and ^{26}Al in the quartz by comparing it to the estimated rates at which they are produced during exposure. On the other hand, if the quartz was covered and protected from further bombardment shortly after it was exposed (maybe by a later ice advance), the relative rates of decay of ^{10}Be and ^{26}Al can be used to determine its age. Somewhat similar methods, known as optically stimulated luminescence and thermoluminescence, also measure time since the sample was exposed to sunlight.

Other Methods

Paleomagnetic measurements, which are discussed further in chapter 5, can be used to determine the orientation of the magnetic pole at the time glacial deposits formed. Deposits in which the north and south poles are reversed are older than 780,000 years. Additional methods are based on the decay of amino acids and on fossil remains. These and other less widely used methods have permitted determination of the ages of glacial events throughout the 2.5-million-year Pleistocene epoch.

3.3.2. *Wisconsin Episode Drift Was Deposited by Ice Streams from the Laurentide Ice Sheet*

The large ice sheet that covered North America during Pleistocene time is known as the Laurentide Ice Sheet. Deposits that it left during its last major advance, the Wisconsin stage, are widespread and well enough preserved to provide an estimate of the volume and extent of the ice sheet at that time. At its maximum, it contained about 18×10^6 km^3 of ice, about 20 times the volume of water in Hudson Bay. It reached a maximum thickness of more than 3 km and covered an area that measured 4200 km east-west and 3700 km north-south.

The Laurentide Ice Sheet grew by accumulation of snow in the central part of northern Canada and was destroyed by ablation at its southern extremes. In a recent summary of the Laurentide Ice Sheet, Martin Margold and others showed that it consisted of three centers of ice accumulation surrounding Hudson Bay, the Labrador center on the east, the Keewatin on the west, and the Foxe on the north.[13] Laurentide ice did not advance uniformly; instead it extended outward as long tongues known as ice streams or ice lobes such as the Des Moines and Lake Erie lobes discussed above (fig. 3.3). At least 117 of these ice streams or lobes have been identified around the margins of the Laurentide Ice Sheet, based largely on the distribution of end and terminal moraines. David Mickelson and Patrick Colgan reviewed the history of the main ice lobes along the southern margin of the Laurentide Ice Sheet in the Great Lakes region and showed that early ice lobes followed river valleys in the preglacial topography. As time passed, however, the lobes modified the topography and created their own pattern of ice streams or lobes. For instance, the ancestral Mississippi River originally made an eastward bend through central Illinois, but it was eventually pushed westward toward its present-day path by strong ice advances and deposits from the Michigan and Huron-Erie ice lobes.

Figure 3.8A shows the distribution of Wisconsin-age ice streams or lobes that moved into the Great Lakes region. In the western part of the region, the Des Moines and James lobes were relatively isolated. Although they moved far to the south, they did not coalesce into a single ice front, and, as we will see shortly, they were a very late stage of the overall Wisconsin advance.

The pattern of ice advances in the east and central parts of the Great Lakes region was much more complex and involved many more lobes, most of which came from the north and northeast and moved down the valleys of the modern Great Lakes. Many lobes coalesced into larger ice masses and formed deposits between them. In the east, one ice stream moving down Lake Huron into the western end of Lake Erie merged with another moving down Lake Ontario into Lake Erie. Between them, smaller lobes moving through Georgian Bay and the Lake Simcoe area were stopped by high ground north of Toronto. The prominent Oak Ridges moraine north of Toronto formed between the Simcoe and Lake Ontario lobes. In Wisconsin, the Michigan and Green Bay ice lobes flowed together to form the long Kettle Moraine that parallels the east side of the state, and in the northwest the Langlade, Wisconsin Valley, Chippewa, and Superior ice lobes formed an almost uniform front (fig. 3.8A), as described by Kent Syverson and Patrick Colgan. Similar deposits formed where the Huron and Saginaw lobes converged in Michigan, but the champion area of dueling ice lobes was in central Minnesota (box 3.3).

BOX 3.3. DUELING ICE STREAMS IN MINNESOTA

The western part of the Great Lakes region was the scene of a battle between ice streams that were vying for the right to cover Minnesota with drift (fig. 3.8B). As the Laurentide Ice Sheet approached its maximum southern limit, local topography allowed portions of the ice to form lobes that advanced and retreated episodically, forming end moraines. Although some of the ice lobes may have been active at the same time, a general sequence has been recognized. According to Howard Mooers, it started with the Rainy ice lobe, which advanced and deposited the Alexandria moraine. Ice then retreated, and the next major phase is marked by the Itasca and St. Croix moraines, formed by the Itasca and Rainy/Superior lobes, respectively. The Rainy ice lobe came from the northeast, moving over the Rainy River district of northern Minnesota, whereas the Superior lobe flowed down Lake Superior. Finally, the Des Moines ice lobe, moving down from Manitoba and the Keewatin ice center, pushed much farther south, depositing the Bemis, Altamont, and Algona moraines. Perhaps to confuse things, the Des Moines lobe also detoured eastward in the St. Louis and Grantsburg lobes, which buried earlier moraine deposits in northern and central Minnesota. One indication of the power of these ice advances can be seen at the Powder Ridge ski area near Kimball, Minnesota, where a borrow pit at the top of the hill exposed till and underlying Cretaceous sediments that had been deformed into thrust sheets. Efforts to unravel the complex history of ice advances in Minnesota were aided by the fact that ice coming through Manitoba eroded Paleozoic limestone, which gave the till a gray color, whereas ice coming down Lake Superior from the east eroded red sandstones and shales from the Midcontinent Rift.

High areas were effective barriers to flow as long as the ice was not too thick. The Lower Peninsula of Michigan was protected by the Niagara Escarpment, mentioned in chapter 2 as the site of so many waterfalls. The escarpment deflected the Green Bay, Michigan, and Huron-Saginaw ice streams and caused them to concentrate their erosion along softer shales and evaporite (salt) layers carved into the low valleys that now contain Lakes Michigan and Huron.[14] Similar highlands impeded the southward flow of the Lake Erie lobe, forcing it to the west. High areas underlain by hard Precambrian rocks in northern Wisconsin also deflected the Chippewa, Langlade, and older ice streams that flowed southward from Lake Superior and probably contributed to protection of the driftless area where Wisconsin, Minnesota, Iowa, and Illinois come together (fig. 3.8A).

The amazing driftless area, which is about the size of the state of New Jersey, appears to have escaped all the Great Lakes ice advances. The topography of its land surface is quite different from that found elsewhere in the Great Lakes region, with obvious river valleys that are typical of unglaciated topography to the south. In fact the driftless area is so hilly that you can locate it simply by looking at road maps of the upper midwest; roads in the driftless area are irregular, very unlike the rectilinear pattern on the surrounding flat, glaciated areas. Although it is tempting to imagine that the driftless area was completely surrounded by a high wall of ice during some part of its history, this was not the case. For one thing, glaciers on land ended in slopes rather than the cliffs that are seen in some tidewater glaciers (like the towering Ross ice shelf in Antarctica). More important, although drift deposits surround the area, they are of many different ages, none of which are continuous enough to indicate a complete circle of ice. Instead, openings to unglaciated areas probably persisted throughout the glacial periods. Because these openings allowed stream erosion to continue in the unglaciated area during the Pleistocene, the present topography is not entirely preglacial.

3.3.3. Detailed Information about Glacial Advances and Retreats Came from Excavations

Although the main highs and lows in the marine isotope record are clear enough, transitions between them are irregular, with numerous second- and third-order highs and lows that were associated with smaller-scale advances and retreats of individual ice streams. Sawtooth-type advances and retreats happened on a global scale, and they can be seen in the Great Lakes region in excavations that cut through the latest glacial deposits.

One of the best such records, at Two Creeks, Wisconsin (fig. 3.7B), along the shore of Lake Michigan, was described by Wallace Broeker and Bill Farrand. At the bottom of the sequence of glacial deposits at Two Creeks is a till that clearly represents a glacial deposit. This is overlain by fine-grained clay with very thin layers known as varves, which was deposited in a proglacial lake that formed in front of glaciers. The presence of a proglacial lake indicates that the glacier had retreated farther north, leaving a lake behind. Above the varved clays is a layer containing remains of a forest with logs, twigs, bark,

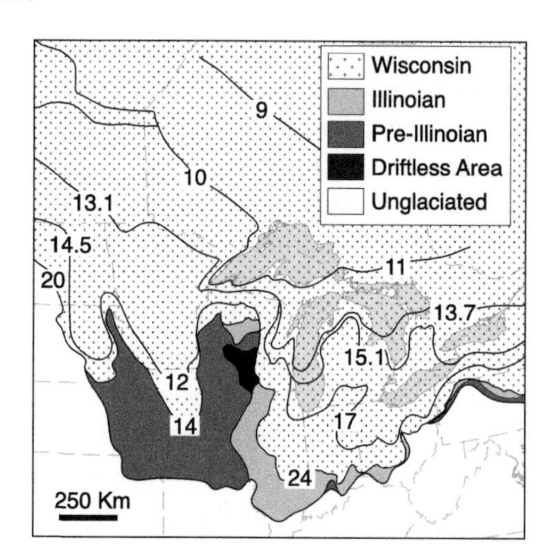

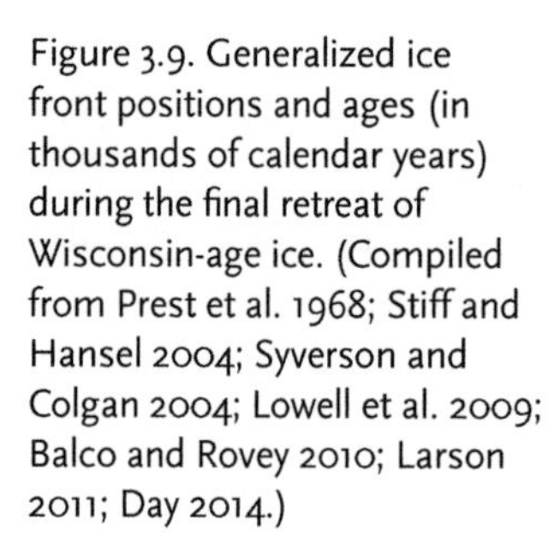

Figure 3.9. Generalized ice front positions and ages (in thousands of calendar years) during the final retreat of Wisconsin-age ice. (Compiled from Prest et al. 1968; Stiff and Hansel 2004; Syverson and Colgan 2004; Lowell et al. 2009; Balco and Rovey 2010; Larson 2011; Day 2014.)

and even cones of spruce and pine trees. Carbon-14 analyses of these remains give an age of about 13,600 years, indicating that by this time the ice had retreated far enough north to allow the lake to drain and a forest to form. The forest bed is overlain by a layer of bedded sand deposits, indicating the return of a proglacial lake, and these deposits are overlain by a second till that was deposited when the glacier returned to the area.

The carbon-14 age places this advance-retreat-advance cycle in the latter part of the Wisconsin episode. Numerous observations of this type, including other buried forests (fig. 3.7B, box 3.4), form the basis for figure 3.9, which shows the timing of the final Wisconsin-age glacial retreat from the Great Lakes region. The large end moraines and heads of outwash marked the times when the ice front paused during its retreat. In some areas, such as the western end of Lake Erie, the end moraines are widely separated and easily identified. In other areas, such as central Michigan and northern Wisconsin, the glacier margin did not move as far between stops, leaving several overlapping moraines that are harder to distinguish. You can see that the ice front in the eastern and central part of the region was near the Ohio River about 24,000 years ago but that it had retreated out of the Lower Peninsula of Michigan by about 14,000 years ago. To the west, however, the Des Moines and James lobes were still far to the south at that time. By about 10,000 years ago, the ice had left the Great Lakes region and was retreating toward the James Bay area. As the ice retreated, it was followed by the proglacial lakes, which formed by meltwater collecting along the ice margin.

BOX 3.4. ANCIENT FORESTS AND HUNTERS

During retreats of the glaciers, the climate warmed, and forests grew across the area. When the ice advanced again, the forests were buried beneath lake and drift deposits. Several drowned forests have been found by divers exploring the lakes, and others have been unearthed beneath sediment on land (figs. 3.7B, 3.10A, B [in color section]). These forests were particularly widespread during a period known as the Stanley low-stand, when water levels in the ancestral Great Lakes were about 100 m lower than today. Early humans hunted in these forests, as indicated by submerged stone features that have been found in Lake Huron. These consist of long rows of stones similar to the ones known to have been used by ancient hunters on land to channel caribou (fig. 3.10C in color section). As explained by John O'Shea and Guy Meadows, they are located along a ridge of high ground beneath Lake Huron offshore from Alpena that would have been exposed during the Stanley low-stand.

3.4. Lakes Were an Important Part of the Glacial Legacy

3.4.1. The Evolution of the Lakes Was Controlled by the Bedrock That Was Eroded, Isostatic Uplift, and the Level of Outlets

Most glacial lakes form as water accumulates in the irregular topography left by retreat of a glacier. The irregular topography forms in two main ways: ice damming and ice scour. Ice-dammed lakes, which are called proglacial lakes because of their position at the front of the ice, form because the weight of the glacier depresses the underlying crust, causing the land surface to slope toward the front of the glacier. Water released by the melting of the glacier ponds against the front of the glacier and adjacent moraines until it reaches a level high enough to find an outlet through which it can flow away. These lakes move along with the glacier as it recedes and the land rises.

Ice scour forms longer-lived lakes because it excavates depressions that are deep enough to accumulate water long after the glaciers are gone. The Great Lakes were formed by this type of ice scour when the large ice streams flowing off the southern margin of the Laurentide Ice Sheet cut into the bedrock, making basins that retained water.[15] The degree of scour is controlled in part by the ease with which the glacier can cut through the rock. This depends both on the hardness of the rock and the direction of flow of the glacier relative to the orientation of the underlying rocks. As we will see in the next chapter, these factors account for the fact that we have six lakes with very different depths and orientations.

Once a glacial lake is formed, its subsequent history depends on the supply of water from precipitation or melting, as well as two other factors. One is isostatic rebound, which refers to the rise of land level in response to removal of the weight of the ice sheet. This is a major factor. The weight of Laurentide ice depressed the land beneath the central part of the ice sheet by hundreds of meters and by smaller but still very significant amounts to the south. As the ice was removed during the late Pleistocene the land began to rebound, a process that continues today. Evidence of the rebound can be seen in the raised shorelines of Hudson Bay, as well as in the Great Lakes, as discussed further below. The other factor is the outlets, or sill levels, through which water can flow out of the lake.

These two factors operate at completely different rates. Isostatic rebound involves the flow of rock in Earth's mantle that was originally displaced by the weight of the ice. Flow is very slow, usually producing elevation changes of millimeters per year. In contrast, sills can be blocked or unblocked by flow and melting of ice, which is much faster, with rates of tens to hundreds of meters per year. Thus, changes in sill levels account for most of the short-term changes in lake levels whereas isostatic rebound is more effective over the long term.

The impact of isostatic uplift and sill levels on the evolution of the Great Lakes can be mapped on the ground by locating the ancient shorelines of the ancestral Great Lakes (box 3.5). Studies like the ones reported by Michael Lewis, Scott Drzyzga, and others for the ancestral lakes can be aggregated to determine the present levels of ancient beaches and the way they have changed through time. In the small inset in figure 3.11, you can see that beaches for two of the ancestral Great Lakes, Algonquin (now Lake Michigan) and Whittlesey (now Lake Erie), rise to higher elevation as you move northward. Something is wrong here, of course, because a beach should have the same level all the way around a lake. But, because the northern parts of the two lakes have moved upward more than the southern parts have, the beaches increase in elevation as you move northward.

The main part of figure 3.11 shows how lake level histories can be reconstructed from these observations. This is a complex diagram that does not "speak" to you immediately. It does show two important features, however. First, the gradual upward slope of the lake levels from right to left (i.e., from the past toward the present) shows the slow isostatic uplift of the area that resulted from removal of the ice. Second, the jagged ups and downs of the

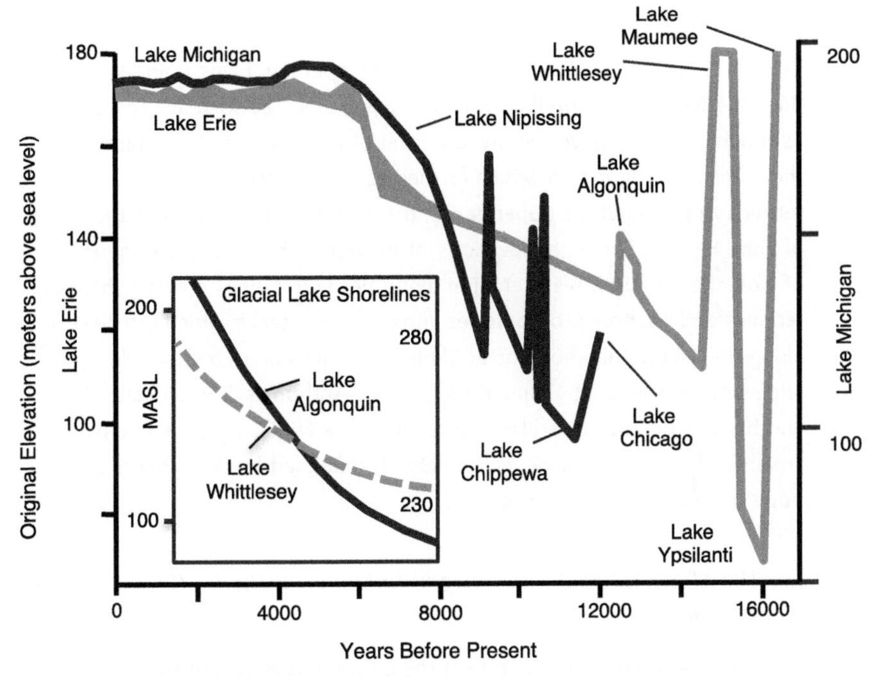

Figure 3.11. Changes through time in lake levels of ancestral Lakes Erie and Michigan. These curves show original water levels that have been corrected for isostatic rebound. The inset shows change in elevation of shorelines of ancestral Lake Erie (*left scale*) and Lake Michigan (*right scale*) moving northward and northeastward, respectively. (Modified from Larson 1987; Barnett 1992; Lewis et al. 2005, 2012; Fisher et al. 2012.)

lake levels show the rapid and abrupt changes that took place in water levels as different sill levels became available to control outflow of water.

BOX 3.5. SHORELINES OF GLACIAL LAKES

Beaches and terraces marking the ancient shorelines are widespread in the Great Lakes (fig. 3.12). Some of the most interesting shoreline features are found on Mackinac Island in the northern part of Lake Huron, and they have been described by Ron Sage in his 2006 guidebook to the geology of Mackinac Island. The island is at the far northwest end of Lake Huron (which has a water level of about 177 m [580 ft]), and it consists of limestone bedrock with glacial striations covered by a thin layer of till. Most of the limestone bedrock consists of a rubble of collapsed rock known as the

Mackinac Breccia. You can see it on the island at Eagle Point Cave where the breccia was eroded into a small cavelike opening. Eagle Point is part of the Lake Nipissing shoreline, which formed about 4000 years ago and is as much as 17 m above the present level of Lake Huron. The shoreline of the older Lake Algonquin, which formed about 12,000 to 11,000 years ago, forms a ring around the upper part of the island up to 70 m above the level of Lake Huron. The high elevations of these beaches today are the result of isostatic rebound. Wave erosion along both shorelines created shelflike terraces and offshore promontories known as sea stacks, which can be seen along the Algonquin shoreline at Skull Cave and Sugar Loaf and along the Nipissing shoreline at Pulpit Rock and St. Anthony's Rock. Castle Rock, another prominent, towerlike feature along the Nipissing shoreline on the "mainland" near St. Ignace, is not actually detached from the shore and is not a true sea stack (fig. 3.12B).

3.4.2. The Modern Great Lakes Began to Evolve as the Ice Retreated for the Last Time

The Great Lakes we see today reflect the combined effects of these processes during the last major ice retreat. At least 30 different ancestral ice-margin lakes formed in the Great Lakes region, as described most recently by Kevin Kincare, Grahame Larson, and Randall Schaetzl. They reflect an amazing sequence of glacial advances and retreats, with lakes forming, merging, separating, and being overrun by glaciers as they advanced again. An important legacy of the lakes is the sediments that cover their original beds. The lake deposits are composed of fine-grained material that forms unusually flat plains, in contrast to the drift deposits or glacially eroded rocks that surround them. Much of the good farmland around the margins of Lakes Huron, St. Clair, and Erie is on the flat plains of outwash and lake sediment from the older ancestral lakes.

Among the earliest of the late Wisconsin-age lakes were glacial Lakes Chicago and Maumee at the ends of present-day Lakes Michigan and Erie, respectively (fig. 3.13A). Glacial Lake Maumee grew in size and invaded the Huron Basin where it reached an outlet that crossed the Lower Peninsula of Michigan known as the glacial Grand River. This allowed it to drain into the southern end of Lake Michigan (then glacial Lake Chicago) and eventually into the Mississippi River system (fig. 3.13B). Ice retreat after glacial Lake Maumee time opened a lower outlet into the Lake Ontario Basin and the St. Lawrence system, causing a huge drop in lake levels, forming glacial

Figure 3.12. Preserved beach features: (**A**) Wave-cut terrace, southern Lake Superior; (**B**) Castle Rock, Michigan; (**C**) Terraces, Mackinac Island, Lake Huron, and (**D**) Little Traverse Bay, Lake Michigan.

Lake Ypsilanti and a related lake in the Michigan Basin. The upper and lower glacial Great Lakes were isolated from each other at this point, with the upper lakes flowing through Georgian Bay into the St. Lawrence River and the lower lakes flowing eastward through the Niagara River (fig. 3.13C). When those outlets were closed by another ice advance, lake levels increased again, forming glacial Lake Whittlesey in the Erie Basin. It drained northward through the Ubly Outlet, which cut across the thumb of Michigan, into glacial Lake

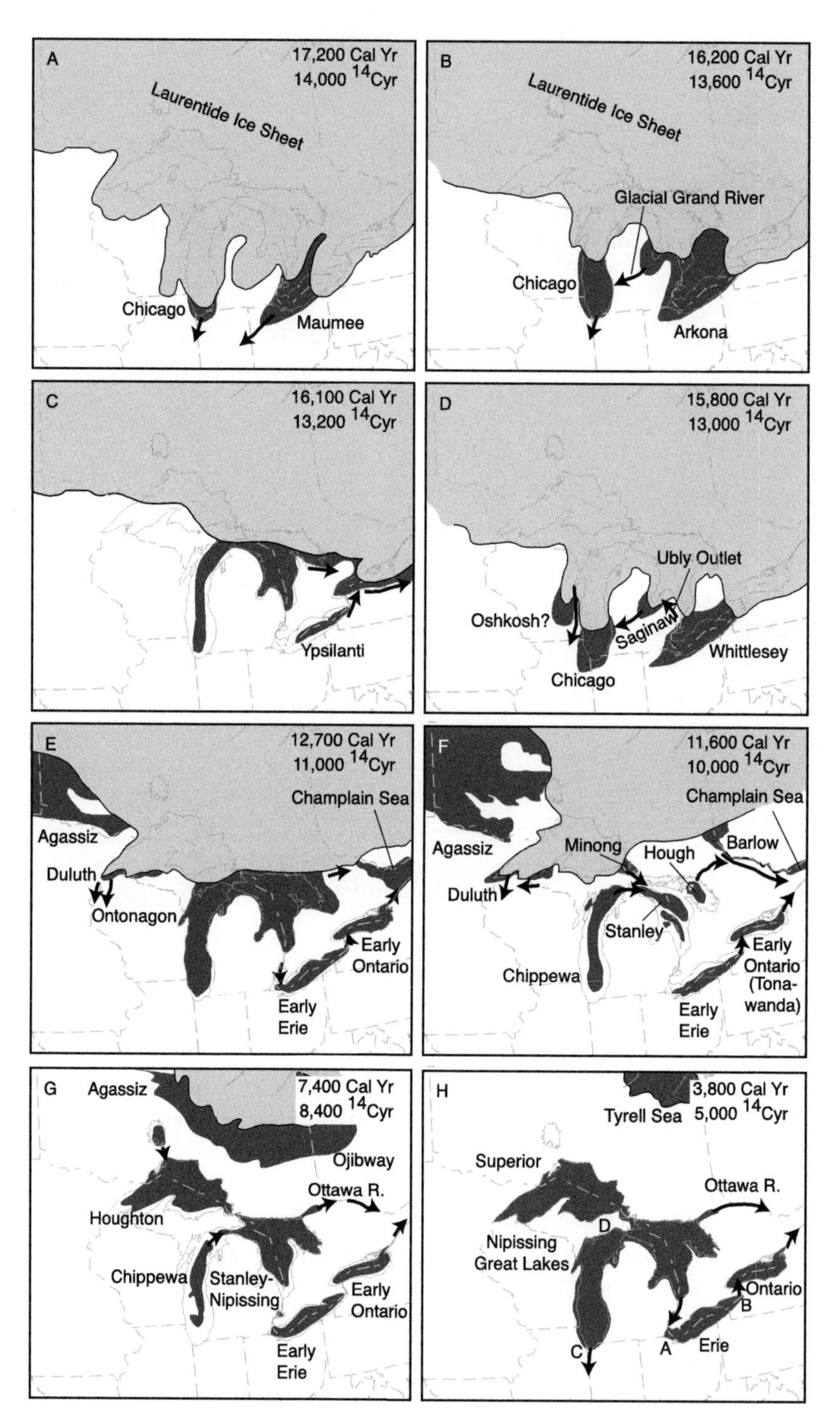

Figure 3.13. Evolution of the ancestral Great Lakes and other proglacial lakes. A-B and C-D in panel H show the locations of sections in figure 3.11. (Modified from Larson 1987; Sado and Carswell 1987; Barnett 1992; Larson and Schaetzl 2001; Kincare and Larson 2009; Breckenridge and Johnson 2009; Anderson and Lewis 2012.)

Saginaw, which drained through the glacial Grand River across lower Michigan into glacial Lake Chicago in the Michigan Basin and finally into the Mississippi system (fig. 3.13D).

By this time, the ice had mostly retreated out of the lower Great Lakes basins. It was still just to the north, however, and water still ponded against it to form glacial Lake Algonquin, which covered much of the Huron and Michigan Basins and drained seaward through both the lower Great Lakes and the Ottawa River to the north (fig. 3.13E). As the ice retreated farther northward, a lower outlet became available through the deeply depressed crust around Georgian Bay and the down-faulted graben of the Ottawa River, causing another big drop in lake levels in the Michigan Basin (glacial Lake Chippewa) and Huron Basin (glacial Lake Stanley). Glacial Lake Chippewa drained into glacial Lake Stanley through a 100-km-long drowned canyon in the Straits of Mackinac known as the Mackinac Channel. The water level was so low that the upper Great Lakes became isolated from the lower Great Lakes again, with the lower glacial lakes draining through the St. Lawrence system.

Glacial Lake Superior was a late player in the Great Lakes hydrodrama simply because it was still largely ice covered until about 10,000 years ago when the other lakes were ice free (fig. 3.13F). At the southwestern end of the Lake Superior Basin, glacial Lake Grantsburg, which was ponded north of the Grantsburg lobe, and glacial Lake Duluth drained southward through what became the valley of the St. Croix River. Today's St. Croix Valley is much too large to have been eroded by the present-day river, and drainage from these glacial lakes must have been much larger. One indication of the power of glacial lake drainage is seen at Taylors Falls where potholes as much as 30 m deep were cut in basalts of the Midcontinent Rift (fig. 2.12D in color section).

Isostatic uplift has been particularly important to the more recent evolution of the Great Lakes. By about 8400 years ago, glaciers had retreated from the Great Lakes region, but the land was still depressed and the upper and lower Great Lakes remained separated. Although both drained to the St. Lawrence, the upper Great Lakes drained through glacial Lakes Stanley and Nipissing into the Ottawa River and the St. Lawrence, whereas the lower Great Lakes drained through Lake Ontario into the St. Lawrence (fig. 3.13G). About 6000 years ago rebound raised glacial Lake Stanley (ancestral Lake Huron) to a level that allowed it to flow to the south through the St. Lawrence and the Mississippi (fig. 3.13H). About 5000 years ago, continued isostatic uplift closed the Ottawa outlet, forcing water from the upper lakes to exit through to the south. Glacial lakes in the Michigan and Huron basins merged to form postglacial Lake Nipissing, which drained to the south. Competition between the two outlets for control of lake drainage involved which could cut downward faster as rebound raised the land. The Erie–Ontario–St. Lawrence outlet won because it was underlain by glacial sediments that were more easily eroded than the Paleozoic sediments that blocked flow out of the south end of Lake Michigan into the Mississippi.

Continued drainage through the Erie-Ontario system lowered Lake Nipissing to form Lakes Huron and Michigan. Finally, Lake Superior was isolated from Lake Nipissing when the area around Sault Ste. Marie was lifted enough by isostatic rebound to form the St. Marys River.

3.4.3. *Glacial Lake Agassiz, Which Formed to the North, Probably Affected Global Climates*

Just north of the Great Lakes region was the enormous glacial Lake Agassiz, by far the largest proglacial lake in North America. As you can see in figure 3.14, it covered large parts of Manitoba, Minnesota, Ontario, North and South Dakota, and Saskatchewan and dwarfed the Great Lakes. David Leverington and others showed that the lake probably had an area of about 400,000 km^2, similar to the area of California, when it merged with Lake Ojibway in Quebec, and an area of about 150,000 to 250,000 km^2 during later phases. The Agassiz-Ojibway lake complex was active for about 5000 years, starting about 13,000 years ago, and it followed the front of the Laurentide Ice Sheet as it retreated northward. The limits of the lake can be determined by the distribution of lake sediments and shorelines, which consist of a combination of wave-cut terraces and beaches. In northern Ontario, sediments that formed in these lakes form the 450,000 km^2 Lesser and Greater Clay Belts, which contain the best arable land in northern Ontario and Quebec. They were farmed in the early days, as discussed by G.V. Burke, and modern farming continues, especially in Quebec, as you can see by roaming around the area with Google Earth.

The level of water in glacial Lake Agassiz varied through time, and changes appear to have happened relatively rapidly, particularly lowering of the water level. The amount of water in Lake Agassiz was so enormous that any rapid lowering could have caused a continent-scale catastrophe. These releases happened when a lower outlet became available, by means of either ice retreat or catastrophic erosion. One of the most important early outlets for Lake Agassiz was through Glacial River Warren, which flowed southward through what is now the valley of the Minnesota River into the ancestral Mississippi River and eventually the Gulf of Mexico. Charles Matsch showed that

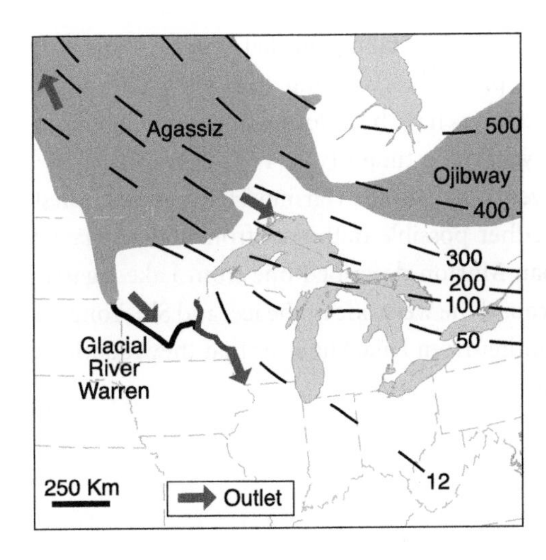

Figure 3.14. Lake Agassiz, showing its maximum extent, outlet zones, and contours of crustal rebound caused by removal of the ice. The southern outflow passed through Glacial River Warren into the Mississippi River. (Compiled from Lewis et al. 2005; Jennings and Johnson 2011; Teller 2013.)

the present Minnesota River is an underfit river, meaning that it is too small in relation to the size of its valley, which was eroded by the much larger and more powerful Glacial River Warren. Keep this river in mind; we will see it again in subsequent chapters.

As the glacier receded northward, Lake Agassiz followed and found more northerly outlets, both southeastward through the Great Lakes into the St. Lawrence River and the North Atlantic Ocean and to the northwest around the Keewatin ice cap and into the Mackenzie River and the Arctic Ocean. The transition to northerly outlets occurred about 12,900 years ago when the level of Lake Agassiz dropped abruptly and stayed low for about 1000 years, a period known as the Moorhead low-water stand. During this event, the lake lost an amazing amount of water, about 4900 km^3, equivalent to ten times the volume of Lake Erie, and it decreased in area by about 36 percent.[16]

The flow of this enormous amount of cold sediment-laden water into the ocean could have changed ocean circulation patterns and the global climate. In fact some researchers have suggested that the outflow of Lake Agassiz water caused the Younger Dryas,[17] an unusually cold period in northern areas, by putting so much cold water into the North Atlantic that the thermohaline circulation system was shut down. In the Atlantic Ocean, the thermohaline system consists of warm water such as the current Gulf Stream, which flows northward at the surface, providing warmth to Europe. Entry of cold water into the North Atlantic and possibly into the Arctic Ocean would have disturbed this flow, cooling Northern Europe, just where the Younger Dryas cooling was most pronounced. Drill cores in the Atlantic contain layers of coarse-grained sediment of continental origin that require some sort of unusual transport mechanism, probably ice rafting, to reach points so far from land. At least seven of these layers have been found ranging in age from about 60,000 to about 12,000 years, the time of the Lake Agassiz event.[18] They are known as Heinrich Events, named for Harmut Heinrich, who first recognized their significance. Other possible outbursts from the proglacial lakes have been suggested. Julian Murton described one from Lake Agassiz through the Mackenzie River around the west side of the ice, and Shi-Yong Yu and his colleagues described another from Lake Superior that they identified with a 9300-year-old cold event.

3.5. What Caused It All?

Global glacial events cry out for an explanation. As we will see in later chapters, they appear to have happened intermittently around the globe for at least the last 2.5 billion years. In fact they are so recurrent that Earth history has been divided into the greenhouse and less common icehouse phases mentioned in chapter 1.

If we are going to explain global glaciation, the best place to start is the Pleistocene, where we have much more information on Earth history before, during, and after glaciation. This gives us a better chance of coming up with an explanation. Unfortunately, we have too many. It's one thing to show that the ice sheets, once they formed, were influenced by Milanković cycles. It's entirely another thing to explain why global glacial events are so scarce when Milanković cycles have been going on for billions of years.[19] In fact, we are trying to explain several different things: why glaciation started at all; and, once it started, why it went through several large and many small cycles of advances and retreats; and why it stopped.

If you want to start a glacial period, it helps to have a relatively large continent at high latitude with a properly located source of moisture for precipitation—sort of like North America and Europe during the Pleistocene. This requires some cooperation from plate tectonics and continental drift, and it might even help to change ocean currents. Then you need relatively cool summers to preserve the snow and ice that accumulates, and that brings you back to global climates and long-term geologic processes like weathering of a newly uplifted mountain range that draw CO_2 out of the atmosphere. You definitely do not want large amounts of CO_2 entering the atmosphere. So minimal sea floor spreading would be desirable.

Even with all these assists, many model calculations require some sort of additional kick, such as the 1991 Pinatubo volcanic eruption, which cooled Earth for several years by erupting millions of tons of sulfur dioxide and ash into the atmosphere. Once glaciation is under way, you need shorter-term processes to lead your glacier cycle through advances and retreats. This can be done in part with Milanković cycles with help from the carbon cycle, especially cyclical plant and reef growth as temperatures change.

With respect to Pleistocene glaciation, one of these many possible factors is relatively clear. Global temperatures have been in almost continuous decline starting about 55 million years ago at the Paleocene-Eocene Thermal Maximum, a period when average Earth temperatures were at least 8°C higher than today (the decline was interrupted by an increase in termperature during the mid-Pliocene about 3 million years ago as described by Deepak Chandan and Richard Peltier). Estimates of atmospheric CO_2 concentrations during the decline indicate that it dropped from about 1500 parts per million (ppm) to modern preindustrial levels.[20] Gretta Bartoli and others have mapped this decline during the last 5 million years, leading up to the start of glaciation. Uplift and weathering of the Himalayas is one of the most commonly cited reasons for this long-term decline in temperatures. As explained by Carmala Garzione, this uplift can also be correlated with the appearance and growth of the Antarctic ice sheet. Some other mechanism might be needed to push this process over the edge, and one suggestion is an increase in the burial of organic carbon caused by fertilization of the oceans with iron and other nutrients resulting from weathering. At this point, options are numerous, and the search continues to identify the magic glacial bullet (or bullets).[21]

3.6. The Main Mineral Resource in Glacial Deposits Is Sand and Gravel

By far the most important mineral resources bequeathed to us by the glaciers are the sands and gravels that are mined from the drift deposits. In the United States, at least 30 percent of the 850 million tons of sand and gravel produced each year comes from states with Pleistocene drift deposits. In Canada total production of about 300,000 tons also comes largely from glacial deposits. Almost all this material goes into the construction of offices, homes, and infrastructure ranging from roads to airports, much of it combined with cement to make concrete. Although the total value of this production amounts to a little more than 8 billion dollars, the value per ton of the product is low. Because large tonnages of material are needed for construction projects, transportation costs to bring the sand and gravel to the project can be an important part of overall costs. Thus, sand and gravel are usually mined as close as possible to the area of construction. And because most construction takes place in populated areas, that means the mining takes place near lots of people. All types of deposits, whether sand or gravel and whether drift or outwash, are used. In the states of Michigan, Wisconsin, and Minnesota, there are more than 1400 active mines. Fortunately, individual sand and gravel mines are relatively small and do not require blasting, which reduces their environmental and social impact.

The other mineral resource of interest is peat, which consists of partly decayed vegetation that collected in a swamp or lake setting. Peat is the first step in the formation of coal. If it remains just below the surface rather than being buried, however, it can be a resource. This has happened in many parts of the world that underwent Pleistocene glaciation. One of the largest such areas is Lake Agassiz, where much of the sediment that accumulated in the lake is now peat. Most of this material has been raised by isostatic uplift where it interacts with the atmosphere and hydrosphere. Although peat might be considered a fossil fuel resource, most interest today is in its role in global climate change. Because it consists largely of plant material, peat is an important sink for CO_2. According to a recent summary by David Bello, peatlands make up about 3 percent of Earth's surface and store as much as 2 trillion tons of CO_2. However, when the peatlands are uplifted naturally or drained anthropogenically, the plant material reacts with the atmosphere to release CO_2.

A final aspect of glacial geology that we can consider a mineral resource is wind energy. This is because glacial deposits in the Great Lakes region control wind patterns, and this is what controls the location of wind turbines. An excellent example of this control is Buffalo Ridge, which is underlain by the Bemis moraine, the outermost moraine in the Des Moines lobe (fig. 3.3B). At Lake Benton, Minnesota, there are more than 600 wind turbines, and many hundreds more occupy a narrow corridor along the moraine from near Clear Lake, Minnesota, to Lake Park, Iowa. Elsewhere in the Great Lakes region, the Kettle moraine in eastern Wisconsin also hosts a long corridor of wind turbines. Other large concentrations of wind turbines along the eastern shore of Lake Huron and the northern shore of Lake Erie in Ontario take advantage of relatively low glacial lake sediments that allow unobstructed wind fetch.

3.7. Pleistocene Life Involved Humans Facing Large Animals and a Harsh Climate

Life in the Great Lakes region during the Pliocene and Pleistocene was dominated by large mammals, collectively known as the American megafauna. These included giant cats, wolves, lions, bears, and their prey, including bison, beavers, peccaries, tapirs, and sloths, as well as mastodons and mammoths. Then humans showed up. In a recent review, James Dixon showed that they could have taken two different routes into the Great Lakes region. The route that receives the most attention is along the western coast of North America, where large areas of land were exposed by the 100- to 120-m lowering of sea level caused by the accumulation of ocean water in the ice sheets. Recession of mountain glaciers that fed into the ocean probably formed an ice-free beach by about 16,000 years ago. The path that would be of more interest to the Great Lakes region is the so-called western interior route through the mountains. Surprisingly large areas of the western cordillera of North America did not undergo glaciation, and it is generally accepted that a throughgoing route opened about 15,000 years ago.

However they arrived, aboriginal populations were residing in the Great Lakes region at least 13,000 years ago and probably even earlier. This suggests that they were in residence when the previously mentioned Younger Dryas cool event began. Christopher Ellis and others have pointed out that the archaeological record in the Great Lakes region documents a decline in mobility and more abandoned sites during this time. Despite this setback, aboriginal populations increased in numbers and sophistication, as shown by submerged barriers thought to have been used in hunting.

This raises the question of just how good early humans were as hunters. Cached bones found in former wetlands, described in the previous chapter, have stimulated debate about the role they might have played in the extinction of the large animals that roamed North America during the Pleistocene. This question is part of a much larger controversy about the interaction of early humans with megafauna throughout the world. Before about 50,000 years ago, large animals were common on all the continents, but after that and until about 10,000 years ago, they gradually disappeared from all continents except Africa. This same period, from 50,000 to 10,000 years ago, coincides with the rise of humans, and this has led to hypotheses that they played a role in the extinctions.[22]

Explanations for this extinction event range along a spectrum, as summarized by Paul Koch and Anthony Barnosky. At one end of the spectrum is the "proposed overkill hypothesis," which holds that increased hunting by humans, both numerically and in terms of efficiency, simply killed the animals. It was proposed originally by Paul Martin and has been supported more recently by the work of Christopher Sandom and colleagues. The other end of the spectrum holds that ecosystem collapse caused by natural processes such as climate change, volcanic eruptions, fire, or other natural processes did the job, a position held by Adrian Lister and colleagues.[23] Between these poles are suggestions that various degrees of human impact on the environment, including fire, disease, and competition for food, pressured the animals into extinction.

In North America, the issue is more focused than elsewhere because many important extinctions generally coincided with the appearance of humans on the continent, whereas in Europe, Asia, and Australia humans were present long before the extinctions were complete. Cached bones are particularly numerous in North America because they were preserved in glacial lakes and wetlands, which formed during the retreat of the glaciers just as humans appeared. Studies of many of these bone caches provide information on the condition of animals that were killed, as well as the timing and conditions of extinction and preservation. The tusks of mammoths and mastodons, for instance, have annual growth zones that can be used to determine not only the condition of the animal and the nature of its food but also the climate in which it was living. In a review of this sort of detailed evidence, Dan Fisher concluded that hunting was the main cause of extinction of mastodons in the Great Lakes region. That seems a reasonable conclusion. After all, the mastodons had persisted through the harsh climates and changing landscapes of the "ice ages." As you will see in the next chapter, that was a tough time to be alive.

Notes

1. Another term used to describe glacial deposits of this type is *diamicton*, which refers to unsorted sediment deposited by any process but most commonly by glaciation.

2. The two key observations that distinguish ice-deposited till from water-deposited outwash are sorting and layering. *Sorting* refers to the range of size of sediment grains. Sediment with a small range of sizes (such as a beach sand deposit) was formed in water, which separates (sorts) grains by size. *Layering* refers to the presence of internal layers or strata, which form in water-deposited sediment. Sediment deposited when ice melts has no internal layering and consists of sediment of all sizes (unsorted).

3. *Ablation* refers to all processes by which a glacier loses snow and ice, including melting and evaporation.

4. Terminal moraines are special types of end moraines that form at the maximum extent of a glacial advance. Moraines that form along the sides of glacial lobes are known as lateral moraines.

5. *Till plain* is a general term that includes features referred to as ground moraine or lodgment till, both of which can form when a thin sheet of ice deposits its sediment.

6. Eskers take the place of streams in areas such as northern Ontario where a throughgoing regional stream system has not had time to develop on the surface left by the glaciers. Because sediment in the esker sediment was eroded upstream, it provides a regional-scale sample of rocks in the upstream area. Joe Brummer and others have described how samples of sediment from the Munro esker near Kirkland Lake contained minerals that led to discovery of a kimberlite intrusion (discussed in chapter 7) containing diamonds upstream from the sample site; subsequent esker surveys helped in the discovery of several diamond-bearing kimberlites in the James Bay region.

7. Icehouse conditions can vary from those that form continental glaciers to climate periods such as our present-day interglacial warm period. Conditions and causes of icehouse and greenhouse states are reviewed in Tabor (2016).

8. "Ma" and "Ga" refer to geologic ages of millions and billions of years, respectively. They are used throughout this book to refer to geologic ages greater than thousands of years (Ka). The terms Gyr, Myr and Kyr refer to periods of billions, millions, and thousands of years, respectively.

9. Oxygen consists almost entirely of two isotopes with slightly different masses (weights). Oxygen-18 (^{18}O) contains two more neutrons in its nucleus than does oxygen-16 (^{16}O). This makes ^{18}O slightly heavier and causes any compounds enriched in ^{18}O to react more slowly in common processes such as evaporation. During the normal water cycle, evaporation preferentially removes H_2O containing lighter ^{16}O from the ocean. This means that rain and snow will be enriched in ^{16}O relative to its major source (the ocean). As long as the water is quickly cycled from ocean to atmosphere to precipitation and runoff back to the ocean, the average oceanic ratio of $^{18}O/^{16}O$ remains relatively constant. However, when this moisture is "captured" and retained on land as snow and ice during the growth of large ice sheets, the average $^{18}O/^{16}O$ ratio in the ocean increases, as does the average ratio in precipitation from the ocean. Analysis of the $^{18}O/^{16}O$ ratio in successive layers of sediment or ice that were deposited during the last few million years of Earth history shows that global ice volume has undergone repeated cycles of growth and decay.

Since evaporation is most common during warm periods, this process increases the $^{18}O/^{16}O$ ratio of the ocean, and the shells of marine organisms that grow in the ocean reflect this changed composition. The $^{18}O/^{16}O$ ratio of the ocean and shells that grow in it will be highest when evaporation is greatest during warm periods. The opposite response is seen in ice that accumulates in glaciers on land during cold periods. This ice forms from precipitation of water that evaporated from the ocean, and therefore it is enriched with ^{16}O. So the $^{18}O/^{16}O$ ratio of ice is lowest during cool periods. Analysis of the $^{18}O/^{16}O$ ratio in successive layers of sediment or ice that were deposited during the last few million years of Earth history show that temperatures have undergone repeated cycles between warm and cool periods (fig 3.15).

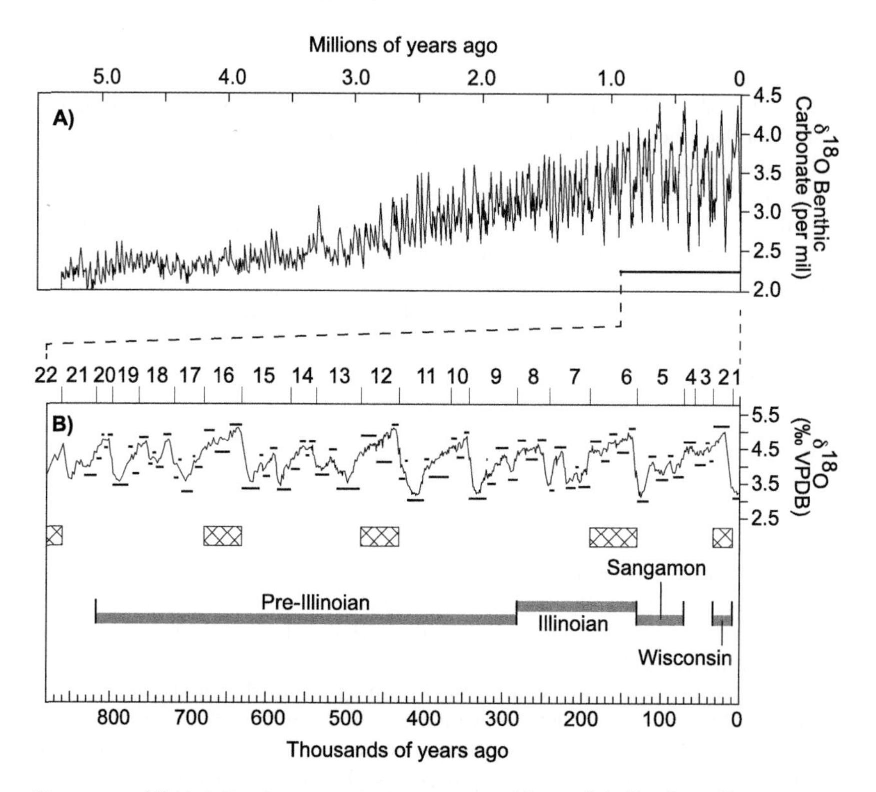

Figure 3.15. (**A**) Variation in oxygen isotope compositions of shells of small marine organisms (foraminifera) during the last 5 million years, showing gradually decreasing global temperatures; (**B**) Isotope record for the last 800,000 years showing the main Marine Isotope Stages. Patterned rectangles show glacial periods represented by deposits in Illinois. (Compiled from Prest et al., 1962; Soller, 2001; Mickelson and Colgan, 2003; Lisiecki and Raymo, 2005; Raymo and Huybers, 2008; Curry et al., 2011; Larson, 2011; Lang and Wolff, 2011.)

10. The most obvious marine isotope cycle has a 41,000-year (41Kyr) period and is seen in the record between about 3 and 1 million years ago. Since about 1 million years ago, a 100,000-year cycle is more obvious. The 41 Kyr period corresponds to cyclic changes in the tilt of Earth's rotation (known as obliquity) and the 100 Kyr period corresponds to cycles in the shape of Earth's elliptical orbit (eccentricity). Both control the degree to which solar radiation reaches Earth and therefore its capacity to accumulate enough ice to form and maintain an ice sheet.

11. In Illinois, which has glacial deposits ranging in age over at least a million years, Brendan Curry and his coworkers have correlated deposits with MISs 2, 6, 12, 16, and 22. Deposits from MIS 6, which represent the Illinoian episode, include at least four different cycles of glacial advances and retreats. Those from MIS 2, the Wisconsin episode, comprise five separate cycles. The Illinoian glacial deposits are overlain by an ancient soil, or paleosol, which formed during the warm Sangamon interglacial. The Wisconsin-age glacial deposits are from the latest part of the Wisconsin episode; older deposits are

not present. The same pattern is present in the Illinoian deposits, which come from only the latest part of the Illinoian episode. And the pre-Illinoian deposits that have been found in Illinois so far represent only about half the known glacial periods that have been recognized in the marine isotope record. All this confirms that the continental glacial deposit record in any one place is fragmentary and that we must look at a large region to understand how the ice sheets behaved. By far the best deposits to study for this are the most recent ones from the Wisconsin episode, which are best preserved.

12. See Ivy-Ochs and Kober (2008) for information on surface exposure dating methods.

13. Omars, which are erratics consisting of unusually distinctive sandstone with calcareous concretions from the Belcher Islands on the east side of Hudson Bay, were especially helpful in reconstructing the glacial history of the southern part of the Laurentide Ice Sheet where it entered the Great Lakes region. V.K. Prest and others reviewed the role of omars in the Laurentide Ice Sheet story and produced maps of the amazing dispersal of these curious features.

14. Glacial erosion was concentrated along the Michigan breccias, as discussed in chapter 4.

15. Fjords, which form when U-shaped valleys are flooded, are not present in the Great Lakes region, although hilly terrane at the east side of Lake Superior in Ontario, has deep, steep-sided glacial valleys that almost qualify. To the casual observer, the narrow arms of Traverse Bay in Lake Michigan resemble fjords, although they are actually tunnel valleys. The Finger Lakes in New York are larger tunnel valleys, and similar small ones are found south of Mille Lacs in Minnesota. Tunnel valleys are formed by rapidly flowing meltwater moving beneath glacial ice; they are the outlets for the large volumes of meltwater that form beneath some glaciers.

16. Additional information on the geology of Lake Agassiz and its freshwater outbursts is in Teller and Clayton (1983); and Teller et al. (2002).

17. The Younger Dryas, a major cool period in the northern hemisphere was named for a flower that is common in glacial areas. It extended from about 12,900 to 11,700 BCE and caused significant cooling, especially in Europe (Carlson 2013). See Condron and Winson (2012) for further discussion of its relation to meltwater.

18. For a discussion of Heinrich Events and global climate, see Hemming (2004); Naafs et al. (2013).

19. See Berger and Loutre (2005) for a discussion of the relationship between Milanković cycles and paleoclimates.

20. We will see in chapter 9 that most models for glaciation require low global CO_2 levels similar to those in preindustrial times.

21. Recent research remains fragmented among the many possible processes that could start or mediate glaciation. Despite its age, the early summary by Chester Beaty (1978) is a useful introduction to the problem.

22. Proof that we are capable of large-scale extermination is seen in New Zealand, where humans killed all the great auks, and in Madagascar, where we did the same to the elephant bird. Continental-scale extermination is a larger assignment, but many researchers feel that we met the challenge (MacPhee and Marx 1997).

23. See Dixon (2013) for a discussion of late Pleistocene colonization in North America. For additional arguments related to natural causes of megafaunal extinction, see Lister and Stuart (2008); Yansa and Adams (2012); Cooper et al. (2015).

CHAPTER 4

Flooding the Continent

Paleozoic-Mesozoic Sediments and the Michigan Basin

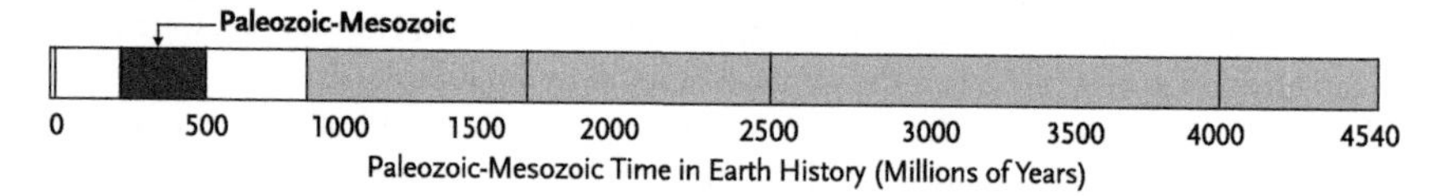

4.1. Paleozoic and Mesozoic Flooding of the Continents Was a Global Event

Our first step into deep time in the Great Lakes region goes back to the Cambrian Period at the start of the Paleozoic Era over 500 million years ago (fig. 1.4 in color section). Upon reaching this Cambrian world, our time traveler would have had an immediate problem: too much ocean. Instead of dry land, most of the Great Lakes region was submerged beneath a shallow sea (fig. 4.1). You can get an idea of the predicament in figure 4.2 (in color section). The top photo shows Cambrian-age Mt. Simon Sandstone resting on weathered Pre-

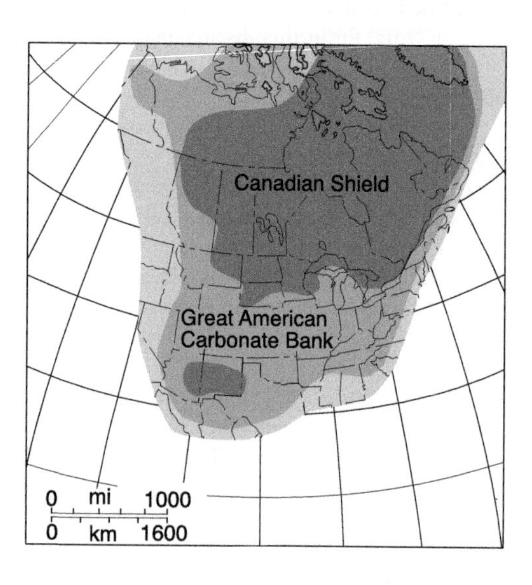

Figure 4.1. Paleogeography showing extent of Laurentia during Early Ordovician time. Archean and Proterozoic rocks of the Canadian Shield were exposed most of the time, but to the south they were covered with limestone and dolomite that had been deposited in a shallow sea known as the Great American Carbonate Bank. (Modified from Derby et al. 2014).

Cambrian granite. Similar sandstones were being deposited over most of the Great Lakes region in Cambrian time, leaving little dry land. About the only place to land a time machine would have been small islands of Precambrian rocks that poked up through the Cambrian seas, and they probably looked uninviting. The bottom photo shows the shoreline of one of these islands in the Baraboo Hills in Wisconsin. There, boulders of Precambrian rock were deposited in the sea at the foot of a cliff. These boulders are well rounded and Robert Dott and Charles Byers have estimated that it took waves as high as 8 m to do the job.[1]

Eventually, our time traveler would have found dry land in the northern part of the Great Lakes region on the Canadian Shield where older Precambrian rocks were exposed. This part of Laurentia, as the North American continent was called at the time, was at least partially emergent (exposed to the atmosphere) during most of Paleozoic and Mesozoic time.[2] It was deeply weathered and had a gently rolling surface known as a peneplain that had developed over hundreds of millions of years, although the lack of land plants and animals made it look something like a wet Mars.

Our time traveler was not experiencing a short-term problem. In fact, Laurentia was flooded off and on for much of Paleozoic and Mesozoic time. How could this happen? How could the ocean have covered so much of the land? Making ice cover land is easy enough; it just has to get cold and wet. But making water cover continents requires much more. One obvious question is whether it happened only in North America or on a wider scale. The answer is that flooding of the continents was a global phenomenon; continents all over the world were covered with ocean during Paleozoic and Mesozoic times. But they cycled between flooded and emergent conditions several times during this long period. So we have to find a way for continents to alternate between submerged and emergent through time and that involves eustasy and glaciation.

4.2. Global Cycling of the Continents between Flooded and Emergent Conditions Was Caused Largely by Eustasy

Flooding of a continent can happen if the continent sinks into the sea. Or the continent might remain fixed while the ocean gets bigger (sea level rises). Eustasy refers to this second process. Eustatic sea level changes are global effects; they are very different from local changes in sea level, which are referred to as relative changes or dynamic topography.[3]

Eustatic changes in sea level can happen if the actual size (volume) of the ocean basin decreases or if the volume of water in the ocean increases. Changing the size of the ocean basin sounds like a big job, but a very effective mechanism is right at hand: plate tectonics (fig. 1.5). Seafloor spreading zones at mid-ocean ridges are underlain by hot mantle, which expands the ocean floor, causing it to rise. This takes up space in the ocean basin, decreasing its volume, forcing water onto the continents, and increasing sea level by up to several hundred meters. Another few tens of meters of sea level can be gained by depositing sediment into the ocean, which takes up space that would have been occupied by water, or by making mountain belts that also take up space in the ocean. Reversal of any of these processes causes a corresponding decrease in sea level.

Changing the volume of water in the ocean also sounds like a challenge, but it, too, has a ready agent: ice. We have already seen that the Pleistocene ice sheets removed enough water from the ocean to cause a huge drop in sea level and that sea level rose again when the ice melted. According to Kenneth Miller, Richard Alley, and others, changes in the volume of ice can vary sea level by a hundred meters or more; for instance melting of the Greenland and Antarctic ice sheets would raise modern sea level by about 70 m. Smaller changes in sea level of only a few meters can be caused by changing the temperature of the ocean or by adding or removing water, such as the water that flowed into the Atlantic Ocean from glacial Lake Agassiz, as discussed in chapter 3.[4]

These two primary eustatic mechanisms, plate tectonics and glaciation, are sometimes referred to as glacio-eustatic and tectono-eustatic, respectively, and they operate at different rates. Glacio-eustatic changes take place over periods of several to thousands of years, whereas tectono-eustatic changes require millions of years. For example, Pierre Deschamps and his colleagues have estimated that glacio-eustatic processes caused sea level to rise about 20 m in less than 500 years during the Bølling warming 14,600 years ago. In contrast, tectono-eustatic changes in ocean basin volume caused by uplift of the Himalayas took millions of years. By operating at different rates, the two mechanisms produced the complex variations in sea level that we see in Paleozoic and Mesozoic sedimentary rocks of the Great Lakes region (box 4.1).

BOX 4.1. WHAT IS SEA LEVEL?

It should be obvious at this point that sea level is not constant or the same everywhere. We already know that it is changing by small amounts today as polar ice caps melt. But it changed by much more in the ancient past. Figure 4.3 shows that sea level has ranged from hundreds of meters above its present level to a few tens of meters below throughout Phanerozoic time.[5] Viewed in this way, it is clear that Earth's normal state during Phanerozoic time was for the continents to be partially flooded. Our present sea level is unusually low. Similarly low levels happened in the past only during the latest Paleozoic and briefly during the Middle Mesozoic. This is not very encouraging news for coastal property owners and, in the long term, for anyone living at an elevation of less than a few hundred meters above present sea level.

Another curious feature of sea level is its variation from place to place. The ocean is pulled toward any massive object—something we know because of the tides. But there are smaller pulls from adjacent bodies of ice and land. In one of the ironies of global warming, melting of the Antarctic ice sheet is expected to lead to lower sea levels in the southern polar seas because the mass of the ice sheet will no longer be pulling water upward toward it. Instead, the extra water will show up as increased sea level in equatorial regions.[6]

Eustatic sea level changes are closely linked to the two most important "cycles" that operate on the planet, the supercontinent cycle (fig. 1.6) and the carbon cycle (box 1.2). When supercontinents break up, the number of spreading ridges increases, causing the continents to flood. Conversely, during supercontinent aggregation, spreading ridges are less numerous and sea levels are lower. Increased volcanism releases CO_2, which can warm the world and melt glaciers, causing sea level to rise. The increased CO_2 can increase weathering rates, sending dissolved calcium and carbonate to the oceans to be deposited as limestone. Or it can enhance plant growth, possibly leading to increased burial of organic matter or even eutrophication of the ocean. As these processes are proceeding, plate tectonic migration can place continents in polar or equatorial regions. This will have an effect on the types of sediments that are deposited, which can range from clastic to chemical and biogenic.[7] Continents in polar regions develop glaciers with abundant clastic sediments, whereas those in equatorial regions have warm seas with abundant limestones and evaporites. Plate tectonics and the larger supercontinent cycle can also form mountain belts, which increase weathering rates and change atmospheric and oceanic circulation patterns, further affecting global climate and levels of CO_2. Add to this the Milanković orbital cycles discussed in the previous chapter and you can see that change is the only constant when it comes to deep time global climate and sedimentation.

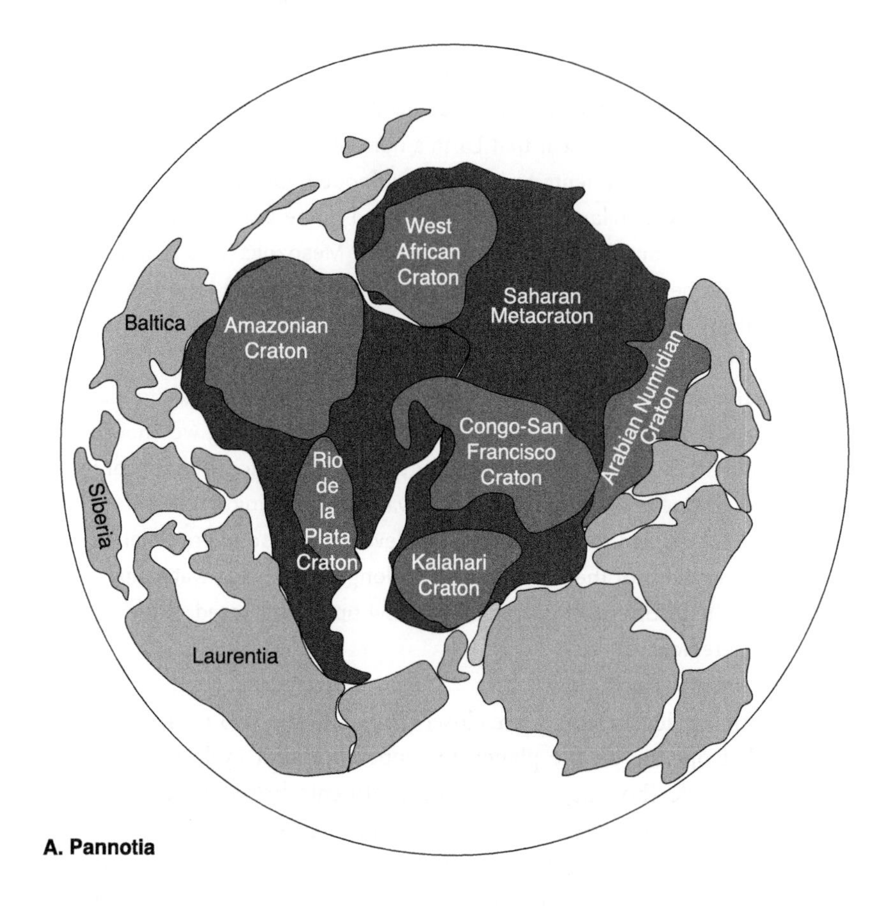

A. Pannotia

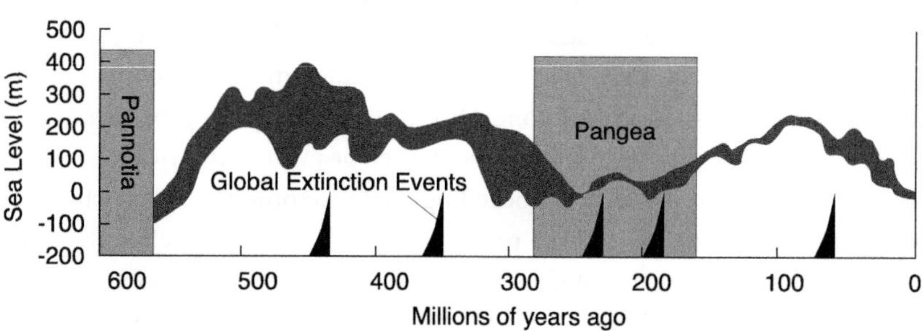

B. Pannotia-Pangea

Figure 4.3. (**A**) Reconstruction of the supercontinent Pannotia centered on the south pole (note that the Great Lakes region of Laurentia was near the equator at this time); (**B**) relations among Pannotia and Pangea supercontinents, changes in sea level, and global extinction events during Paleozoic time. (Modified from Raup and Sepkoski 1982; Hallam 1992; Dalziel 1997; Haq and El-Qahtani 2005; Haq and Schutter 2008; McElwain and Punyasena 2007; Nance and Murphy 2013; Nance et al. 1988, 2014.)

The supercontinent framework that partly controlled the Paleozoic and Mesozoic geologic history of the Great Lakes region began to develop in latest Precambrian time when several continent fragments merged into a large supercontinent known as Pannotia, which you can see in figure 4.3. According to Chris Scotese and others, these fragments included Laurentia, which forms the core of present-day North America, as well as parts of the Baltic, Siberian, Amazon, and Congo continents.[8] They had been freed when the older supercontinent, Rodinia, began to come apart about 800 to 700 million years ago, forming the Panthallasic Ocean. After several hundred million years and wander paths of up to 5000 km, the newly released fragments rejoined to make the supercontinent Pannotia near the beginning of Paleozoic time. Shortly after it formed, Pannotia began to break up, releasing Laurentia to resume its solitary existence.

The early Paleozoic breakup of Pannotia involved increased seafloor spreading, which caused sea level to rise (fig. 4.3B). Because Laurentia remained close to the equator during much of Paleozoic time, the seas were warm, encouraging deposition of limestones and evaporites. By Devonian time, however, Laurentia had recombined with Baltica to form a somewhat larger continent known as Euramerica (also known as Laurussia), and by Permian time it had joined all the major continents, including Gondwana, to form the supercontinent known as Pangea (fig. 1.6). In later sections, we will see how this choreography of continents controlled the type and distribution of sediments on Laurentia and its mineral resources. For the moment, we can summarize by noting that the collisions opened and closed the ancestral Atlantic Ocean (Iapetus Ocean) and resulted in the Taconic (Late Ordovician), Acadian (Late Devonian), and Alleghenian (Permian) orogenies,[9] which formed the Appalachian Mountains along the eastern side of Laurentia. The Appalachians were big mountains during these orogenies, and they had an impact on our Paleozoic story even far to the west in the Great Lakes region. But first we need to spend a moment on the concept of sedimentary basins, where all this Paleozoic and Mesozoic sediment collected.

4.3. Paleozoic and Mesozoic Sediments Were Deposited in Two Types of Sedimentary Basins

Areas where large thicknesses of sediment were deposited are known as sedimentary basins, and there were several of these in and around the Great Lakes region, as you can see in figure 4.4. The Michigan Basin, which is the

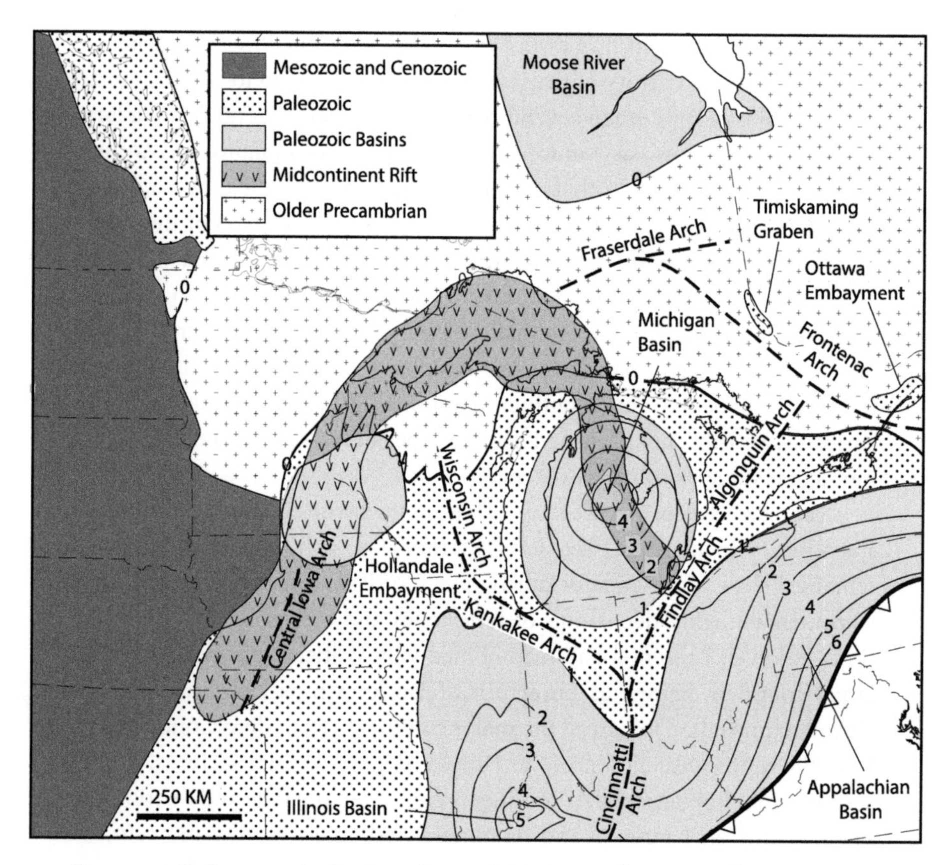

Figure 4.4. Sedimentary basins in and around the Great Lakes region showing the thickness of sediments in thousands of meters. Arches, which separate the basins, are high zones in the underlying Precambrian rocks. The Midcontinent Rift (see chapter 5) underlies the Michigan Basin and Hollandale Embayment, and probably caused their initial subsidence.

largest complete basin in the Great Lakes region, contains a sequence of sedimentary layers with a maximum thickness of over 4000 m. The only other sedimentary basin that is completely within the Great Lakes region is the Hollandale Embayment in Minnesota, which contains only about 1000 m of sediments.[10] Larger sediment thicknesses are present in the Illinois and Appalachian basins, which are on southern and eastern margins of the Great Lakes region, respectively. To the north, the Moose River Basin under James Bay is the southern continuation of the larger Hudson Bay Basin. Small enclaves of

Paleozoic sediment are also present in the Timiskaming Graben and Ottawa Embayment.

As you might imagine, sedimentary basins can form in all sorts of plate tectonic settings. For our purposes, we can condense these into two main types of sedimentary basins: craton margin and intracratonic basins.[11] Craton margin basins form on the edges of cratons or continents. They are sometimes called passive margin basins because the margins along which they form are passive; subduction is not happening at the margin. Instead it is a rifted margin, where the craton pulled apart to form two smaller pieces that moved away from each other. The Appalachian Basin is a craton margin basin; it formed off the edge of Laurentia when it rifted apart to form the Iapetus Ocean.

The rest of the basins are intracratonic basins, and they are rather strange lot. The term *intracratonic* means that the craton or continent began to subside somewhere in its middle. Why would a perfectly good craton do that? The answer, of course, is that it probably wasn't a perfectly good craton, that it had some sort of weakness. This is the situation for the Michigan Basin, which is one of the world's iconic intracratonic basins. Figure 4.4 shows that the craton beneath the Michigan Basin is underlain by the Proterozoic-age Midcontinent Rift, which formed millions of years before Paleozoic time. In the next chapter, we will see that the Midcontinent Rift formed when the craton rifted (or faulted) apart and that it contains rocks that are denser than the rest of the craton. This combination of dense rocks bounded by big fractures or faults would not have made for a very stable craton.

Paul Howell and Ben van der Pluijm have suggested that just this situation formed the Michigan Basin, when heavy Midcontinent Rift rocks were jostled by formation of the Appalachian Mountains along the east coast of Laurentia. The heavy rift rocks began to subside, causing sediment to collect on top, and the weight of new sediment led to continued subsidence, more sedimentation, and growth of the basin. The probable catalytic role of the Midcontinent Rift is supported by the fact that the Hollandale Embayment in Minnesota is underlain by the western extension of the Midcontinent Rift, as you can see in figure 4.4. Crust beneath the Illinois and Moose River basins has similar weaknesses, suggesting that they had a similar origin. Not everyone agrees on this fault-related origin, however. For instance, Norman Sleep has recently proposed that the Michigan Basin formed largely because the underlying crust cooled and contracted.

4.4. Sedimentation in the Basins Was Controlled by Eustatic Sea Level Changes That Can Be Divided into Six Cycles Separated by Unconformities

So far, we have been looking at the process of flooding and emergence from the standpoint of the ocean. Early geologists actually looked at it from the standpoint of the continents. They spoke of transgression when the sea advanced across the continents to flood them and regression when the sea retreated into its basins. To understand how and why they took this perspective, we need to look briefly at the history of geologic mapping in the Great Lakes region.

In the early days of Great Lakes work, geologists mapped the distribution of various types of sedimentary rocks, such as sandstones and limestones, and used this information, along with their fossil content, to divide the rocks into formations that represented sediments deposited at about the same time over a wide region. Related formations were combined into units called groups. In most cases, they named formations and groups for areas where the rocks were best exposed and most typical, such as the Devonian-age Detroit River Group along the Detroit River and the Amherstburg Formation just across the river in Ontario. Figure 4.5 summarizes some of the important formations and groups in the Great Lakes region, and we will return to it frequently in discussing the history of the sedimentary basins.[12]

Comparison of formations and groups of the same age throughout the region showed that sediments of different types were deposited at the same time in different areas. Figure 4.6 shows that Cambrian and Ordovician sediments in the Hollandale Embayment changed from sand and mud near the shore to limestones farther from shore. This variation in rock types of the same age from place to place is referred to as facies change, and it helps us locate shorelines and determine water depths. The movement of the shoreline through time can be determined by looking at the sequence of formations and groups at a single location (the stratigraphic sequence). If the stratigraphic sequence changes upward from formations containing sandstone (shorelines) to formations containing shale and limestone (deep water), the sea was deepening. Geologists viewed this as a transgression of the sea. Conversely, stratigraphic sequences that changed upward from limestone to shale to sandstone (as seen in the diagram in fig. 4.6) indicated that the sea was retreating, which is termed regression.

Most of the sediments that were deposited during these episodes of transgression and regression are separated by features called unconformities, which are a very important part of the story. Unconformities represent periods of time when no sediment was being deposited or previously deposited sediments were removed by erosion. They are periods of time for which we have no information.[13] For example, an enormous unconformity separates the Paleozoic-Mesozoic sedimentary rocks in the Great Lakes from the Pleistocene glacial deposits that cover them. This unconformity represents about 200 million years of time for which we have very little information in the Great Lakes region and must look elsewhere to see what was happening.

Unconformities in the Paleozoic stratigraphic sequence provide a useful alternative framework for understanding the Paleozoic history of Laurentia and the Great Lakes region. It turns out that the entire Paleozoic-Mesozoic sequence can be divided into packages, called stratigraphic sequences, that are separated by unconformities, as first explained by L. L. Sloss.[14] Sloss recognized six stratigraphic sequences that represent major transgression-regression events during which the sea advanced across the continent, covering it with sediment, and then retreated, leaving the emergent continent exposed to erosion. The oldest of these stratigraphic sequences, the Sauk sequence, began in Cambrian time when the sea advanced across a widespread Precambrian erosion surface. It was succeeded by the Tippecanoe sequence, which began in Middle Ordovician time, and then the Kaskaskia, Absaroka, Zuni, and Tejas sequences (fig. 4.7). The early Paleozoic sequences are based largely on rocks in the Great Lakes region. For instance, the Sauk sequence is named for exposures in Sauk County, Wisconsin, and the Tippecanoe sequence is named for Tippecanoe County, Indiana.

Stratigraphic sequences can be linked to the supercontinent cycle, periods of mass extinction (fig. 1.3), and other global processes. For instance, the widespread unconformity at the base of the Sauk sequence clearly resulted from prolonged erosion of the uplifted Pannotian supercontinent after regression of the sea. Flooding of the continent during Cambrian transgression started when the Pannotia supercontinent began to break up. Stephen Meyers and Shanan Peters have shown that the entire series of sequences has a 56-million-year rhythm that is probably linked to long-term tectonic or mantle-scale processes They note further that this cyclic change correlates with the change in number of genera in the marine fossil record, suggesting that global-scale sedimentary sequences are significant even to life on the planet.

It's one thing to talk about these sedimentary sequences and their global significance. It's a completely different thing to recognize them on the ground. So in the next section we will spend a little time reviewing the geology of Paleozoic and Mesozoic sedimentary rocks in the Great Lakes region and see-

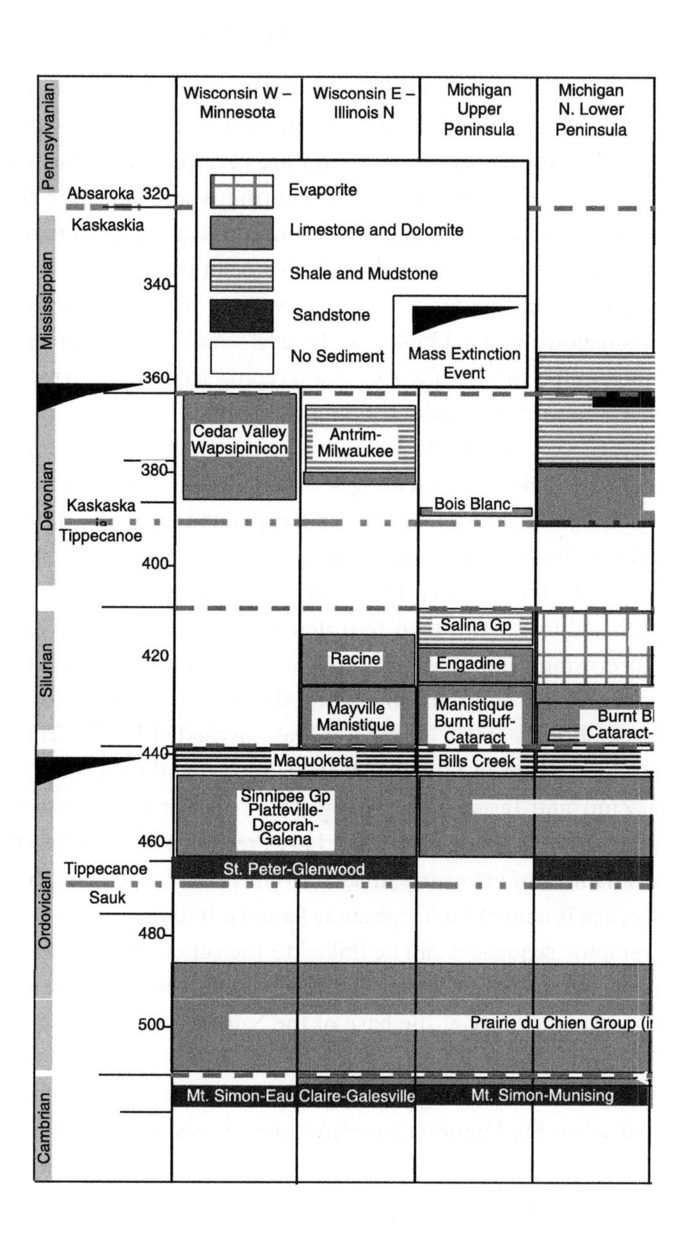
Pennsylvanian
Mississippian
Devonian
Silurian
Ordovician
Cambrian
Wisconsin W – Minnesota
Wisconsin E – Illinois N
Michigan Upper Peninsula
Michigan N. Lower Peninsula
Absaroka
Kaskaskia
Kaskaska
ia
Tippecanoe
Tippecanoe
Sauk
320
340
360
380
400
420
440
460
480
500
Evaporite
Limestone and Dolomite
Shale and Mudstone
Sandstone
No Sediment
Mass Extinction Event
Cedar Valley Wapsipinicon
Antrim-Milwaukee
Bois Blanc
Salina Gp
Racine
Engadine
Mayville Manistique
Manistique Burnt Bluff-Cataract
Maquoketa
Bills Creek
Sinnipee Gp Platteville-Decorah-Galena
St. Peter-Glenwood
Mt. Simon-Eau Claire-Galesville
Mt. Simon-Munising

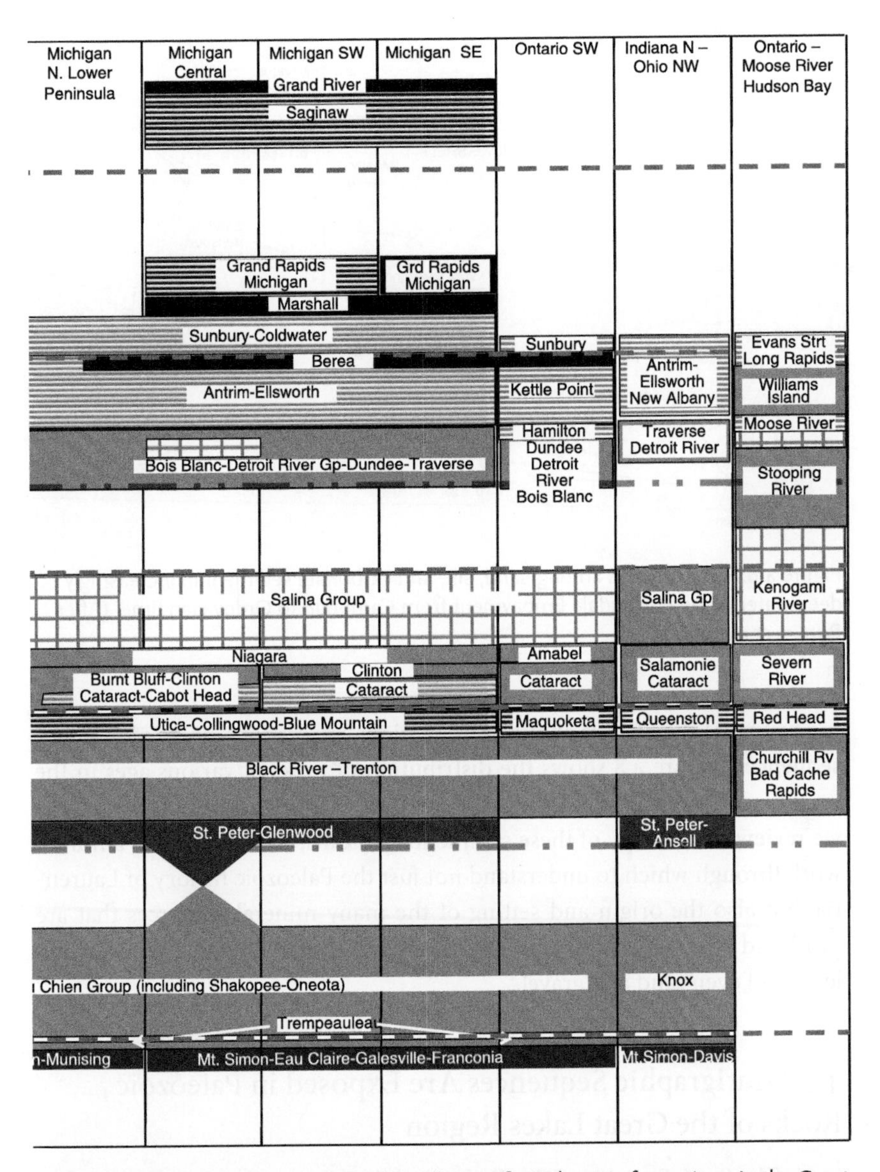

Figure 4.5. Representative stratigraphic columns for Paleozoic formations in the Great Lakes region showing their relation to the Sauk, Tippecanoe, Kaskaskia, and Absaroka stratigraphic sequences and Ordovician and Devonian mass extinction events. The column for the central part of the Michigan Basin shows that sedimentation was continuous across the Sauk-Tippecanoe boundary. (Modified from Fisher et al. 1988; Catacasinos and Daniels 1991; Witzke et al. 1996; Catacosinos et al. 2000; Armstrong and Dodge 2007; Mossler 2008; Wisconsin Geological Survey 2011; MacLeod 2014; Swezey et al. 2015.)

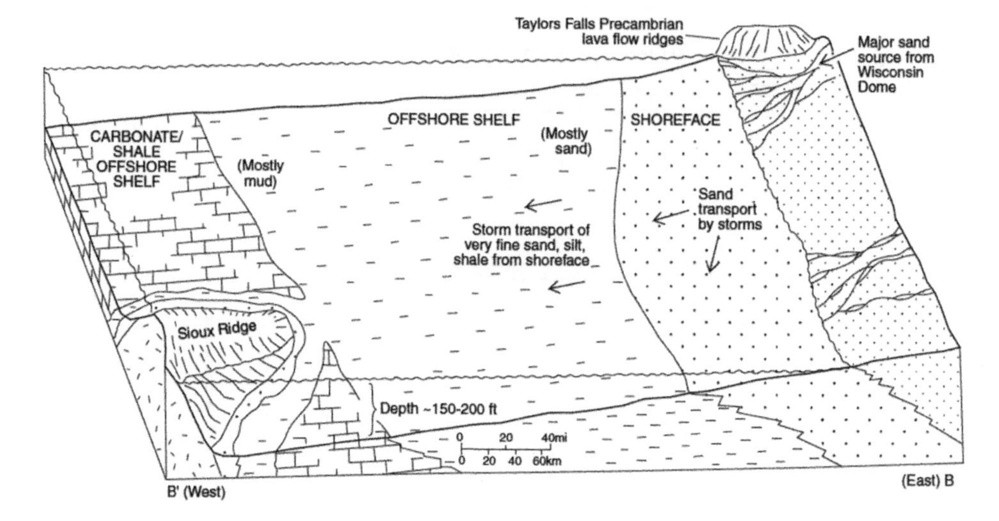

Figure 4.6. Relationships among sand, silt, and carbonate sediment in increasingly deep water in the Hollandale Embayment from Cambrian to Ordovician time. (After Runkel et al. 1998.)

ing just what they tell us about transgressions, regressions, and sedimentary sequences. Figure 4.8 shows the distribution of rocks of various ages in the region, and figure 4.5 will help you keep track of formations and groups.[15] As we review the geology of these sedimentary basins, we are building a framework through which to understand not just the Paleozoic history of Laurentia, but also the origin and setting of the many mineral resources that are produced from its sedimentary rocks, including oil and gas, coal, salt, and lead, and even sand and gravel.

4.5. Stratigraphic Sequences Are Exposed in Paleozoic Rocks of the Great Lakes Region

4.5.1. The Sauk Sequence Was Deposited during Cambrian–Lower Ordovician Time

If it followed the rule book, the Sauk stratigraphic sequence should have started with a marine transgression as the Pannotia supercontinent broke up, and that is exactly what we see in the Great Lakes region. The oldest Paleozoic sediments, which are Late Cambrian in age (about 500 Ma), are sandstones and minor conglomerates that formed as beaches and other near-

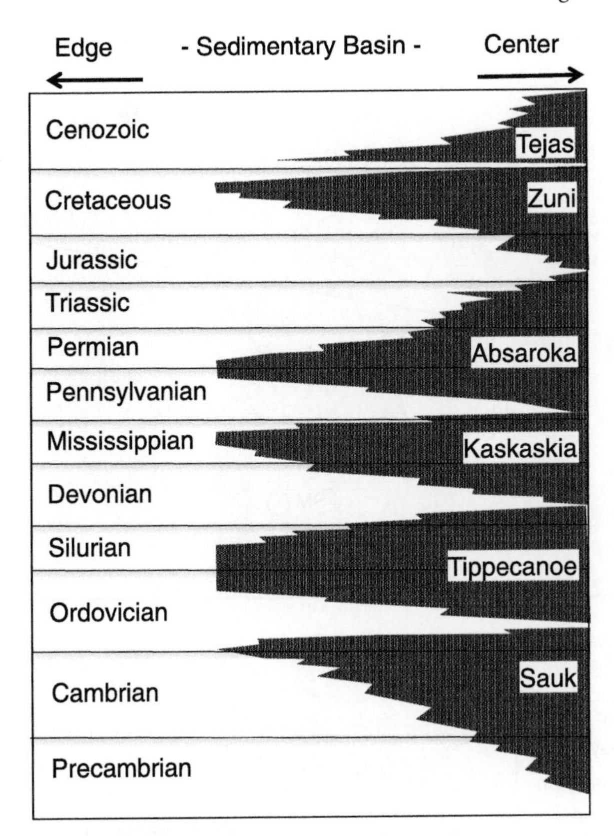

Figure 4.7. North American (Laurentian) stratigraphic sequences showing that sedimentation was continuous in the center (deepest part) of basins (including the margins of continents) and discontinuous on basin margins. (Modified from Sloss 1963.)

shore deposits (fig. 4.5). They are exposed today around the margins of the basins. In Minnesota and Wisconsin, the oldest formation is the Mt. Simon Sandstone, which we saw in figure 4.1A. To the north, in the Pictured Rocks National Lakeshore, the oldest unit is the Munising Formation, which you can see in figure 4.9A (in color section). According to Kenneth Hamblin, the sand in the formation came from Precambrian rocks to the north. The unconformity that these sandstones rest on had been developing for several hundred million years and was relatively flat; all the big Precambrian mountains had been eroded down to small hills. The transgression that deposited these Late Cambrian sandstones in the Great Lakes region is actually the western extension of transgression that started in early Cambrian time along the east side of Laurentia as the Appalachian Basin began to form. The rate at which the sea advanced was very slow, at least in human terms, probably only kilometers per million years—plenty of time to sell oceanfront homes.

The sequence of Cambrian sediments in the Great Lakes region actually shows us that the transgression was not continuous. Instead, it included

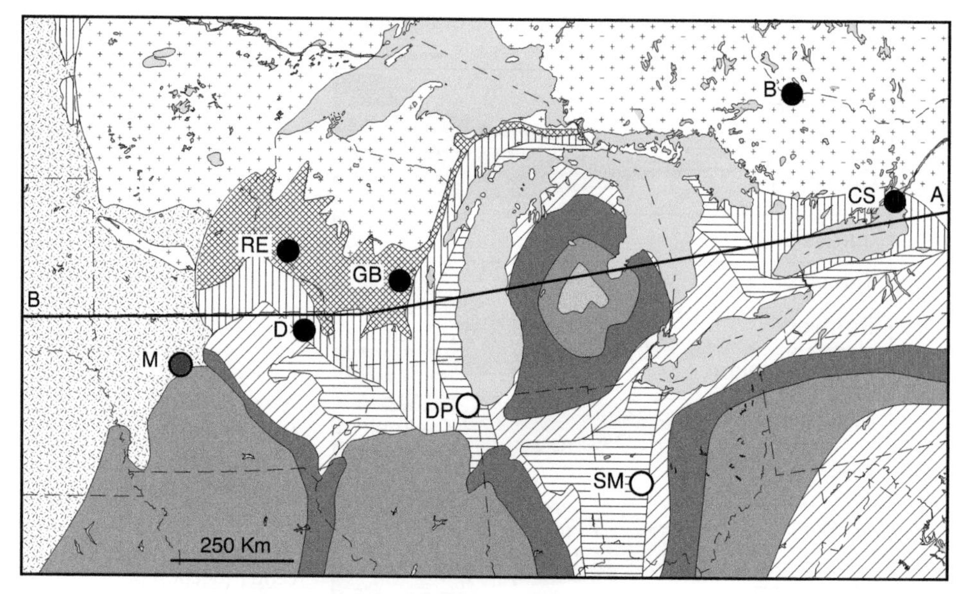

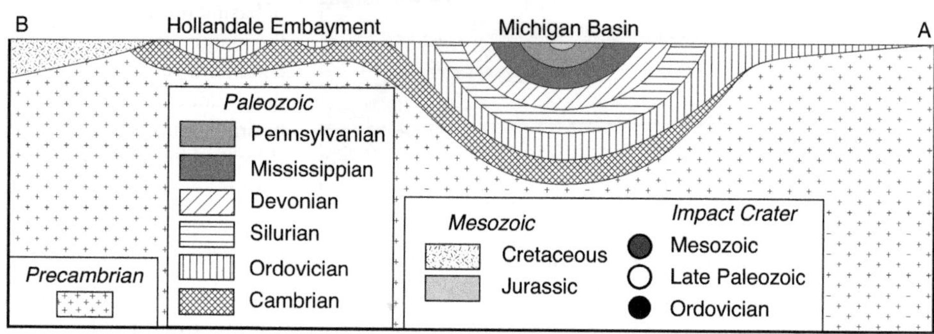

Figure 4.8. (**A**) Geologic map showing sedimentary basins and meteorite impact craters in the Great Lakes region (B – Brent, CS – Charity Shoal, D - Decorah, DP – Des Plaines, GB – Glover Bluff, M – Manson, RE – Rock Elm, SM – Serpent Mound). (**B**) Simplified cross section along line B-A in the map.)

three smaller transgressions and regressions that you can see in the formations listed in figure 4.5. The first transgression deposited sands of the Mt. Simon–Munising Sandstones, which grade upward into the finer-grained silty sediments of the Eau Claire Formation. After the shoreline retreated, a second marine transgression deposited sands of the Galesville and Wonenoc formations and then silty deposits of the Franconia Formation and Tunnel City Group (fig. 4.10A in color section). These are overlain by carbonates of the St. Lawrence Formation in Wisconsin and Minnesota, the Trempealeau

Formation in the Michigan Basin, and the Potosi Dolomite in Illinois, all representing deeper water conditions. Finally, a third transgression deposited the Jordan Sandstone and Trempealeau Formation.

The third transgression continued into early Ordovician time (about 485 to 470 Ma), when carbonates of the Prairie du Chien Group (named for Wisconsin's second-oldest community) were deposited. There is no unconformity between Cambrian and Lower Ordovician formations in the central part of the basin in the Lower Peninsula of Michigan.[16] In peripheral areas, however, an unconformity is present, and unconformities are even seen between other layers in the Lower Ordovician sequence. In Wisconsin and Illinois, for instance, Prairie du Chien equivalent sediments are divided into the Oneota and Shakopee formations (fig. 4.5), which are separated by an unconformity that represents another, more local retreat of the sea.[17]

The thickness of sedimentary layers can be used to estimate the degree of development of sedimentary basins, and thicknesses of Sauk sequence sediments in the Great Lakes region provide information on early subsidence in the Michigan Basin and Hollandale Embayment. A contour map with lines of equal thickness, known as isopachs, for sediments deposited during this period shows that the Michigan Basin was open to the southwest and merged with the Illinois Basin (fig. 4.11A). Thus, although more than 1000 m of sediment were deposited in the Michigan Basin during this time, it was not yet independent of the Illinois Basin. Sediment thickness in the Hollandale Embayment at this time was considerably smaller, reflecting more limited subsidence.

4.5.2. The Tippecanoe Sequence Was Deposited during Middle Ordovician–Silurian Time

When the Tippecanoe sequence began in Middle Ordovician time at about 470 Ma, Laurentia was at the equator and the much larger Gondwana continent had migrated to the South Pole. Glaciation on Gondwana caused a global drop in sea level. The resulting land surface in the Great Lakes region was highest along the Transcontinental Arch in the west and north, as you can see on the paleogeography map in figure 4.11B. Farther to the east and south, emergence and weathering of the Prairie du Chien carbonates formed caves and karst topography that were to be buried by later sediments.

The first sediment to appear during the subsequent Tippecanoe transgression as the Gondwanan glaciers melted was the Middle Ordovician St. Peter Sandstone (fig. 4.10B in color section), which was deposited on beaches, tidal flats, and offshore bars. It is an unusually pure sandstone and consists in large part of quartz grains eroded from older sandstones that had been deposited along the Transcontinental Arch. St. Peter sand grains are also frosted (or pitted), suggesting that they were once part of sand dunes. The St. Peter is overlain by shales of the Glenwood Formation, indicating transgression and deepening water.

By early Tippecanoe time, the Michigan Basin had attained its independent existence, as you can see in isopach map of figure 4.11C, which shows a strong bull's-eye pattern. In the central part of the basin, the Prairie du Chien Group grades upward into the St. Peter Sandstone with no unconformity, as described by David Barnes, Bill Harrison, and Tom Shaw and indicated in figure 4.5 with the merging triangles. This is an important feature. It shows that sedimentation continued in the middle of the basin while erosion was taking place on the margins. This is exactly what Sloss predicted, as shown in figure 4.8: continuous sedimentation in the basin and intermittent sedimentation on its margins. In other words, the deep basin remained full of water while the seas rose and fell along the margins, just as we saw for the preceding Sauk sequence.

The St. Peter and Glenwood are overlain by a thick sequence of Middle Ordovician carbonate sediments, indicating further transgression and deepening of the seas. These include the Black River and Trenton formations in Michigan and the Platteville and Galena formations (fig. 4.10D in color section) to the west in Wisconsin and Minnesota. The shallow sea in which they were deposited is known as the Great American Carbonate Bank (fig. 4.1). The Black River–Trenton carbonates were originally deposited as limestones but were converted to dolomites with the addition of magnesium from formation waters, especially in peripheral areas in Wisconsin, Illinois, and Indiana. Alissa Kendall has explained that this conversion destroyed the limestone raw material required to make cement, and this means that limestone for cement-making must come from states to the east where limestones are present.[18] To the north, sedimentation in the Moose River–Hudson Bay Basin (fig. 4.11B) began at this time, a sign that transgression had finally reached into the center of Laurentia.

Shale and sandstone in these sediments were derived from erosion of the Taconic Mountains, which were forming along the eastern side of Laurentia as the Iapetus Ocean closed. The isopach map of these rocks in figure 4.11E shows a tilt to the east rather than the Michigan Basin bull's eye seen earlier. Bernard Coakley and Mike Gurnis suggested that this was caused by the Taconic orogeny, which was deforming the Appalachian Basin at that time.[19]

Near the end of the Ordovician, the sky began to fall . . . and this went on for quite a while. First came a major meteorite shower, probably related to

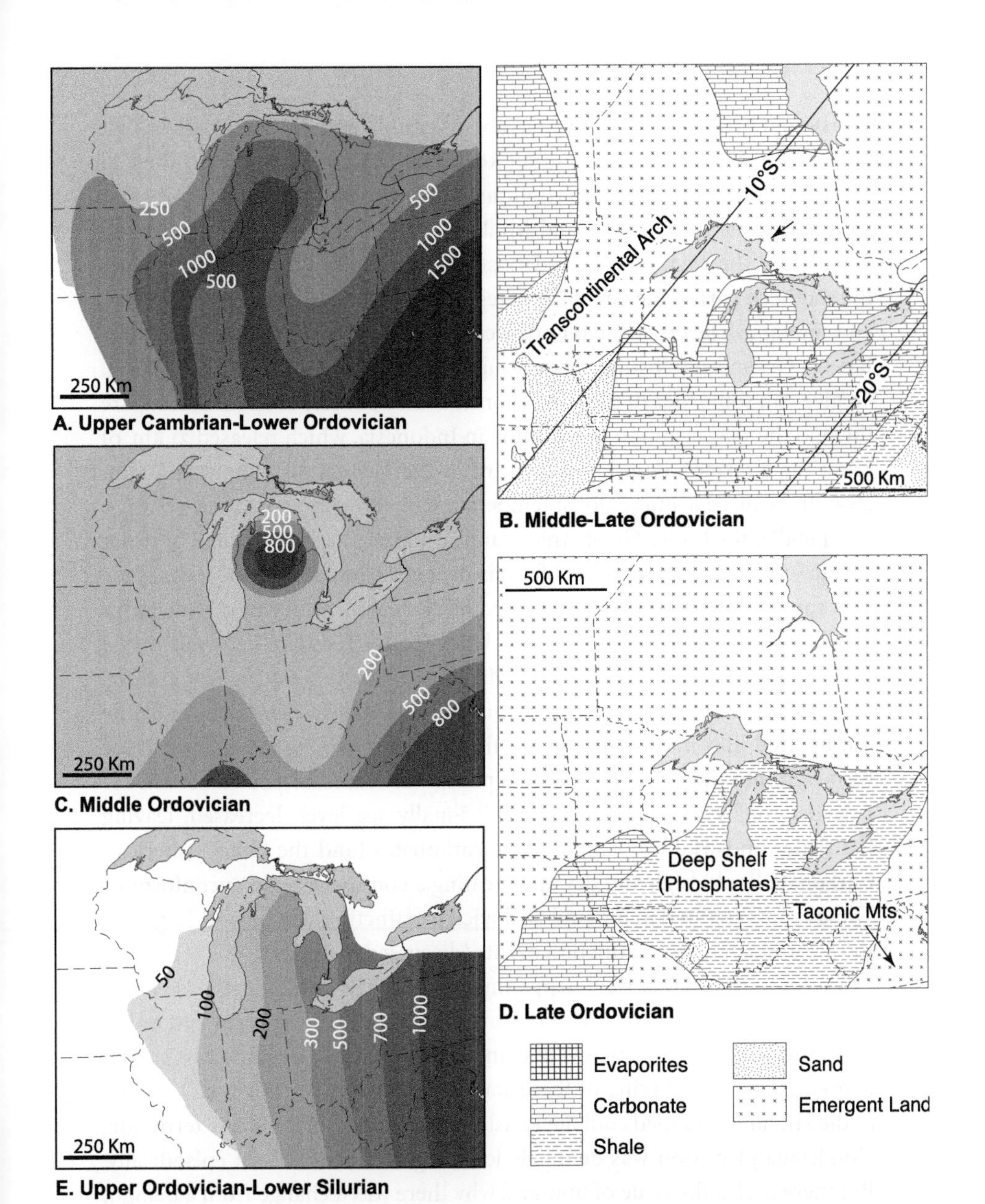

Figure 4.11. Cambrian and Ordovician sediments: (**A**) Thickness (isopachs) of Upper Cambrian–Lower Ordovician sediments; (**B**) paleogeography during deposition of Middle Ordovician Black River–Trenton sediments; (**C**) isopachs of Middle Ordovician Black River–Trenton sediments; (**D**) paleogeography during deposition of Late Ordovician Queenston-Utica-Maquoketa sediments; (**E**) isopachs of the shales and overlying Lower Silurian sediments. (Compiled from Sloan 1987; Witzke and Kolata 1989; Howell and van der Pluijm 1990; Friedman and Kopaska-Merkel 1991; Gutschick and Sandberg 1991; Sonnenfeld and Al-Asam 1991; Mossler 1992, 2008; Swezey et al. 2008.)

an asteroid breakup. It happened at about 470 Ma and generated debris that impacted in several areas, including Rock Elm, Wisconsin (fig. 4.8). Then, at about 457 Ma, came volcanic eruptions associated with the Taconic orogeny, which deposited the Diecke and Millbrig volcanic ash beds in the Trenton Formation. The ash layers covered most of the east side of Laurentia, including the southern part of the Great Lakes region, from New York to South Dakota, and similar ash layers covered northwestern Europe (which was nearby as the Iapetus Ocean was closing). These eruptions released about 2500 km^3 of ash, To give you some idea of the truly enormous size of these eruptions, compare them to the 1883 eruption of Krakatoa in Indonesia, which released 21 km^3 of ash. (The Krakatoa eruption was heard in Perth, Australia, about 2000 km away and caused average global temperatures to drop by more than 1°C.[20])

Finally, the lovely Great American Carbonate Bank, the home of widespread and varied Ordovician life, was overrun by oxygen-poor seas. These seas snuffed out the life and deposited black shales that go by a wide range of names, including Queenston and Utica in the east, Collingwood in the Michigan Basin, and Decorah and Maquoketa to the west (fig. 4.5 and fig. 4.10C). They were part of the large Queenston Delta system that was fed by the uplifted Taconic Mountains on the east side of Laurentia. They grade westward into carbonates containing phosphate lenses, as you can see in the paleogeography map in figure 4.11D.[21] Finally sea level decreased, leaving the shales and what remained of the carbonates (and the Great American Carbonate Bank) high and dry and creating a continent-scale unconformity. These events resulted in the End Ordovician extinction, the second-largest of the great Phanerozoic extinction events (discussed below).

A lot of things that were happening at or near the end of the Ordovician probably played a role in this series of events. On the long-term end of the scale, the Taconic orogeny was pushing mountains up along the east side of Laurentia, and weathering of these was probably drawing down CO_2, much as the Himalayas helped cool the Pleistocene climate. On the short-term side, Gondwana glaciation was definitely lowering sea level, at least episodically. But there is also the issue of how and why there was a change from a benign, oxygenated shallow sea to an anoxic sea flooded with dead organic matter, a change that seems to require a stratified ocean. Just how all these events were related is an area of active research.[22]

The world started over at the beginning of Silurian time, at about 444 Ma, when the sea advanced across this unconformity, depositing shales and then widespread carbonates nourished by the near-equatorial position of the continent. By Middle Silurian time, carbonate reefs appeared. These are generally referred to as Niagaran and include formations and groups named Lockport, Guelph, and Amabel in the east and Cataract, Burnt Bluff, Manistique, and Engadine in the central and western part of the region. One of these, the Guelph Formation, can be seen outside a quarry near Wiarton, Ontario where there is a statue of Wiarton Willie, a now-deceased competitor of the groundhog Punxsutawney Phil. Where the Niagaran carbonate rocks reach the surface today (after consolidation and erosion of the Michigan Basin), they form a ridge that starts on the west in the Door and Garden peninsulas and continues eastward through the Upper and Bruce peninsulas into the Niagara area of Ontario and New York. This prominent cuesta, known as the Niagara Escarpment, is the precipice for many waterfalls, including Niagara Falls, as discussed in chapters 2 and 9.

During Middle and Late Silurian time, the Niagaran carbonates and overlying Salina Group formed an extensive two-part reef system around the Michigan Basin (fig. 4.12A). The reef system began with an outer complex that caused an imbalance between inflowing and evaporating water, leading to the deposition of hundreds of meters of evaporite deposits and interlayered carbonates of the Salina Group (fig. 4.5). The main mineral in the evaporites is halite (salt), although an unusually thick layer of sylvite (potash) is present in the A-1 evaporite layer described recently by Bill Harrison and discussed further in the section on mineral deposits. The evaporation crisis was caused in part by the relatively southerly position of the Great Lakes region at the time, about 20 degrees south of the equator. Five episodes of evaporite deposition took place, apparently representing episodic eustatic changes in sea level.[23] As the Salina Group accumulated, pinnacle reefs (box 4.3) with a towerlike form developed in a belt on the deep-water side of the main reef (fig. 4.12E).[24] Isopach maps of the Silurian sediments (fig. 4.12B) show a return of the bull's-eye pattern, indicating independent subsidence in the Michigan Basin.

BOX 4.2. HOW TO DEPOSIT HUNDREDS OF METERS OF SALT IN A SHALLOW BASIN AND SURROUND IT WITH PINNACLE REEFS

Evaporation of seawater in most sedimentary basins typically takes place in relatively shallow water, only a few meters to tens of meters deep. You might ask, how such a shallow sea could deposit layers of salt that are hundreds of meters thick, such as we find in the Salina Group. The problem gets even more perplexing if you consider that seawater contains only about 3.5 percent dissolved salt. So complete evaporation of all the seawater in a basin 100 m deep would yield a layer of salt only about 3.5 m thick. The only solution to this problem is for the basin to subside continually as evaporites are deposited. The rate of subsidence must keep pace with the rate of water flow into the basin and with the rate of evaporation, so that the depth of the basin remains constant. Pinnacle reefs around the evaporite basin provide additional evidence for this remarkable period of continuous subsidence (fig. 4.12E). Individual pinnacle reefs are only about 0.5 km^2 in area, but they are as much as 200 m in height, very different from conventional reefs with lateral extents of many kilometers. Their shape reflects the fact that reefs can grow only where sunlight reaches them, in the shallow photic zone. Apparently, the rate of subsidence was a little faster than the reef organisms could grow over a large area, causing them to crowd together into a small area along the top of the reefs, forming towers or pinnacles.

The history of the Moose River and Hudson Bay basins during Silurian time is generally similar to that of the Michigan Basin. Limestones of the Severn River Formation built reefs that led to closed basin conditions and deposition of several hundred meters of halite evaporites of the Kenogami River Formation in the Hudson Bay Basin to the north (fig. 4.5). In the Moose River Basin, correlative sediments are largely sands and siltstones. The presence of these sediments suggests that much of the area between the Moose River and Michigan basins was also covered by a thin layer of Paleozoic sediments that has since been eroded.

4.5.3. The Kaskaskia Sequence Was Deposited during Devonian–Lower Carboniferous Time

By the start of Devonian time at around 419 Ma, Laurentia had joined with Greenland and Europe to form Euramerica (Laurussia). Erosion removed Silurian sediments that had been deposited in the western part of the Great Lakes region and formed the basal Kaskaskia unconformity (fig. 4.5).[25] As the sea advanced westward into the Great Lakes region, it did not form a widespread sandstone. Instead we see carbonates of the Detroit River and Traverse Groups. Extensive reefs again created intermittent closed basin conditions and deposition of several hundred meters of evaporites in the Detroit River Group (fig. 4.9E in color section). These reefs contain the corals that form Petoskey stones, the state stone of Michigan. The stratigraphic

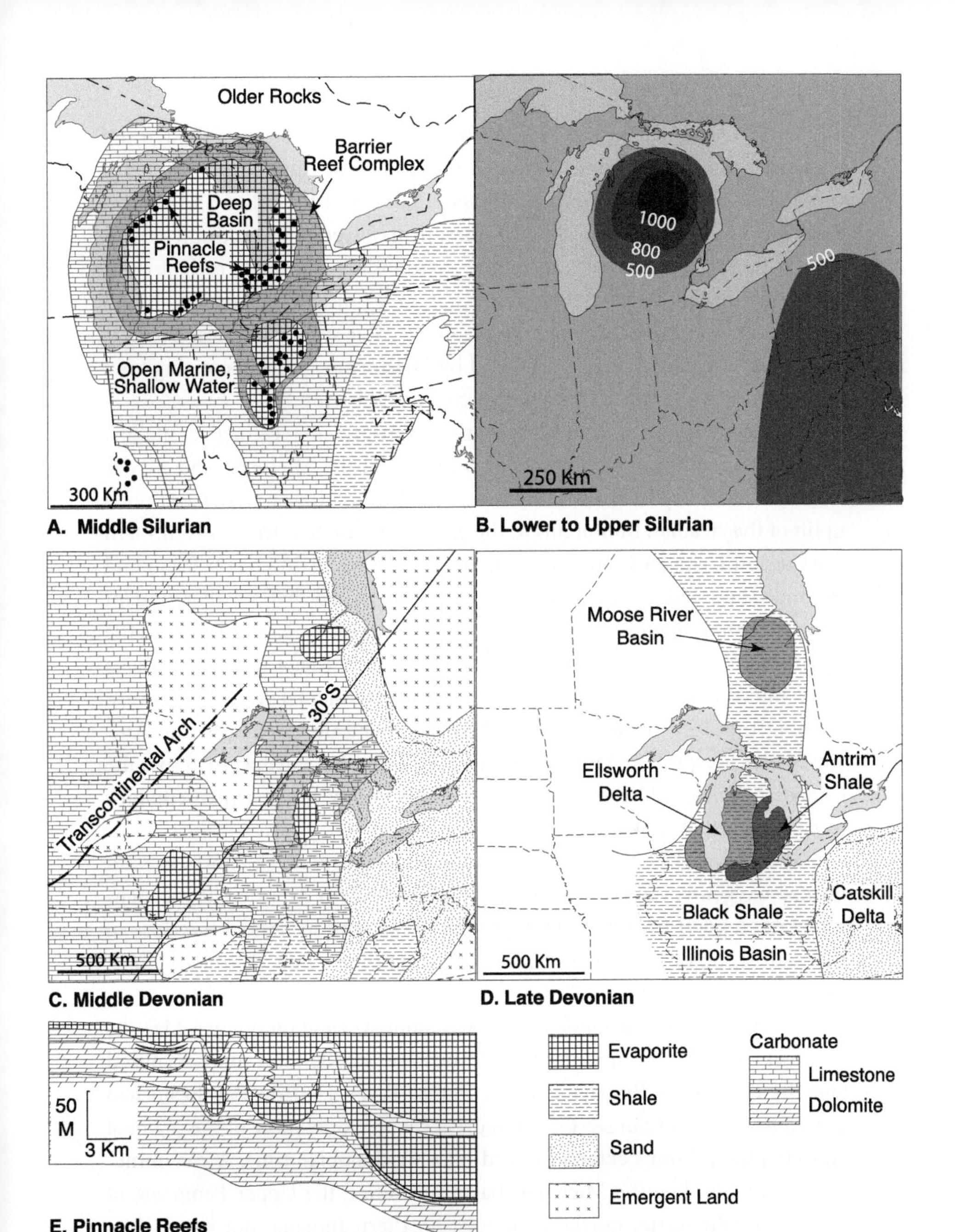

Figure 4.12. Silurian and Devonian sediments: (**A**) Middle Silurian paleogeography during deposition of Salina Group carbonate reefs and evaporites; (**B**) thickness (isopachs) of Lower to Upper Silurian sediments; (**C**) Middle Devonian paleogeography during deposition of Detroit River evaporites; (**D**) Late Devonian paleogeography during deposition of the Antrim and Ellsworth shales; (**E**) cross section through the pinnacle reef trend in A, showing distribution of limestone, dolomite, and evaporite. (A through D compiled from sources listed in figs. 4.5 and 4.11.)

sequence was similar in the Moose River–Hudson Bay Basin, although the evaporites are much thinner (fig. 4.5). Some of these salt deposits were not preserved. By Middle Devonian time, salt deposits of the Salina and Detroit River Groups that were near the surface were dissolved by ground and basinal waters. Removal of the salt caused the surrounding carbonate rocks to collapse, forming the Mackinac breccias that are exposed on Mackinac Island and around St. Ignace, Michigan (fig. 4.9F in color section). (These breccias formed the weak rock that was eroded by Pleistocene glaciers, creating low-lying areas now occupied by Lakes Michigan and Huron.)

By Late Devonian time, the marine transgression formed a shallow seaway from south of the Illinois Basin through the Michigan and Moose River–Hudson Bay basins and into northern Canada (fig. 4.12D). Continued uplift of the Acadian Mountains to the east shed sediment westward through the Catskill Delta and across the Findlay Arch into the Michigan Basin and beyond. These sediments show up now as fine-grained shales with abundant organic matter, including the Antrim, Ellsworth, Ohio, New Albany, Kettle Point (Figure 4.9B in color section), and Milwaukee (fig. 4.7). Sediment also came into the Michigan Basin from the Transcontinental Arch to the west, forming the delta of the Ellsworth Shale that interfingered with the Antrim Shale (fig. 4.12D). Similar shale deposits extended northward across the Canadian Shield into the Moose River–Hudson Bay Basin, which contains organic-rich shale of the Long Rapids Formation.

These remarkable shales mark a time when stagnant, shallow seas with limited oxygen covered much of midcontinent North America. It was definitely not a good time to be living in the ocean, and it marked the close of the flourishing of Devonian life and the onset of the End Devonian global extinction (discussed below).

By the start of Carboniferous (Mississippian) time (about 359 Ma) the Michigan and Appalachian basins had become smaller and more localized and the Hollandale Embayment and Moose River–Hudson Bay Basin had ceased to exist. You can see this in figure 4.5 where only three of the regional stratigraphic columns extend upward into Mississippian time. Areas immediately surrounding the Michigan Basin, including the Upper Peninsula of Michigan, Wisconsin, northern Illinois, northern Indiana, northern Ohio, and southwestern Ontario, were exposed and being eroded. This erosion produced sediment that formed the Berea Sandstones, as well as shales of the Sunbury and Coldwater formations, followed by the Marshall Sandstone (fig. 4.5), which is an important groundwater aquifer in Michigan because of its high porosity and permeability and its setting at the top of the basin immediately below the Pleistocene glacial deposits. According to James Harrell and his coworkers, all the clastic sediments were deposited in a shallow marine environment, with stream and river (fluvial) deposits only in the eastern part of the basin. So we see that the Michigan Basin was beginning to stabilize, with less and less subsidence.

4.5.4. The Absaroka Sequence Was Deposited during Pennsylvanian Time

Near the end of Mississippian time, glaciation in the newly formed supercontinent Gondwana (which was to combine with Euramerica to form Pangea at the end of the Paleozoic) led to a major decline in sea level and formation of a widespread unconformity that marked the end of the main Kaskaskia stage and nearly the end of the Michigan Basin. When sea level rose again in Pennsylvanian time, starting at about 323 Ma, only the central part of the Michigan Basin and the main part of the Appalachian Basin were flooded. Exactly how the Michigan and Illinois basins were connected to the sea is not clear because no sediments remain to show its path. The main Pennsylvanian sediments in the region are deltaic sandstones and shales (fig. 4.9G,H in color section) along with coal (box 4.4) and some limestones of the Saginaw and Grand River formations.

BOX 4.3. PALEOZOIC COAL DEPOSITS OF LAURENTIA

Plants moved out of the seas and onto land in Middle to Late Ordovician time, but they did not become large and abundant enough to form coal deposits until Late Carboniferous (Pennsylvanian) time. Coal deposits in the Appalachian, Illinois, and Michigan basins formed in swamps associated with deltas of rivers that were flowing westward from the Alleghenian Mountains into shallow seas that were rising to cover the Absaroka erosion surface (fig. 4.8). The coal layers are part of a repeating sequence of marine and nonmarine layers known as cyclothems, which probably formed during cyclic changes in sea level related to variations in the intensity of glaciation in Gondwana.

Individual coal layers and swamps in the Illinois and Appalachian basins were enormous; the famous Pittsburgh coal seam in the Appalachian Basin covers an area of about 21,000 km^2, is up to 1.5 m thick, and contained an original resource of about 34 billion short tons of coal. Continuous coal layers of this type can be mined with large-scale, mechanized, underground or open pit methods, depending on the thickness of overlying rocks, and both the Illinois and Appalachian basins have been important coal producers. Coal deposits in the Michigan Basin are smaller in terms of the number, thickness, and extent of coal layers. From 1835 to 1950, underground mining managed to produce 46 million tons of coal in Michigan, a very small amount compared to the more than 4 billion tons produced in Pennsylvania between 1960 and 2014.

Although younger Paleozoic sediments are not known in the Great Lakes region, there is good reason to suspect that they were present originally and have since been eroded. It all relates to the coal deposits. It is well known that coal is deposited as peat and that it changes from peat to lignite to bituminous rank as it is heated by burial beneath younger sediments. It turns out that burial to produce the thermal maturity of Michigan coal requires more sediment than is present today. Karen Cercone and Henry Pollack showed that about 2 km of overlying sediments are required to account for the present rank of coal in the Michigan Basin. These sediments are not there. Furthermore, the lack of thermal effects in overlying Mesozoic sediments in the area (described below) shows that the coals were heated before these sediments were deposited. Thus, it is likely that the original thickness of Pennsylvanian or Permian sediments was greater than what we see now and some were eroded. These vanished sediments probably came from the east where collision and uplift of the Alleghenian Mountains marked the final assembly of the supercontinent Pangea.

4.6. Mesozoic Rocks Are Found in Only a Few Locations

At the start of Mesozoic time around 252 Ma, the Great Lakes region was near the center of the Pangea supercontinent, and the Michigan Basin had stabilized enough to stop subsiding and receiving sediment. It reactivated briefly in Middle Jurassic time, for unknown reasons, when the Ionia Formation was deposited in the center of the basin (fig. 4.8). The Ionia consists of about 100 m of sand and silt with layers of gypsum that accumulated in dry, desertlike stream beds. Similar sediments are found in north-central Iowa (including the Fort Dodge Gypsum) and in a zone extending from northwestern Iowa into Manitoba, indicating that desert conditions were widespread.

During an otherwise uneventful Jurassic period, at least 27 kimberlite pipes were intruded in the Upper Peninsula of Michigan and adjacent

Wisconsin. Kimberlites, which are discussed further in chapter 7, are igneous intrusions that come from the mantle. In some places, they contain diamonds, although economically interesting amounts have not been found in the Upper Peninsula kimberlites. Farther to the north, however, kimberlites were intruded in the Timiskaming Graben at about 155 to 134 Ma and in the James Bay lowlands (including the Victor diamond mine, mentioned in chapter 7) between about 180 and 170 Ma. These kimberlites may have been related to the breakup of Laurasia (which separated Laurentia and Europe) or to the passage of a hot spot (mantle plume).[26]

By the beginning of Late Cretaceous time, at about 145 Ma, most of the Great Lakes region was above sea level, although the shallow western interior seaway invaded the far western margin of the Great Lakes region. The surface that this sea invaded had been largely emergent since Devonian time and had a subdued relief of up to 400 m. Marine sandstones and shales in the western and northern parts of Minnesota change southward into delta and lake sediments in southern Minnesota and Iowa.

Meteorites kept falling. The peaceful western interior seaway was disturbed near the end of Cretaceous time at about 74 Ma by the Manson meteorite impact in Iowa (fig. 4.8). The huge crater, which is about 38 km in diameter, is one of the largest in North America, and ejecta from it has been found in adjoining states. It is so large, in fact, that the Manson impact crater was originally considered a candidate for the end-Cretaceous dinosaur extinction event. However, detailed age studies by Glen Izett and his colleagues showed that it is about 10 million years too old. Although it might not have exterminated all the dinosaurs, the Manson impact did cause a big disturbance; sediment at 17 localities in the surrounding region are thought to have been deposited by a tsunami generated by the impact.[27]

4.7. Paleozoic Rocks Have a Large Oil and Gas Endowment and a Wide Variety of Hard Mineral Resources

4.7.1. The Great Lakes Region Has a Long History of Oil and Gas Production

Oil and natural gas is produced in the Great Lakes region from the Michigan Basin, as well as the margins of the Illinois and Appalachian basins (fig. 4.13A). In terms of historical oil production, the Illinois Basin is first, followed by the Michigan Basin and then our part of the Appalachian Basin. For natural gas, the Appalachian Basin is the hands-down winner because of recent production from the Marcellus Shale, followed by the Michigan Basin with production from the Antrim Shale.

The first well that produced oil in North America was a 3-m hole dug near Oil Springs, Ontario, in 1858 (to be followed in 1859 by the better-known Drake well in Pennsylvania [fig. 4.13A]). Early efforts were attracted to Oil Springs and nearby Petrolia because the ground was saturated with tar, which formed when oil migrated upward to the basin margin. The most important of these marginal fields began with a well at Findlay, Ohio, in 1884, which led to discovery of the Lima-Peru-Trenton trend (fig. 4.13A,B). As methods improved, the Deerfield and Port Huron fields in Michigan were discovered in the basin farther down from the Lima-Peru-Trenton trend and the Oil Springs fields.

4.7.2. Separate Hydrocarbon Systems Formed in Different Parts of the Basins

Oil and gas in and around the Michigan Basin comes from geologic features called hydrocarbon systems, as described by Christopher Swezey and others. The hydrocarbon systems consist of a source rock from which the oil was derived by reactions that took place during burial; a porous, permeable reservoir rock in which the oil collected as it seeped out of the source rock; and an impermeable cap rock that prevented the oil from escaping.[28] Hydrocarbon systems can be linked to the rock sequences reviewed above. Within a single transgressive-regressive system, shales serve as source rocks and sandstones and reef carbonates serve as reservoir rocks. A key part of these hydrocarbon systems is the source rock shale, which needs lots of organic matter and must have been buried deeply enough to reach temperatures and pressures at which oil or gas are generated.

In terms of total production, the fabulous Middle Ordovician Trenton–Black River system is the most important hydrocarbon system in the Great Lakes region (fig. 4.13B). Reservoirs for this system are in the Trenton–Black River carbonates, and the source rocks are probably shale layers within the carbonates or the overlying Collingwood Shale. Oil was probably released during Late Devonian subsidence of the Michigan Basin associated with uplift of the Acadian Appalachians. Late burial of the Collingwood during Pennsylvanian-Permian time probably generated additional gas.

The Trenton–Black River system hosts important fields within the Michigan Basin and along its southern margin. The single most important field in the basin is the large Albion-Scipio field, which has produced more than 125 million barrels of oil and 250 billion cubic feet of natural gas. Albion-Scipio occupies a fracture zone in the Trenton–Black River carbonates where the limestone has been converted to dolomite and partially dissolved to produced cavities called vugs. Cap rock for the fracture zone is low-porosity limestone and shale of the upper Trenton and the overlying Collingwood Shale. Several other, similar fields trend roughly parallel to Albion-Scipio including the most recent discovery, the Napoleon field, just 25 km to the east. Other fields formed along the southern margin of the basin, notably the Lima-Peru trend, which is also in vuggy Trenton Dolomite. The Lima-Peru trend, including its extension to the west into the Trenton field, is credited with production of 500 million barrels of oil and 1 trillion cubic feet of gas, which started in 1884 and peaked in 1896.[29] The Lima-Peru fields are linked to the Albion-Scipio field by a swarm of smaller fields, including the Deerfield, which contain oil and gas that was apparently intercepted as it migrated upward toward the margin of the basin (box 4.5, fig. 4.13B). Things are not as clear for the Trenton field at the west end of the Lima-Peru-Trenton trend, which straddles the Kankakee arch and may have come at least in part from the Illinois Basin.

BOX 4.4. WHOSE OIL?

Early production of oil and gas involved all sorts of chicanery. One favorite practice was to pump as fast as possible from a well along a property line, trying to drain oil from the adjacent property. In recognition of this, the industry developed the practice of unitization or pooling, in which all property owners were allotted a fair share of total production from a field. The concept was not extended to the source rock, however. Does the owner of property over the source rock have any right to the value of oil and gas that was derived from the source rock? Could unitization be taken one step further to include the entire basin, both source rock and traps? (In fact, this does happen when source rocks are fracked today, as in the Antrim fields discussed below, but it was not done in older fields.) We can probably identify the source of oils. Jürgen Rullkötter, Philip Meyers, and others have shown that oils of the Michigan Basin can be distinguished in part by the use of isotopic and chemical tests. But can we convince owners of the reservoirs to share their bounty? Residents of Michigan might wish for this to happen, however, if it could be shown that the enormous amount of oil and gas produced from the Lima-Peru-Trenton fields in Ohio was actually sourced from their Michigan Basin.

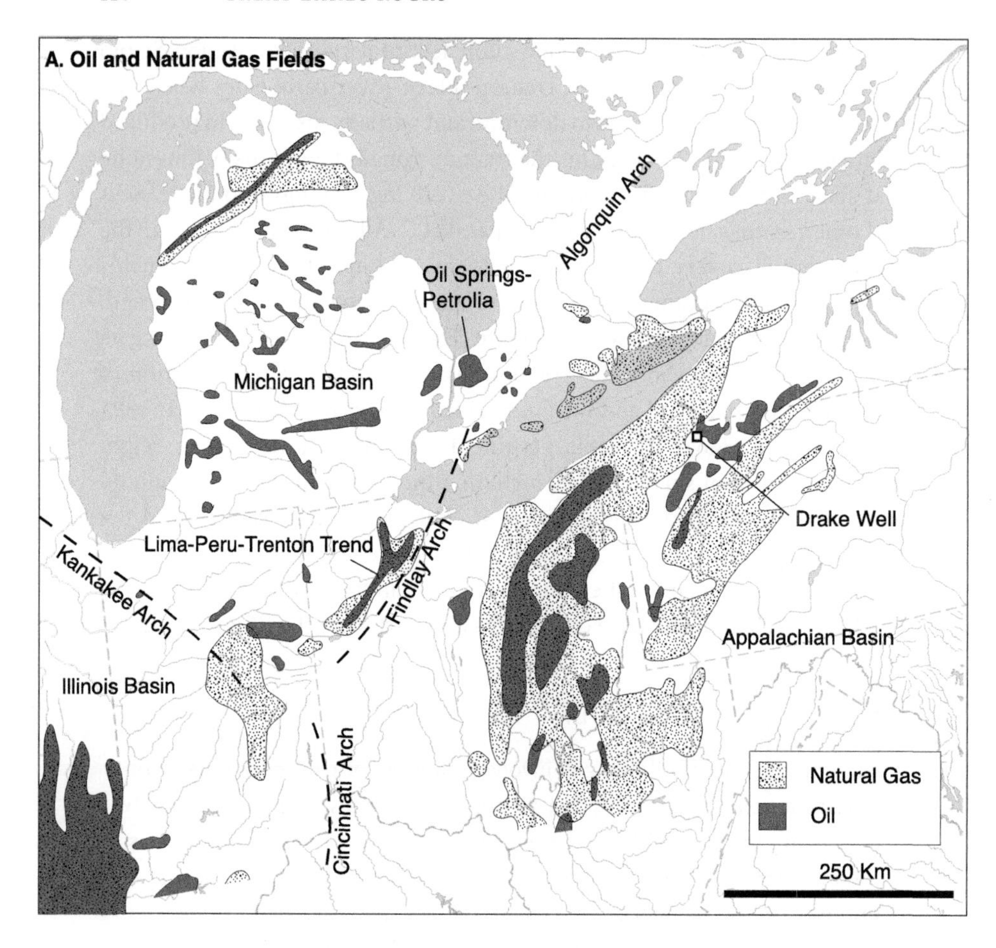

The Upper Devonian Antrim gas system (fig. 4.13C) in the Michigan Basin is a remarkable, almost unique hydrocarbon system that consists solely of source rock with no reservoir rock. Gas is produced directly from the Antrim Shale (which is the source rock) by fracking the rock using the same process used to produce oil and gas from other important source rocks in North America, including the Marcellus Shale in the Appalachian Basin of Pennsylvania and the Bakken Formation in the Williston Basin of North Dakota. But there is a big difference. In Pennsylvania and North Dakota, the source rocks have been buried deeply enough in their basins for the organic matter to start forming oil and gas, but it has not had time to seep far enough out of the source rocks to reach a reservoir rock. In contrast, the Antrim Shale has never been buried deeply enough for the organic matter to be converted to gas. Something else formed the gas. Anna Martini, Lynn Walter, and others

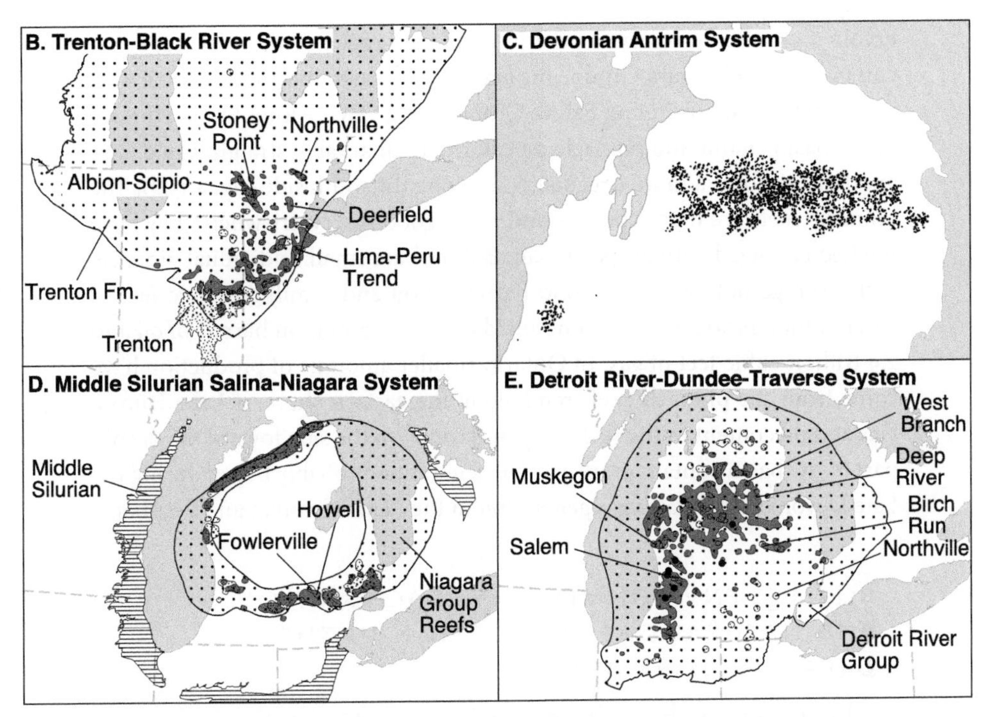

Figure 4.13. Distribution of oil and gas fields in (**A**) the Great Lakes and adjacent regions; (**B**) Middle Ordovician Trenton–Black River system; (**C**) Upper Devonian Antrim system; (**D**) Middle Silurian Salina-Niagara system. (Modified from Swezey et al. 2015.)

have shown that the gas was produced when microbes ate organic matter in the shale, not unlike what happens in modern landfills that produce gas. The microbes entered the shale when meltwater from the Pleistocene glaciers seeped into fractures in the near-surface part of the Antrim. This makes the Antrim one of the most unusual gas producers in the world. So far it has produced 3.1 trillion cubic feet of natural gas from a tremendous number of closely spaced wells that are clustered just a few hundred meters down-dip from the outcrop of the Antrim in the northern part of the basin. (As you can see in fig. 4.13C, the wells are so closely spaced that they cannot be shown separately.) Of all the hydrocarbon systems in the region, this one has the most obvious environmental legacy (box 4.6).

The Salina-Niagara system is another peculiar hydrocarbon system. For one thing, the source rocks for the system are thought to be carbonate rocks and shales in the Salina and Niagara groups, indicating that the oil and gas did not travel far from its source. Even more unusual are some of the reservoir rocks. Some reservoirs are in carbonate rocks of the first stage reef complex in the Niagara Group, but many more are found in individual pinnacle reefs of the overlying Salina Group (fig. 4.12A). Along the north side of the basin, about 800 pinnacle reefs form a remarkably straight trend, and another 300 form a more irregular zone along the south side (fig. 4.13D). The pinnacle reefs are highly porous and make good reservoirs where they have not been plugged with evaporite salt. But they are small. Individual pinnacle reefs average only about 300,000 barrels of oil and 2 million cubic feet of gas, but total production has amounted to over 500 million barrels of oil and 2.9 trillion cubic feet of gas. In Ontario smaller amounts of production have come from the pinnacle reef trend along the eastern shore of Lake Huron. Gas and oil content in the reefs vary, and those on the side toward the deeper part of the basin contain more gas. Oil generation probably began during the Late Devonian with later gas generation in the Pennsylvanian and Permian.

BOX 4.5. ENVIRONMENTAL LEGACIES OF OIL AND NATURAL GAS PRODUCTION IN THE GREAT LAKES REGION

Oil and gas production in the Great Lakes region has a long and continuing environmental impact. It started in 1860, when a well drilled near Oil Springs, Ontario, became North America's first uncontrolled gusher, releasing thousands of barrels of oil a day, which eventually reached Lake St. Clair. Since then spills have been relatively small, and interest has centered on two environmental issues. The first is the possibility of producing oil and gas from reservoirs under the Great Lakes. Potential reservoir rocks for oil and natural gas continue beneath all the Great Lakes except possibly Lake Superior, and the US Geological Survey has estimated that they contain up to 430 million barrels of oil and 5.2 trillion cubic feet of gas. The only offshore production that has taken place so far is on the Ontario side of Lake Erie, where it began in 1913. Since then at least 2000 wells have been drilled from platforms in the lake, and about 480 have produced gas, all from Middle Silurian Clinton sands at a depth of only 100 to 200 m. Oil production is not allowed from offshore platforms in Canadian waters, although directional drilling from shore can be used to produce oil from beneath the water. In the United States, Michigan has allowed directional drilling beneath Lake Michigan, and production still comes from some of these wells, although the practice was ended in 2002. The second issue centers on the use of water in fracking of wells, especially in the Collingwood and other deep oil or gas shales. Fracking of this type requires injection of large amounts of water, much greater than the 50,000 gallons or less that is used for fracking of wells in the Antrim Shale. Much of this water is returned to the surface by backflow after fracking, and it is then disposed of by injection into deep formations that contain saline waters (brines) of no current economic interest. Although the total volume of water used in fracking is small compared to the total hydrologic cycle, this practice is of concern because it essentially removes water from the hydrosphere.[30]

Detroit River–Dundee–Traverse (Devonian) system fields (fig. 4.13E) are near the top of the Michigan Basin and were among the earliest to be discovered. Reservoirs in these rocks consist largely of vuggy carbonates with caps of evaporite, shale, and low-porosity limestone. Many of the traps are subtle anticlines that formed during the Late Mississippian or later. The anticlines are not really the result of conventional compressive folding; some follow reactivated faults in underlying Precambrian rocks, and others formed by dissolution of underlying Salina evaporites. Some anticlines host production from several horizons; the anticline at the Northville field produced originally from the Silurian Salina and Niagara groups and then progressed deeper to production from the Devonian Dundee Limestone and finally the Trenton–Black River formations. In Ontario a few fields along the north side of the Algonquin Arch appear to be the up-dip extension of the Michigan Basin system. Oil and gas in all these fields probably came from the Amherstburg and Lucas formations, both of which contain organic-rich carbonates and are in the oil window in the middle of the basin.[31]

4.8. The Paleozoic Sedimentary Rocks also Contain Hard Mineral Deposits

4.8.1. The Most Important Hard Minerals Are Called Industrial Minerals

Mining of the Paleozoic sedimentary rocks yields sand, limestone, dolomite, gypsum, and salt, which are known as industrial minerals because of their markets (fig. 4.14). Some sand from the Cambrian sediments in the Great Lakes region is used in construction and as an abrasive, but most is used in fracking during oil and gas drilling, where it is injected into fractures to prop them open, allowing fluid to flow to a well. Proppant or "frac" sand is mined from the St. Peter, Jordan, Wonewoc, and Mt. Simon formations, especially in Wisconsin. The St. Peter Sandstone is particularly good because it was derived from erosion of the Late Cambrian sandstones along the Transcontinental Arch, where the second phase of erosion and transport made the sand grains more spherical and therefore easier to inject during fracking.

Carbonate rocks are mined in the Great Lakes area for several products. The Mississippian Bedford (Salem) Limestone in Indiana is used widely as a building stone (and is the state stone). Devonian and Silurian limestones along the shores of Lake Huron and the eastern shore of Lake Michigan are mined for crushed stone, cement, and chemical markets. Many of these operations use barge transport to major markets, taking advantage of their shoreline location, as discussed by Alissa Kendall and others. Along the western side of Lake Michigan, carbonate rocks have been converted to dolomite, which is mined largely for crushed stone.

Salt deposits in the Salina Group evaporites are mined by underground operations where they come near the surface along the St. Clair and Detroit rivers on both sides of the US-Canada border. Detroit River Group evaporites are "mined" from drill holes near Midland, Michigan, where they are dissolved and the brines pumped out. Elsewhere, mining is carried out underground. The largest salt mine in the world, near Goderich, Ontario, extends more than 7 km beneath Lake Huron.

A very thick zone of sylvite in the Salina Group evaporites in the Michigan Basin is a candidate for possible future solution mining to produce potassium-bearing (potash) fertilizer. Gypsum has been produced from the Michigan Formation in open pit mines near Alabaster and Tawas, Michigan, and by underground methods in the western part of Michigan near Grand Rapids. It was deposited originally as gypsum, was converted to anhydrite during burial,[32] and finally changed back to gypsum during uplift and contact with groundwater.

4.8.2. Other Hard Mineral Deposits Contain Lead-Zinc-Fluorine-Barium and Iron

Hard mineral deposits containing galena (PbS, the Wisconsin state mineral), sphalerite (ZnS), fluorite (CaF_2, the Illinois state mineral), barite ($BaSO_4$), and celestite ($SrSO_4$) fill fractures and holes in limestone and dolomite. They are known as Mississippi Valley type (MVT) deposits because of their abundance in this part of the world. The minerals were precipitated from brines that flowed out of the basins and encountered H_2S or some other form of dissolved sulfur in the porous limestones and dolomites. The most important MVT deposits in the Great Lakes region are in the Upper Mississippi Valley

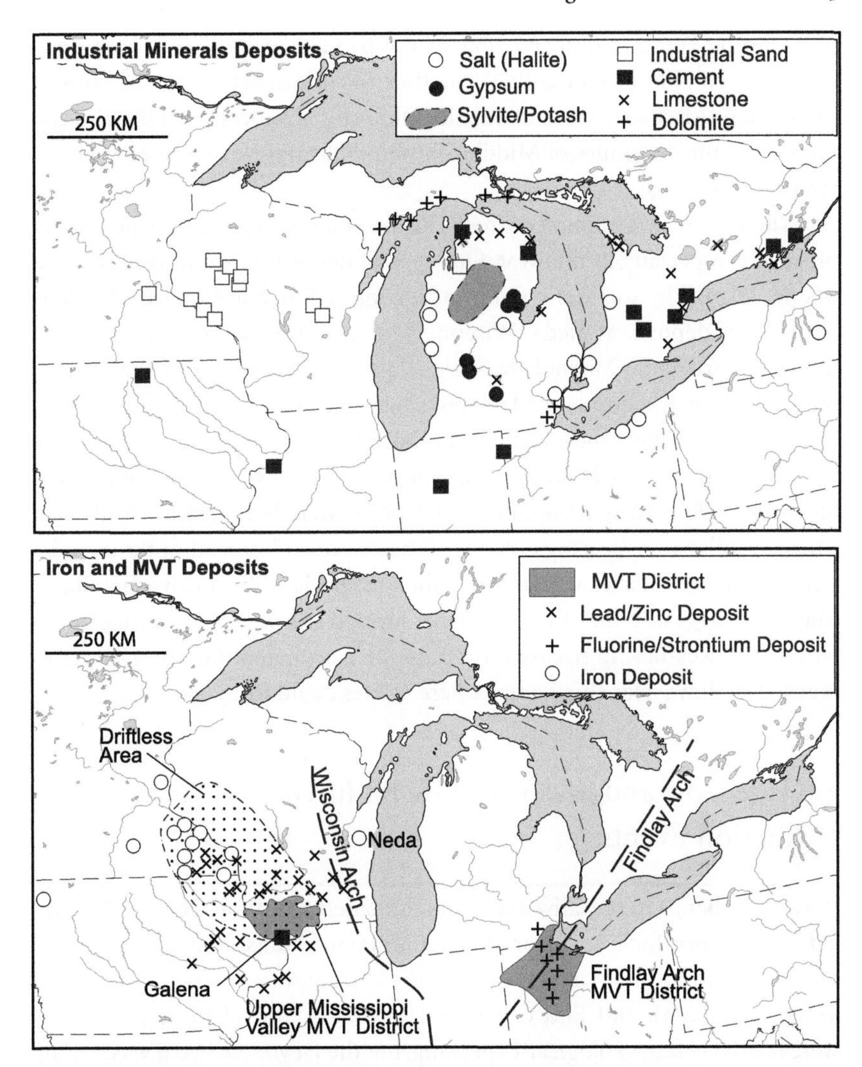

Figure 4.14. Distribution of industrial mineral, iron, and Mississippi Valley type (MVT) lead-zinc-fluorite-strontium deposits in the Great Lakes region. (Modified from Heyl 1968; Kean 1981; Harrison 2014.)

district in the southern part of the driftless area (fig. 4.14), where the town of Galena, Illinois, hints at the historical importance of mining. Brines that formed these deposits probably came from the Illinois Basin and migrated toward its margins along the Mt. Simon or St. Peter Sandstone and then up into the carbonate rocks above. Joyce Brannon and others have shown that the sphalerite has an age of 269 Ma, indicating that fluid migration took place during Permian time, possibly driven by plate tectonic collisions along the southern side of Laurentia.[33] Another MVT district along the crest of the Findlay Arch southwest of Toledo, Ohio (fig. 4.14) contains fluorite and celestite in the dolomites of Middle Silurian (Niagaran) to Middle Devonian (Detroit River) age.

The Paleozoic sedimentary rocks also contain several iron deposits and prospects, especially in the west. One type of deposit, in southwestern Minnesota, formed by weathering of iron-bearing sedimentary rocks.[34] A different type of deposit, at Neda, Wisconsin, is in the Neda Formation, which directly overlies the Maquoketa Shale. Here, the upper part of the Neda is made up of small spheres called oolites, consisting of iron hydroxide (goethite, FeO[OH]), and the uppermost 0.3 m is hard hematite (Fe_3O_4). This hematite is Permian in age, much younger than the Ordovician sediments, and probably formed when the goethite-bearing deposits were buried after Ordovician time.[35] Similar iron-bearing oolitic sediments are widespread in the Appalachians and Great Lakes region,[36] and their origin is still controversial. Barry Maynard suggested that these deposits formed during very unusual times when deep weathering released iron into the ocean, and the iron-rich water reached shallow, near shore areas where oolites could form.

4.9 Paleozoic and Mesozoic Life Endured Five Major Extinction Events

It was a challenge to stay alive throughout Phanerozoic time. The fossil record shows evidence for five major global extinction events, known as the End Ordovician (440 Ma), Late Devonian (365 Ma), End Permian (252 Ma), End Triassic (201 Ma), and End Cretaceous (66 Ma). Four of these extinctions were relatively fast, geologically speaking, but the Devonian event was so long that it cannot be referred to accurately as End Devonian. Great Lakes rocks record two of these extinctions, the End Ordovician and Late Devonian. Permian sediments are missing in the region; the end Triassic extinction preceded deposition of the small Jurassic sequence in the Michigan Basin, and Cretaceous sediments along the far western margin of the area are too old to record the End Cretaceous event.

These were major events in which as many as 96 percent of known species are said to have disappeared from the fossil record. Although Steve Stanley has shown recently that these estimates might be a bit high, there is no doubt that the extinction events were global in extent and catastrophic to life, as reviewed recently by Peter Brennan in *The Ends of the World*.[37]

The repeat offenders for the extinction events were the carbon and supercontinent cycles and their distortion of global climates. For the End Ordovician extinction, there is growing evidence that weathering of the ancestral Appalachian mountains formed during the Taconic orogeny helped cool Gondwana enough to form glaciers that lowered sea level. Werner Buggisch and others have suggested that cooling started with the enormous Deicke and Millbrig volcanic eruptions at about 457 Ma. The meteorite bombardment, which occurred 10 million years earlier than the eruptions, is a less likely perpetrator, although Birger Schmitz and others have noted that it coincided with the Great Ordovician Biodiversity Event, discussed below. For the Late Devonian extinction, the jury is still deliberating, and recent suggestions are surprisingly scarce. It poses problems in part because it consists of several separate extinction events that took place over a period of 10 to 15 million years. Sea level was lowering because of glaciation, and Tom Algeo and others have called attention to the role of newly evolved vascular land plants, which could have increased soil formation and nutrient supply to the ocean, leading to marine anoxia. Anoxia, of course, would have been especially destructive to life on a shallow shelf like the Great American Carbonate Bank.

Large-scale volcanic eruptions of a different kind have emerged as potentially pivotal agents of later extinction events. These large-scale eruptions are not the sort of explosive, felsic eruptions that formed the Ordovician ash deposits, which would have come from conical stratovolcanoes found in island arcs such as Japan and the Aleutians. Instead they are large plateau basalts similar to those of Hawaii and the Midcontinent Rift, which is the subject of the next chapter. All volcanic eruptions release CO_2, and these plateau basalts are important sources simply because of their enormous volume. The End Permian event is correlated with eruption of the Siberian traps (another name for basalt) in eastern Siberia, which released an enormous amount of lava (as well as CO_2) that could have covered North America to a depth of at least 100 m. The End Cretaceous event, though widely ascribed to devastation from the Chicxulub meteorite impact, also coincides with the Deccan traps, which erupted about 1 million km^3 of basalt. Recent high-precision age measurements have shown that an important phase of the Deccan eruption coincided with the age of Chicxulub, leading to the suggestion that the eruption was triggered by the impact.[38]

The Paleozoic extinctions were a time of dying, but in the wake of these events other oganisms evolved to take the place of those that exited the scene. And some of the new life showed up in what appeared to be rapid explosions. The first of these was the so-called Cambrian explosion at the start of Cambrian time when most important phyla appeared in the fossil record,[39]

Early Cambrian sediments are missing in the Great Lakes region, and the Late Cambrian sediments that are here caught only the last part of this explosion (fig. 4.15A in color section). The best known of the Cambrian fauna is the trilobite (fig. 4.15C in color section). The sudden appearance of an organism with such a complex form, including an external skeleton and segmented body, is one of the factors that encouraged early speculation about the Cambrian explosion. Trilobites persisted in the fossil record for most of Paleozoic time; a few orders disappeared in the End Ordovician extinction, many more in the Late Devonian, and one continued until the major Permian extinction event. Other common Cambrian fossils include those of unusual spongelike organisms known as archaeocyathids, cephalopods that ate trilobites, and brachiopods, which littered the seafloor.[40]

The Cambrian explosion takes a backseat to the Great Ordovician Biodiversification Event, which Thomas Servais and others have called "the most important and sustained increase of marine biodiversity in Earth's history." According to Colin Edwards and others, genus-level diversity quadrupled, possibly due to a cooling climate and an increase in oxygen in the atmosphere. The Ordovician marine fauna were rewarded for their huge diversification by being largely wiped out by the end Ordovician extinction event. Brachiopods and bryozoan were especially hard hit. The new marine fauna that developed in the Silurian and Devonian (fig. 4.15B in color section) included crinoids (fig. 4.15E in color section), echinoderms, articulate brachiopods, two types of corals (rugose and tabulate), and large arthropods such as eurypterids (fig. 4.15D in color section). During the Late Devonian extinction, brachiopods, trilobites, and reef builders such as coral and stromatoporoids were almost completely obliterated.

In addition life began to find its way onto land. Land plants showed up in Ordovician time, as evidenced by the presence of fossil spores, and in Devonian time vascular plants reached large size. Invertebrate animals appeared on land in Silurian time, followed by vertebrates in Late Devonian time.

In Carboniferous time, large forests developed in swamps along the margins of most sedimentary basins in Laurentia. Accumulation of these plants in ancient swamps formed peat deposits that matured into coal by burial beneath younger sediment. The End Permian extinction event had a curiously uneven effect on Paleozoic life. Marine invertebrates were very strongly affected, with almost catastrophic extinction. The greatest impact was on organisms with calcium carbonate shells, possibly because of acidity of the ocean caused by dissolved CO_2. Terrestrial invertebrates and vertebrates also suffered major extinctions. Plants differed from other life forms in showing relatively little change at the family and even the species level in all the major Paleozoic extinction events. A recent study by Shuzhong Shen and others, based on the excellent fossil record in China, suggests that the End Permian event was unusually short, lasting only about 200,000 years, and that it was associated with widespread fires, which released soot and CO_2 into the atmosphere.

This episodic blossoming and extinction of life during Paleozoic and Mesozoic time took place in only the last 10 to 15 percent of Earth history. What was Earth doing and how was life evolving during the almost 4 billion years that preceded the Paleozoic? Fortunately for us, the Great Lakes region contains a lot of rocks that can help answer these questions as we step farther back into truly deep time in the next few chapters.

Notes

1. Don't get the idea that these could be glacial erratics like we saw in the last chapter. The boulders are clearly rounded by wave action, unlike most erratics, and they were deposited in the sea, not on land.

2. Laurentia is the core of North America and part of the supercontinent story that will be discussed later in this chapter. Even the Canadian Shield was covered partly by Paleozoic sediments. Remnants of these sediments are preserved in a few areas that were dropped down by faulting, notably the Timiskaming Graben and Ottawa Embayment along the border between Ontario and Quebec that are discussed further in chapter 9 (fig. 4.4). Also, waters recovered from many of the underground mines in the Precambrian rocks north of the Great Lakes have been interpreted to be Paleozoic-age basinal brines that seeped down from Paleozoic sediments into the basement rocks. See Bottomley et al. (1994) for a description of the basinal brines and Doughty et al. (2010) for a recent description of the Timiskaming Graben.

3. Relative changes in sea level commonly reflect changes in the height of continents, such as might happen when a continent is heated by a rising plume of hot mantle or when a thick ice sheet melts, removing weight from the crust. These changes are also referred to as dynamic topography because they reflect movement in parts of Earth's mantle and crust. Jean Braun (2010) provides a discussion of dynamic topography, and Rebecca Flowers and her coauthors (2012) discuss some of the difficulties in distinguishing dynamic topography from other effects. For recent discussions of eustasy, see Miller et al. (2005, 2015); Rovere and Vacchi (2016); Rowley (2017).

4. See Kopp et al. (2016, including revisions) for estimates of sea level changes caused by anthropogenic warming during the present-day interglacial.

5. The Phanerozoic Eon is the entire span of geologic time since the Precambrian, including the Paleozoic, Mesozoic, and Cenozoic eras. For a review of geologic time, see chapter 1.

6. See Lallensack (2017) for information on how gravity can affect sea level.

7. Sedimentary rocks are divided into three groups: clastic, chemical, and biogenic. Clastic sedimentary rocks deposited by water range from conglomerates consisting of cobbles to sandstones consisting largely of sand-size grains to shales consisting of

smaller silt and clay-size grains. Chemical and biogenic sedimentary rocks consist of material that was originally dissolved in seawater and then deposited from seawater either by simple precipitation or by organisms that used the dissolved material to make shells or skeletons. The most common rocks of this type are limestone and dolomite (or dolostone). Limestone consists of the mineral calcite ($CaCO_3$), ranging from fossils of marine organisms to fine-grained particles of probable biogenic origin. Dolomite [$CaMg(CO_3)$] forms where limestone reacts with magnesium in Mg-rich seawater and basinal brines. In arms of the sea where evaporation exceeds inflow of seawater, evaporation causes precipitation of chemical sediments known as evaporites. When originally deposited, most sediments are unconsolidated; they are converted to sedimentary rocks by burial beneath other sediments and by reactions between grains of sediment and surrounding pore waters, a process collectively known as diagenesis.

8. This description of the formation of Pannotia is based largely on Scotese (2009) and Nance et al. (2014). Earlier descriptions of Pannotia by Torsvik et al. (1996) and Dalziel (1997) differ somewhat but still show Pannotia occupying southerly latitudes. Laurentia included what is known today as the Canadian Shield (see chapters 6 and 7). Rodinia is discussed in chapter 5. Pangea is the large supercontinent that combined almost all the continents at the end of Paleozoic time. Pangea broke into two fragments during the Mesozoic: Gondwana and Laurasia.

9. The Taconic, Acadian, and Alleghenian (also known as Appalachian) orogenies took place over periods of time lasting tens of millions of years.

10. The Hollandale Embayment was an extension or bay on the north side of the larger Forest City basin to the south.

11. Miall (2008) reviewed types of sedimentary basins. Coleman and Cahan (2012) classified sedimentary basins in the United States into four types and provided examples of each, as well as brief discussions of their relation to plate tectonic processes. There are several suggestions for the origin of intracratonic basins. Middleton (2007) suggested that some form above zones of mantle that flow downward (the reverse of a rising mantle plume). More common suggestions involve reactivation of older faults or rifts as suggested by McBride et al. (2007) for the Illinois Basin and Pawlak et al. (2011) for the Hudson Bay–Moose River basins.

12. This compilation shows only the most important formations and groups and divides their lithology into sandstone, shale and mudstones, carbonates (limestone and dolomite), and evaporites. For more details, see Swezey et al. (2015); Mossler (2008); Armstrong and Dodge (2007); Wisconsin Geological Survey (2011); Sanford and Grant (1990). See also reports for specific time periods and locations mentioned in the text and notes.

13. Evidence for the presence of sedimentary rocks in the Great Lakes region that have since been eroded is discussed by Michael Velbel (2009) in his summary of events after the Paleozoic.

14. See Sloss (1963) for an early discussion of sequence stratigraphy, Witzke et al. (1996) and Burgess (2008) for a discussion of its application to North America, and Coe (2002) for a broader discussion of the relationship between the sedimentary record and sea level changes.

15. For more on the Paleozoic stratigraphy of the Great Lakes region, see the many excellent regional syntheses and summaries, including Fisher et al. (1988); Catacasinos and Daniels (1991); Witzke et al. (1996); Mossler (2008); LoDuca (2009); Grammer et al. (2018).

16. Barnes and Harrison (1996) have shown that sediments in the Prairie du Chien Group in the central part of the Michigan Basin are not separated from the underlying Cambrian sediments by an unconformity. Instead sedimentation was continuous.

17. Ordovician limestones and dolomites are exposed at the surface along the western and northern margins of the Michigan Basin (fig. 4.9 in color section), where they form cliffs above the softer Cambrian sandstones and siltstones. These cliffs are the Cambrian-Ordovician cuesta, discussed in chapter 3 as the control for many waterfalls in the Upper Peninsula of Michigan, including Tahquamenon Falls (fig. 2.12A).

18. Budai and Wilson (1991) suggested that this widespread dolomitization was related to late-stage fluid movement, possibly during Late Paleozoic Appalachian uplift, and that the fluids included hydrocarbons, as well as basinal brines containing lead and zinc.

19. This is an example of dynamic topography causing a local (continent-scale) change in sea level (rather than a global scale change. It also provides further support for the proposal that processes along the craton margin caused renewed subsidence of the Midcontinent Rift and that this initiated the Michigan Basin. Ettensohn and Lierman (2015) have shown that deposition of black shales began in the Appalachian Basin and migrated westward as tectonic control of the Michigan Basin was taken over by processes along the eastern margin of the continent.

20. For a description of the Rock Elm feature, see French et al. (2004). Korochatzeva et al. (2007) and Schmitz et al. (2008) discuss the evidence for timing of the event and its relation to life. For descriptions of the Deicke and Millbrig ash beds, see Haynes (1994); Huff et al. (1996). For one side of the controversy about the role of ash in global cooling, see Buggisch et al. (2011).

21. These shales are relatively rich in organic matter and have generated interest as potential oil or gas shales (Daniels and Morton-Thompson, 2010). According to Raatz and Ludvigson (1996), the arm of the Paleozoic sea (known as a seaway) that deposited the Maquoketa Shale in southern Minnesota was up to 200 m deep and contained density-stratified water. Upwelling of water from this deep seaway is thought to have brought the nutrients that allowed deposition of the widespread phosphate deposits in the Maquoketa Shale.

22. Finnegan et al. (2011) showed that Ordovician ice sheets were at least as large as those of the Pleistocene, and Pohl et al. (2017) showed that weathering of newly exposed glacial sediments produced the nutrients that led to anoxia in a sea that was probably stratified by glacial meltwater.

23. According to Sonnenfeld and Al-Aasm (1991), all five of the evaporation episodes in the Michigan Basin reached salt saturation, but only the first deposited sylvite. With the passage of time, the evaporite basin grew larger and transgressed into northern Ohio and Ontario. It stopped when the Salina Group was covered by dolomites of the Bass Island Group at the end of Silurian time.

24. The pinnacle reefs, described in Friedman and Kopaska-Merkel (1991), have played an important role in the oil and gas story of the Michigan Basin, as discussed later in this chapter.

25. The Kaskaskia unconformity does not coincide with widespread glaciation and might be related instead to warping of the Laurentian continent in response to the Taconic collisions that assembled Euramerica.

26. See Cannon and Mudrey (1981) for the geology of the Michigan kimberlites and

Heaman and Kjarsgaard (2000) and Cundari et al (2018) for the distribution and possible origins of kimberlites north of Lake Superior

27. Chapter 6 has more information about features that allow identification of meteorite impact craters.

28. For further information on the generation of oil and gas in sedimentary basins and the morphology of traps, see Kesler and Simon (2015).

29. See Keith and Wickstrom (1992) for information on oil fields of the Lima-Indiana trend. In Ontario the smaller Wigle-Olinda, Rochester, Goldsmith, and Renwick fields are hosted by linear zones of porous (vuggy) dolomite that trend parallel to the Northville field. The cap for these fields is the Blue Mountain Shale, which is correlative with the Collingwood Shale.

30. For a review of fracking and its application in the Michigan Basin, see US Environmental Protection Agency, https://www.epa.gov/sites/production/files/documents/09_Mantell_-_Reuse_508.pdf; and State of Michigan, http://www.michigan.gov/documents/deq/Hydraulic_Fracturing_In_Michigan_423431_7.pdf

31. Production has also come from other less well defined hydrocarbon systems that are both younger and older than the systems described here. They include: (1) the Upper Devonian Berea Sandstone from which the Williams field produced at least 2 million barrels of oil, (2) gas fields in the Upper Mississippian Michigan Formation, (3) Early Ordovician carbonates of the Prairie du Chien Group and the overlying St. Peter Sandstone that have produced natural gas and gas liquids from northwest-trending anticlines up to 10 km in length from fields sealed by the overlying Collingwood Shale. Individual wells are required to have a well spacing of 640 acres; the Reed City field, with five wells, produced 2.4 to 7.3 million cubic feet of gas per day with 2 to 40 barrels of natural liquids. (4) The Cambrian Mt. Simon Formation in Ontario with oil and natural gas fields that have produced up to 1.8 million barrels of oil and 31 billion cubic feet of gas. All fields so far are along the southeast side of the Findlay-Algonquin Arch system, and the oil and gas were probably derived from the Appalachian Basin rather than the Michigan Basin (Lazorek and Carter 2008).

32. Evaporation of seawater starts with deposition of gypsum ($CaSO_4.2H_2O$) followed by (NaCl, salt), sylvite (KCl), and minerals containing Mg, I, Br, and other elements, most of which are produced from brines pumped from the Michigan Basin. Gypsum is converted to anhydrite ($CaSO_4$) by heat and pressure when sediments are buried; the exact temperature at which this happens depends on the presence of the fluid with which it is in contact (Ostroff 1964).

33. Estimates based on thermal alteration of rock materials indicate that the fluids required about 200,000 years to form the Upper Mississippi Valley deposits. Sphalerite in the district has curious repetitive growth zones thought to indicate cyclic advance and retreat of the fluids, possibly related to Milanković cycles (McLimans et al. 1980; Rowan and Goldhaber 1995).

34. These deposits produced about 8 million tons of ore between 1942 and 1968. They consist of iron oxides that resulted from the weathering of iron-bearing sedimentary rocks, especially layers containing siderite ($FeCO_3$) in the Devonian Cedar Valley Formation. The weathering that formed these deposits probably took place during formation of the Cretaceous unconformity (Stauffer and Thiel 1944; Alexander and Wheeler 2015).

35. Oolites are small, spherical grains that usually have a central quartz grain covered

with concentric layers of calcium carbonate. In the Neda Formation, concentric layers in the oolites consist of goethite (Kean 1981; Pattison and Bailey 2016). The deposits at Neda, Wisconsin produced about 2 million tons of ore between 1864 and 1915.

36. Similar rocks have been found in the Manitoulin Dolomite in Illinois and the Cataract Group in Michigan, both of which are near the same Ordovician-Silurian boundary, and iron ores of this type are known in the Clinton Group, also of Lower Silurian age, from New York to Alabama. Observations in the Clinton ores indicate that the ferruginous oolites consist of the iron minerals hematite and chamosite, as well as silica and phosphate (apatite), all surrounded by a matrix of iron-bearing dolomite (Voice and Harrison 2016).

37. Reviews with different perspectives about mass extinction events can be found in Ward (2000); MacLeod (2015); and Benton (2005).

38. For recent studies of the Deccan and Siberian traps and their relation to global extinctions, see Richards et al. (2015); and Burgess (2016).

39. The Cambrian explosion is a hotly debated topic that is highly dependent on the discovery of new fossil evidence. Early opinions in favor of the explosion were influenced by discoveries such as the Burgess Shale in British Columbia. Later opinions favoring a more gradual development of life are based on more recent fossil discoveries, especially those in latest Proterozoic sediments such as the Ediacaran fauna in South Australia, discussed in chapter 5 (Marshall 2006; Briggs 2015).

40. See LoDuca (2009) for a review of Paleozoic life in Michigan. Brachiopods are divided into articulate and inarticulate groups based on the nature of the connection between the two valves (shells).

CHAPTER 5

Rifting the Continent

The Mesoproterozoic Midcontinent Rift and Grenville Province

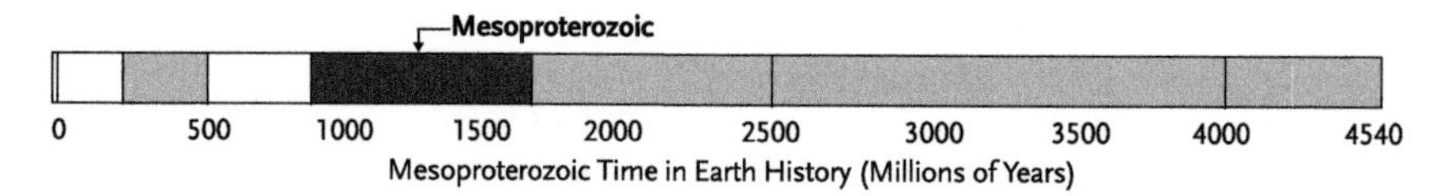

As our time traveler moves farther back through time, geologic processes become increasingly bizarre—and boring. At this point in our story, the time traveler has landed in the middle of the "Boring Billion," a period of geologic time from about 1800 to 800 Ma that geologists elsewhere find boring because there was not much going on. Fortunately for us, the Great Lakes region was one of the places where something was happening. A huge gash in the crust called the Midcontinent Rift was trying to tear Laurentia apart, and in the process it was forming huge deposits of copper, silver, and nickel. There was a faint hint of this process about 1400 million years ago, and it got under way in a big way about 1100 million years ago near the end of Mesoproterozoic time. All this happened long before the Paleozoic Michigan Basin formed, but it created the weakness that started the basin, as we learned in the previous chapter.

The best place to start the Midcontinent Rift story is the Keweenaw Peninsula, which juts into the south side of Lake Superior. The Keweenaw country has always attracted attention because of the native copper that is widespread in its rocks and glacial deposits. Native copper is the elemental form of the metal; it can be picked up off the ground and shaped into complex forms without any further processing. In all other copper-rich areas of the world, copper is combined with elements such as sulfur and requires smelting before it can be liberated and used in industry. Keweenaw copper requires only a little purification for use in industry, and none if used for ornaments and primitive implements. It was the basis for the Old Copper Culture, which included native peoples societies that used and traded copper starting about 6000 years ago.[1] It also attracted the earliest Europeans. First came the missionaries, who found native copper artifacts, decorations, and tools among the local native peoples. They even spoke of a huge boulder of pure copper, the mysterious Ontonagon Boulder mentioned in chapter 1 (fig. 1.2C).

In 1669, the French government sent Louis Jolliet to check on the boulder, but he and his companion, Fr. Jacques Marquette, strayed into the upper reaches of the Mississippi River system instead, as we saw in chapter 1. It was another 100 years before a European trader, Alexander Henry, made two trips to confirm that there was indeed a large boulder of pure copper near the Ontonagon River. Henry had more important challenges than trying to move a 1.6-ton boulder in a canoe. So it wasn't until 1819 that Lewis Cass, the governor of Michigan Territory, arrived to take the boulder back to Detroit. It proved too large to move, however, and remained in place until the 1840s, when Julius Eldred built a temporary rail line to haul the boulder to the Ontonagon River, where it was put on a raft and taken to Lake Superior and then transported by schooner to Detroit. The boulder was eventually claimed by the US government and is now in the Smithsonian Institution.

All this activity attracted Douglass Houghton, a medical doctor and naturalist, who was appointed the first state geologist of Michigan and the first professor of geology at the new university in Ann Arbor. Houghton explored the copper country and wrote the first reports on its geology and copper prospects, which included cautionary notes about the risks involved in trying to open a mine. His final report was never written, however, because he drowned at age 36, in 1845 when his boat was swamped in a storm off the peninsula that had been the focus of his work.[2]

So what was it that Douglass Houghton and his many successors have learned about the Keweenaw and its copper? What is the Midcontinent Rift, how did it form, and how does its copper fit into the story?

5.1. Much of the Midcontinent Rift Is Buried Beneath Younger Rocks and Must Be Found Using Geophysical Surveys

The Midcontinent Rift is an enormous Precambrian gash in the middle of North America filled with lava flows and sediments (fig. 5.1A). The gash formed about 1100 million years ago, when North America's ancestor continent, Laurentia, began to break apart into two smaller continents. We can see rocks that filled the rift along the sides of Lake Superior from Duluth on the west to Sault Ste. Marie on the east (fig. 5.1B), but the rest of the rift is buried

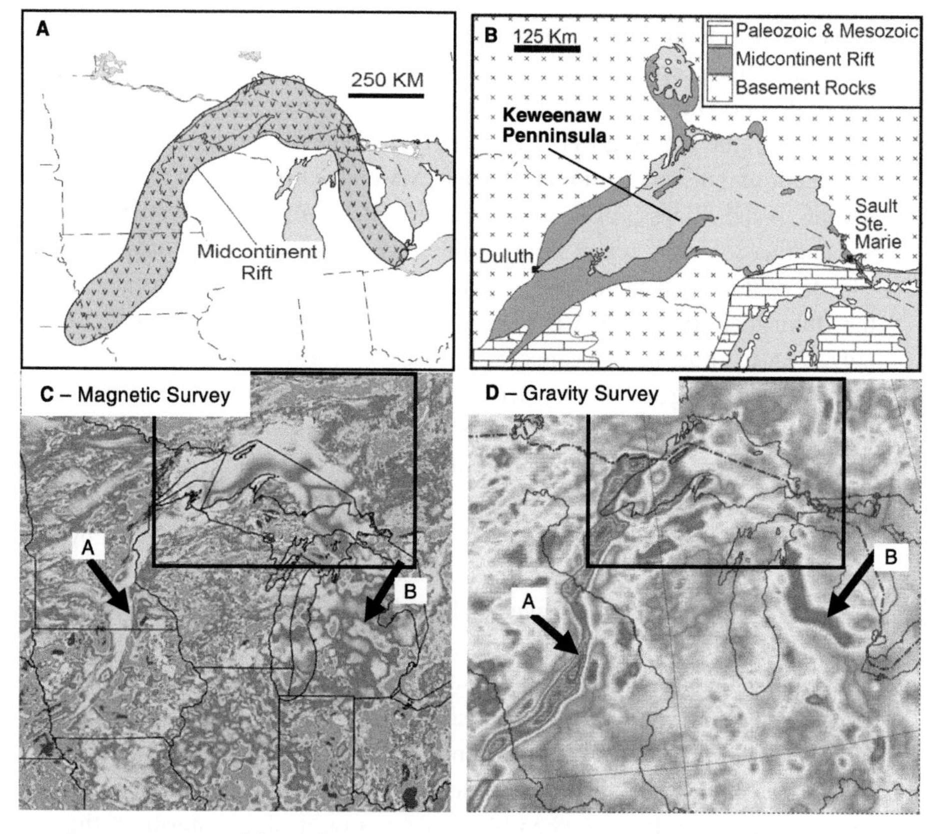

Figure 5.1. Views of the Midcontinent Rift: (**A**) general location of the entire rift, including rocks exposed at the surface and buried beneath younger Paleozoic sediments; (**B**) distribution of exposed rift rocks around Lake Superior (the area labeled "Basement Rocks" includes all Precambrian rocks older than the rift); (**C**) aeromagnetic map and (**D**) isostatic gravity map; rectangle in these maps show area of map in B. Both maps show the rift extending from arrow A to Lake Superior, but only the gravity map shows the rift at arrow B. (Modified from Map 1808A, Isostatic Gravity Anomaly Map of North America, Geological Survey of Canada; http://geogratis.gc.ca/api/en/nrcan-rncan/ess-sst/54d2754e-aa75–5564-a7f6-b2ca0393bfbb.html#distribution)https://mrdata.usgs.gov/magnetic/; Bankey et al. 2002.)

beneath younger rocks at both ends. So just how big is the rift? Did it reach the edges of Laurentia and succeed in splitting the continent, and what role did it play in the Proterozoic supercontinent story?

Geophysical measurements are our best way to see how far the rift extends beneath younger rocks.[3] Start with maps of Earth's magnetic field. Magnetite (Fe_3O_4) is the most important magnetic mineral in the crust, and rocks with magnetite can be readily detected in airborne magnetic surveys,

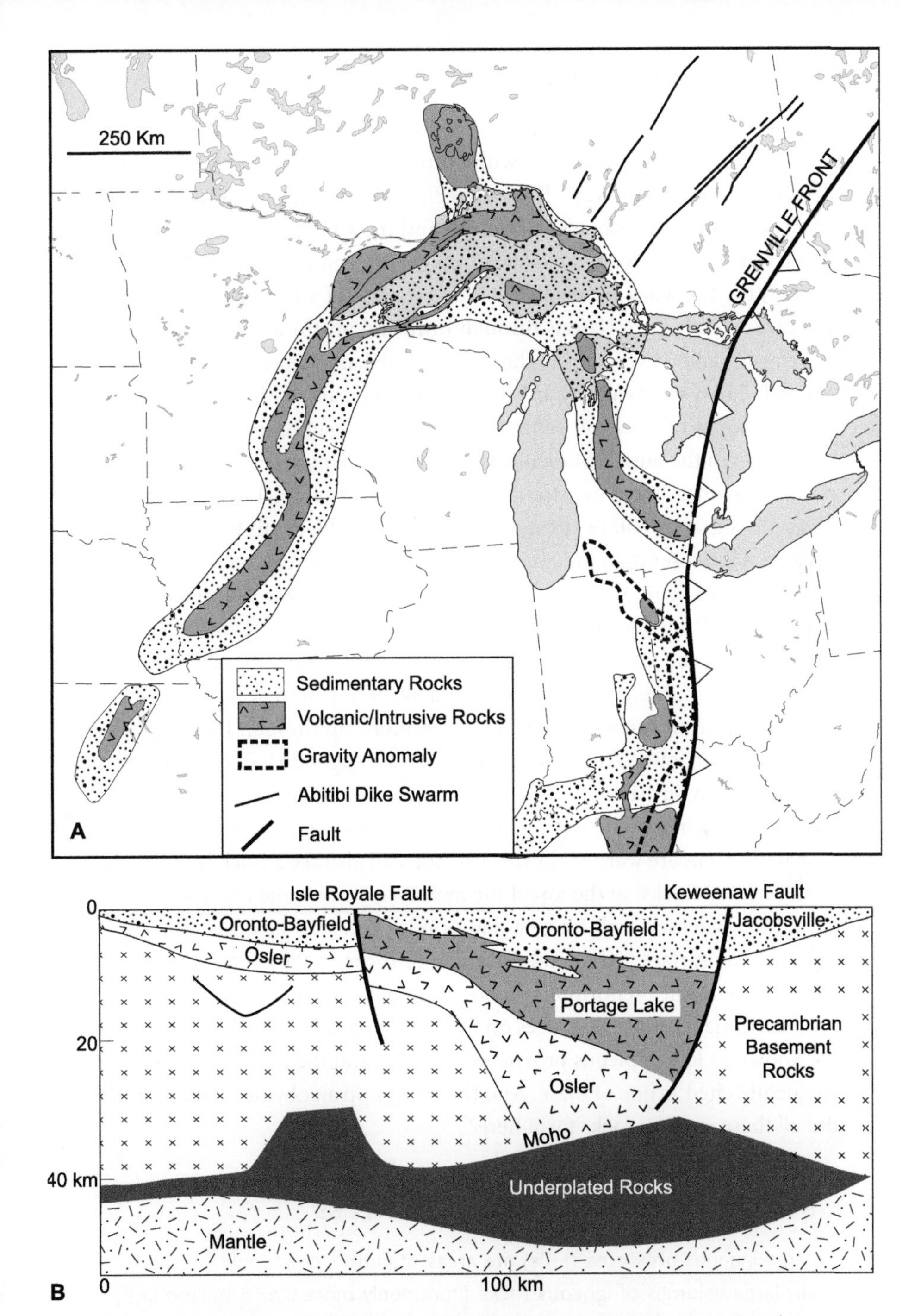

Figure 5.2. (**A**) Midcontinent Rift and other rifts and suspected rifts showing volcanic and sedimentary rocks, and the 1114 Ma Abitibi dike swarm; (**B**) cross section through the Midcontinent Rift beneath Lake Superior showing the probable zone of underplated rocks consisting of magmas that stopped rising at the Moho (crust-mantle boundary). (Modified from Behrendt et al. 1988; Cannon et al. 1989; Trehu et al. 1991; Dickas et al. 1992; Stark 1997; Ojakangas et al. 2001; Heaman et al. 2007; Miller and Nicholson 2013; Stein et al. 2014, 2015, 2016.)

even when they are deeply buried. The main rock in the Midcontinent Rift is basalt, a mafic volcanic rock with abundant magnetite, which is easy to detect in a magnetic survey.[4] This is especially true because the older Precambrian rocks on opposite sides of the rift and the younger sedimentary rocks that cover the rift have much smaller magnetite contents. As you can see in figure 5.1C, the Midcontinent Rift forms a zone of strong magnetic intensity that extends from Kansas through Iowa and northward into Minnesota, Wisconsin, and Michigan. At the east end of Lake Superior, the magnetic anomaly turns southeast toward the Lower Peninsula of Michigan. Here the great thickness of Paleozoic sediments in the Michigan Basin masks the magnetic effect of the buried Midcontinent Rift rocks. However, we can solve this problem by measuring the strength of Earth's gravitational field in the area. The pull of gravity will be greatest over rocks like the basalt that fills the rift because it is denser than the overlying Michigan Basin sedimentary rocks. As you can see in figure 5.1D, the gravity measurements show a strong anomaly in the Lower Peninsula of Michigan, confirming that the Midcontinent Rift is present below the Michigan Basin, as discussed in the previous chapter.

The Midcontinent Rift that is revealed by these geophysical surveys forms a large loop from somewhere in eastern Kansas, up through Lake Superior, and then down toward the west end of Lake Erie (fig. 5.2A). The cross section along line A-A' in Lake Superior (fig. 5.2B) shows that the rift is filled with volcanic rocks below and sedimentary rocks above, which form a syncline in which the rocks are warped downward. You can also see a thick layer labeled "Underplated Rocks" at the top of the mantle just below the rift. This is interpreted to be dense basalt magma that ponded beneath the crust rather than rising to the surface.[5] The total thickness of rocks in the central part of the Midcontinent Rift is enormous, 20 to 25 km of volcanic rocks and another 8 km of sediments. The rare places on Earth that have this much igneous rock are known as large igneous provinces (box 5.1), and the Midcontinent Rift is frequently cited as one of these. And this brings up an obvious question: why did all these igneous rocks form here?

BOX 5.1. LARGE IGNEOUS PROVINCES

Large igneous provinces (LIPs) are areas of Earth's crust in which unusually large volumes of igneous rocks (commonly more than 1 million km^3) are emplaced over a geologically short period (usually a few million years). According to Olav Eldholm and Millard Coffin, LIPs are probably related to plumes of hot rock that rise from depth in the mantle. As the plumes rise, decreasing pressure causes part of the mantle rock to melt (partial melting), generating basalt magmas that rise to the surface. If the mantle plume rises beneath ocean crust, the magma can flow through the overlying crust onto the seafloor where it forms large piles of basalt lava such as Hawaii. If the plume rises beneath continental crust, basalt lava can pile up on top of the continent to form large plateaus called flood basalts, such as the Columbia Plateau in Washington, Oregon, and Idaho and the Deccan and Siberian traps that were discussed in chapter 4. If the plume is active long enough, it can heat the base of the continent, weakening it and causing it to break apart, as happened in the Midcontinent Rift, although this is relatively rare. The heat can also melt the base of the continent generating large volumes of felsic (rhyolite) magmas, as happened in Yellowstone National Park. As we saw in the last chapter, some LIPs are closely linked in time to Phanerozoic mass extinction events, probably through the release of large volumes of CO_2 and SO_2 that were originally dissolved in the magmas as discussed by David Bond and Paul Wignall.[6]

5.2. The Midcontinent Rift Was Formed by a Mantle Plume

It is generally agreed that the Midcontinent Rift formed because the crust was underlain by a hot plume of rock that rose from deep in the mantle. The heat from the plume weakened the crust, causing it to stretch and fracture. As you see in figure 5.3, it all started when the crust began to rift apart and the early plume began to melt, producing basalt magma that rose to the surface forming lava flows that filled the rift. There are three important points to keep in mind here. First, the rift was never a great canyon many kilometers deep, as filling by basalt lavas kept pace with subsidence caused by rifting. Actually, the heat from below made the rocks expand, causing the rift to be a high area with a central valley. Second, the basalt lavas did not form conical volcanoes like Fujiyama. Instead, the volcanoes formed thick plateaus of lava similar to the shield volcanoes in Hawaii. Third, the rising mantle plume melted because of decreasing pressure, known as decompression melting, and this melting was partial, removing only the part of the mantle rock that melted at relatively low temperatures. This low-temperature partial melt was basalt magma.

As you can see in figure 5.3B, during the period of greatest rifting, when the plume was strongest, two large faults, the Douglas and Keweenaw, developed along the sides of the rift. When the plume began to weaken, basalt

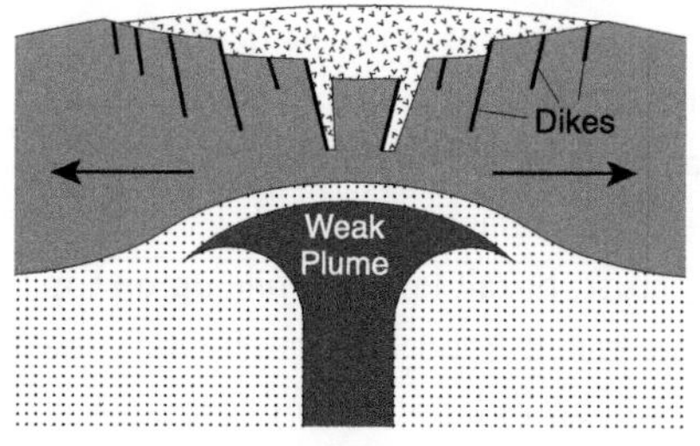

A) Early rifting and extrusion of basalt lavas from partial melting of rock in hot mantle plume

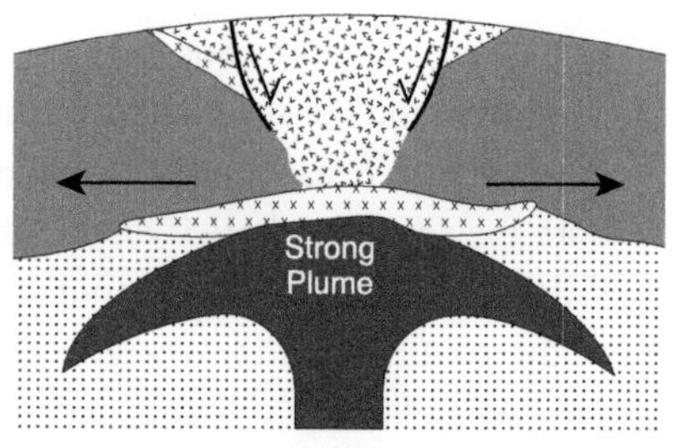

B) Continued rifting with basalt lava and mafic intrusions in the rift and basalt magma ponded below the crust

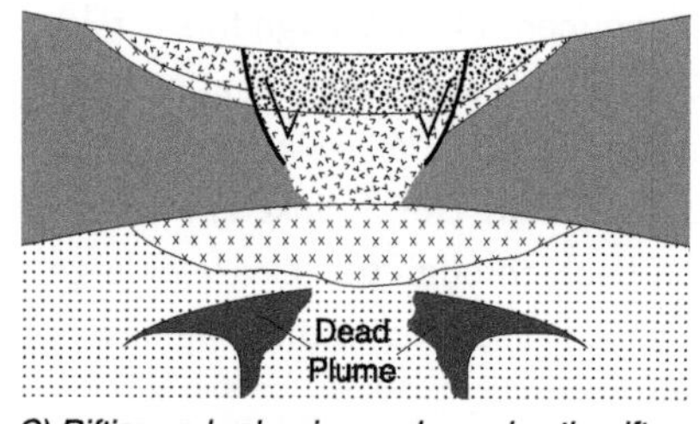

C) Rifting and volcanism end, causing the rift to subside and fill with sediments

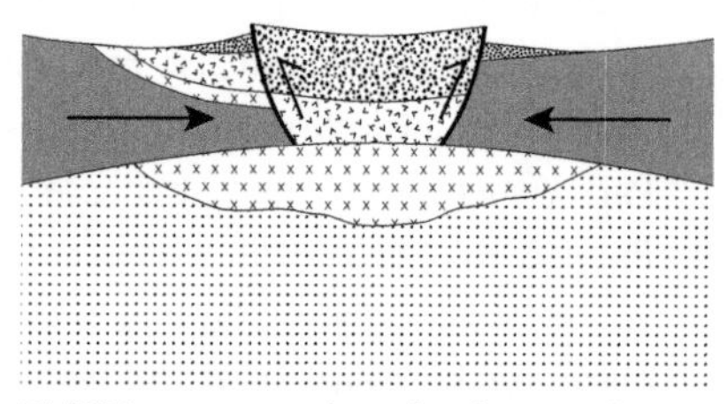

D) Rift is compressed causing the central part to rise and shed late sediments to the sides

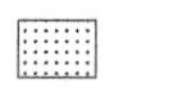

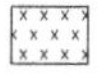

Figure 5.3. Schematic evolution of the Midcontinent Rift. (Modified from Bornhorst and Lankton 2009; Miller and Nicholson, 2013.)

volcanism ceased and the rift began to subside (fig. 5.3C), and this allowed sediments to be washed in from the sides. During formation of the rift, the region was under extension (being pulled apart), but at some point after the sediments were deposited the crust was compressed (fig. 5.3D), causing the central part to pop up along the Douglas and Keweenaw faults and shed sediments to either side.

The Midcontinent Rift ends near the Grenville front (fig. 5.2A), the boundary that separates Laurentia (and the rift) from a long continent fragment known as the Grenville Orogen, discussed later in this chapter. The Grenville Orogen collided with Laurentia at about the same time that rifting and volcanism in the Midcontinent Rift stopped, and Bill Cannon and others have proposed that this collision is the reason that rifting stopped. But gravity surveys and deep drilling described by Joshua Stark, Benjamin Richards and others show that Midcontinent Rift-type basins are present west of the Grenville front beneath Paleozoic sediments in Ohio and Kentucky (fig. 5.2A). And, Carol and Seth Stein, along with others, have suggested that these basins are part of a long, eastern arm of the Midcontinent Rift that extends along the Grenville front as far south as Alabama. This has reopened the question of just how the Midcontinent Rift and Grenville Orogen interacted, and especially their relation to the supercontinent cycle. Before we pursue these questions, though, let's take a look at some of the rocks and the observations that have led to this new controversy.

5.3. The Midcontinent Rift Filled with Volcanic, Intrusive, and Sedimentary Rocks

This section contains brief summaries of the three types of rocks that filled the Midcontinent Rift: volcanic, intrusive, and sedimentary. For all three rock types, the same base map showing Lake Superior (fig. 5.4) is used to show rock units and their locations. Good exposures of Midcontinent Rift rocks are widely scattered around the sides of Lake Superior, including on the Keweenaw Peninsula and Isle Royale in Michigan, the North Shore in Minnesota, and the Thunder Bay (Lakehead) area, Lake Nipigon, and Mamainse (Copper Mine) Point in Ontario. Figure 5.5 shows current estimates of how these rocks are related.

5.3.1. Volcanic Rocks Filled the Early Midcontinent Rift

The Midcontinent Rift, where it is exposed in the Lake Superior region, contains approximately 2 million km^3 of volcanic (extrusive) and plutonic (intrusive) rocks.[7] As first described in detail by John Green, they form thick sequences, called groups, that appear to have been separate centers of volcanism. At the west end of the lake are the North Shore Group in Minnesota and the Osler Group near Thunder Bay, Ontario as well as the Powder Mill Group in Michigan and Wisconsin and the Chengwatana (St. Croix) Group in Minnesota (fig. 5.4A). All these are exposed on only one side of the lake. The Portage Lake Group, however, extends under the lake from the Keweenaw Peninsula to Isle Royale. At the east end of the lake in Ontario are the Michipicoten Island and Mamainse Point Groups. All these, along with the sedimentary rocks described below, make up the Keweenawan Supergroup.

These groups were large lava plateaus fed by fissures (faults) that formed in the old Precambrian crust when it began to rift. Here and there it is possible to see early flows that came out over Precambrian basement rocks (fig.

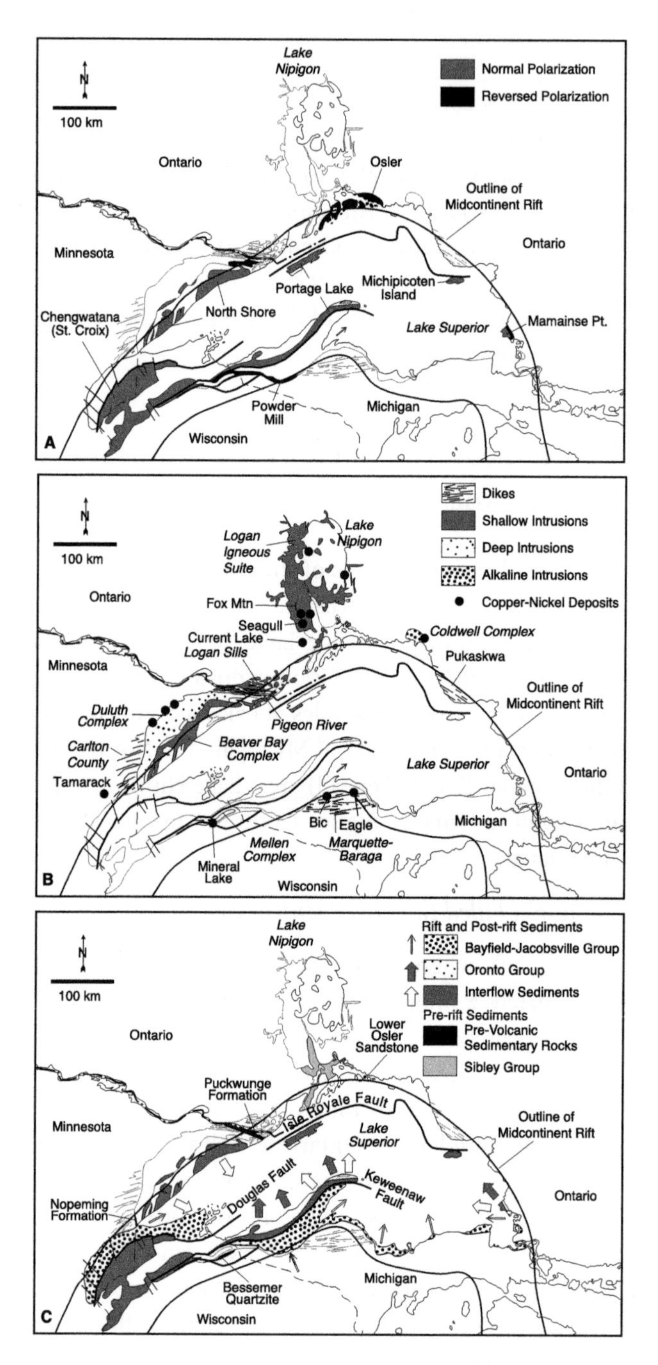

Figure 5.4. Midcontinent Rift rocks and magmatic mineral deposits: (**A**) volcanic rocks showing the major volcanic sequences (plateaus) and their paleomagnetic pole orientations; (**B**) intrusive rocks and magmatic mineral deposits; (**C**) sedimentary rocks showing dominant sediment transport directions. (Modified from Ojakangas and Morey 1982; Swanson-Hysell et al. 2014a,b; Miller and Nicholson 2013; Ripley 2014.)

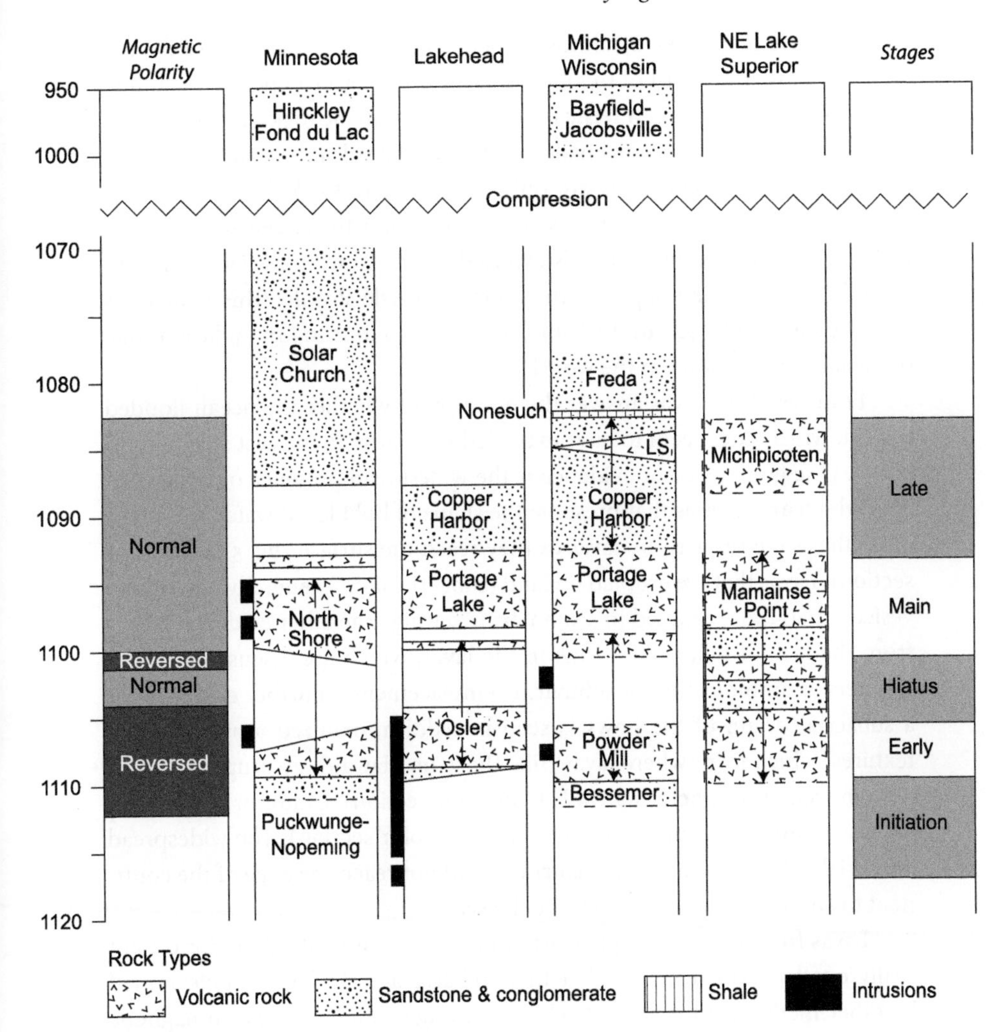

Figure 5.5. Stratigraphic and age relations (left column, millions of years) among Midcontinent Rift volcanic and sedimentary rocks. (Modified from Davis and Green 1997; Ojakangas et al. 2001a,b; Miller and Nicholson 2013; Fairchild et al. 2017.)

5.6A in color section), but most of what we see today are just large stacks of lava flows that are the remnants of the original plateaus, and that are tilted toward the lake because of later subsidence when the rift cooled (fig. 5.6B, C, D in color section). The plateaus consist almost entirely of mafic rocks, largely basalt lava flows and a smaller volume of intrusive rocks. Felsic rocks make up only a small percentage of the volcanic sequences except in the North Shore

Group, where they make up almost 25 percent in some areas. Most of the felsic volcanic rocks consist of rhyolite flows and pyroclastic deposits, including some ignimbrites and a few felsic intrusions.[8] Ultramafic rocks, mostly picrite lava flows, make up an even smaller portion of the volcanic sequences.

Most of the basalt flows are only a few meters thick, but some reach incredible thicknesses and lengths. The champion is the Greenstone Flow in the Portage Lake Group, which is exposed on the Keweenaw Peninsula and extends beneath Lake Superior to resurface in Isle Royale. This enormous flow, which is up to 400 m thick and contains 800 to 1500 km^3 of lava, is one of the largest lava flows in the world.[9]

Features of the lavas answer the question of whether the ocean flooded the rift, which does seem to have extended very close to the east side of Laurentia. One good piece of evidence is the widespread presence of large vesicles, which are cavities that preserve the shape of bubbles of water and other gases that were released by lavas as they reach the surface (fig. 5.6E in color section). Lavas that are erupted on land (subaerial lavas) generally have larger vesicles than those erupted in a submarine setting. The other evidence comes from the lack of pillowlike forms in the lavas, which are discussed in later chapters as an indication of submarine emplacement.[10] Further evidence for a subaerial origin of the flows is seen in reddish, oxidized tops with ropy textures, which form where lavas (like those on Hawaii) are erupted into an oxygen-rich atmosphere (fig. 5.6F, G in color section). Columnar structures are also common in some flows (fig. 5.6H in color section). The widespread subaerial volcanism confirms that rifting did not reach the edge of the continent to allow large-scale invasion by the sea.

Lavas from the various plateaus or groups around Lake Superior do not really overlap, which means that their relative ages had to be determined by other methods. First insights about their ages came from paleomagnetic measurements made by Henry Halls and his colleagues, who showed that lavas near the bottom of some of the Midcontinent Rift plateaus (the oldest lavas in the plateaus or groups) preserved reversed paleomagnetic poles and lavas above them had normal poles (Box 5.2).[11] Later paleomagnetic studies by Nicholas Swanson-Hysell and others added an additional period of reversed and normal poles, found largely in lavas from Mamainse Point, and the paleomagnetic sequence was filled out by isotopic analyses that provided absolute ages showing that volcanism extended from about 1110 to 1083 Ma and producing the correlations shown in figure 5.5.[12]

BOX 5.2. PALEOMAGNETIC MEASUREMENTS IN THE MIDCONTINENT RIFT

Rocks that contain magnetic minerals can preserve the attitude of Earth's magnetic field when they formed. If the age of the rock and its original attitude relative to the horizon are known, this can be used to determine the position of Earth's magnetic poles at that time. Basalt lava flows are by far the best rocks for paleomagnetic measurements because they were emplaced along a roughly horizontal surface and they contain magnetite (Fe_3O_4), which preserves the magnetic field in which they formed.[13] Paleomagnetic poles can also be measured in sedimentary rocks if they contain magnetic minerals that are either part of the original sediment or formed during diagenesis as the sediments were converted to rock.

Paleomagnetic measurements around the world have provided information on two features of Earth's magnetic field. First, we now have maps showing the positions of Earth's magnetic poles for most of geologic time. It is these measurements that have been used to estimate the positions of continent fragments through geologic time and their role in the supercontinent cycle. Second, we now know that the global magnetic field has fluctuated between normal and reversed (North Pole at the South Pole) polarity throughout geologic time. These reversals have happened relatively frequently, and it is common for a thick sequence of lavas like the Midcontinent Rift to show evidence for such a reversal. Recognition of these reversals in thick stacks (sequences) of lava flows that are widely separated helps determine relative ages of the sequences.

5.3.2. *Some Magmas Formed Intrusive Rocks in the Volcanic Sequences and Underlying Rocks*

Some of the magma that rose from the mantle did not make it to the surface and instead formed dikes and sills at shallow depths of a few kilometers.[14] These are collectively known as the Midcontinent Rift Intrusive Supersuite, a true mouthful. The earliest dikes and sills of the supersuite, which make up the Logan Igneous Suite around Lake Nipigon (fig. 5.4B), range in age from about 1117 to 1105 Ma, an age slightly younger than the earliest lava flows in the main part of the rift to the south. Dikes and sills of basalt and ultramafic rocks that are the same age as Midcontinent Rift lavas are found in groups called swarms in basement rocks outside the rift; the largest are the

Marquette-Baraga, Carlton County, and Pukaskwa swarms (fig. 5.4B). These swarms are associated with nickel-copper deposits such as the Eagle deposit in Michigan, which is discussed later.

Deeper intrusions form sheetlike bodies either in the volcanic complex or between the basement rocks and overlying basalt lavas. The largest of these are the Duluth Complex (1109 to 1106 Ma) and Mellen Complex (1101 to 1102 Ma) (fig. 5.4B). They are called complexes because they consist of many small intrusions ranging in composition from gabbro to anorthosite.[15] Some of the gabbro intrusions look almost like a sedimentary rock because they contain (cumulate) layers of minerals that settled from the magma as it cooled (fig. 5.7A in color section). The Duluth Complex was cut by shallow mafic intrusions, including the Beaver Bay Complex (figs. 5.4B, 5.7B [in color section]), which appear to have been feeders for later basalt lavas, as discussed by Jim Miller and Val Chandler.

Alkalic rocks, including the 1108 Ma Coldwell Complex (fig. 5.4B), are also part of the Midcontinent Rift large igneous province. The alkalic complexes include carbonatite, an igneous rock consisting largely of calcite and other carbonate minerals, that often forms during early stages of rifting.

5.3.3. Sedimentary Rocks Filled the Midcontinent Rift as It Cooled and Subsided

Before we deal with sedimentary rocks in the Midcontinent Rift, we should pause for moment and think about what might have been happening immediately before the rift formed. Could the early stages of continental stretching, which precedes actual rifting, have foreshadowed the coming of the rift by forming a shallow depression that collected some pre-rift sediment? By far the largest area of pre-rift sedimentary rocks is in the Sibley Group near Thunder Bay and Lake Nipigon, Ontario (fig. 5.4C), although smaller exposures are found farther south in the Puckwunge, Nopeming, and Bessemer Quartzites. Dick Ojakangas, Glenn Morey and others have shown that these pre-rift sediments were deposited in two different basins at two different times. The first one, which hosted the 1450 Ma Sibley Basin, was oriented north-south, and was probably too much older than Midcontinent Rift to tell us much about pre-rift conditions. The second one, which hosted the later Puckwunge and other quartzites, was oriented parallel to the present Midcontinent Rift and looks like it was a faint warning of the upcoming rift.[16]

More voluminous sedimentary rocks were deposited after volcanism slowed and the Midcontinent Rift began to subside, forming a sedimentary basin (fig. 5.4C).[17] Some sediments formed during volcanism and are interlayered with the lava flows (fig. 5.7C in color section). These interflow sediments account for less than 5 percent of the total thickness of the Portage Lake and North Shore Groups and about 25 percent of the Mamainse Point Group, including the Great Conglomerate, which is 500 m thick. Individual layers consist of fragments of basalt and rhyolite from the volcanic sequence, and transport directions summarized by Dick Ojakangas and others indicate that sediment flowed into the Midcontinent Rift basin from both sides.

The real accumulation of sediment happened after the close of volcanism, however, when the rift began to cool and subside. Oronto Group sediments, which overlie Portage Lake Group volcanic rocks, are by far the most important (fig. 5.5). The stack of Oronto Group sediments is up to 8 km thick and includes three formations. The oldest is the Copper Harbor Conglomerate (fig. 5.7D, E in color section), which formed as alluvial fan deposits shed into the rift from its margins. It is overlain by the Nonesuch Shale (fig. 5.7F in color section), which contains relatively abundant organic matter and pyrite, and the Nonesuch is succeeded upward by sandstone and shale of the Freda Sandstone.

Most geologic evidence indicates that Oronto Group sediments were deposited in a fluvial (stream) or lacustrine (lake) rather than marine (ocean) setting. This aspect of the Oronto Group has been a key part of arguments that the Midcontinent Rift did not reach the edge of Laurentia, where it could have been invaded by the sea, even at this late stage in its evolution. The overlying Nonesuch Shale continues to cause concern about this conclusion, however because it has abundant pyrite (FeS_2), which contains sulfur that could have been derived from sulfate dissolved in seawater.[18]

At this point in the history of the Midcontinent Rift, subsidence ceased and the sedimentary basin was uplifted along the Douglas and Keweenaw faults on the margins of the basin (fig. 5.2). Erosion of the uplifted central part of the basin and of surrounding older Precambrian basement rocks shed sediments that formed the Bayfield Group in Wisconsin and the Jacobsville Sandstone in Michigan and Ontario. These sandstones formed in lakes and dunes, and contain abundant quartz, which was recycled from underlying sandstones. The Bayfield and Jacobsville were used widely as building stone in the area during the late 1800s.[19]

The Oronto Group and overlying Midcontinent Rift sedimentary rocks are obviously younger than the volcanic rocks, but establishing their absolute age has been challenging (box 5.3). We need to know the age in order to understand the full duration of the mantle plume and rift event. The older limit on the age of the sediments comes from a 1087 Ma age of lavas (Lake

Shore traps) that are interlayered with the Copper Harbor Conglomerate. The younger limit comes from an age of 1062 Ma for volcanic rock that cuts the overlying Freda Formation. The best estimate for the age of the compression that uplifted the rift comes from 1060 Ma isotopic ages for minerals in faults that parallel the main Keweenaw fault. The Jacobsville-Bayfield sediments should be about the same age, but there is some confusion because the sediments contain zircons as young as 959 Ma. If these ages are correct, either the uplift continued much longer than originally thought or the Jacobsville-Bayfield sediments are associated with later events in the Great Lakes region.[20]

BOX 5.3. MEASURING/ESTIMATING THE AGE OF PRECAMBRIAN SEDIMENTARY ROCKS

The absence of fossils in most Precambrian sedimentary rocks requires more imaginative methods to determine their age. Most sedimentary rocks do not contain minerals that can be used for conventional isotopic age measurements. This is particularly true for clastic sedimentary rocks, which consist of grains eroded from preexisting older rocks. Relative ages of Precambrian sedimentary rocks can be estimated from the law of superposition mentioned in chapter 1. And the law of crosscutting relations can be used to bracket the age of a sedimentary rock by measuring isotopic ages for older and younger igneous or metamorphic rocks that cut across it (fig. 1.4C in color section). Paleomagnetic methods can also indicate relative and sometimes approximate absolute ages. Absolute ages can also be estimated from strontium and other isotope analyses using the isochron method, although this has challenges, as discussed by Tod Waight. More recently a method based on the isotopic ages of detrital minerals has been used to put absolute limits on the age of sedimentary rocks. The most common version of the method uses U-Pb ages of zircon, a detrital mineral that resists weathering and chemical alteration even at high temperatures (the temperature at which a mineral begins to change composition is called its closure temperature), as discussed further in chapter 6. Obviously, a sedimentary rock cannot be older than the youngest grain of sedimentary zircon that it contains. Using new analytical methods that allow rapid analysis of numerous zircon grains, it is possible to get a good idea of the maximum age of sedimentary rocks.

5.4. Mineral Deposits Formed as the Midcontinent Rift Filled

The Midcontinent Rift was a giant geologic ore-forming engine that has played an important role in the settlement and economy of the Lake Superior region. It started in the early 1800s when prospectors swarmed the Lake Superior country trying to find the copper mother lode. The Ontonagon Boulder turned out to be a glacial erratic that led nowhere, but the discovery of small Old Copper Culture mines led to deposits in the bedrock where mining started.[21] By the 1850s, the need for experienced miners brought immigrants from Europe along with their favorite underground lunch, the pasty, which became the signature food of Michigan's Upper Peninsula. The mining boom that followed extended from the Keweenaw Peninsula to Isle Royale and Mamainse Point and Michipicoten Island in Ontario. Lake Superior became the copper capital of North America. Its copper was known as Lake Copper, and was specially valued because it contained small amounts of silver, which increased its conductivity of electricity and heat. Later exploration found copper sulfide deposits, as well as copper-nickel sulfide deposits. To link all these to the geology of the Midcontinent Rift, we will look first at the native copper and copper sulfide deposits and then at the copper-nickel deposits.

5.4.1. Copper Deposits Were Formed By Hot (Hydrothermal) Solutions Flowing Out of the Midcontinent Rift

The native copper and copper sulfide deposits contain copper in two very different forms even though they formed by the same general geologic process. In the native copper deposits, copper is present in elemental form, whereas in the copper sulfide deposits it is present largely as the mineral chalcocite (Cu_2S). Both were deposited by hot, aqueous solutions that flowed out of the Midcontinent Rift. In terms of historical value, native copper deposits are the big winners, both in total production and in geologic character. There simply are no other places in the world where such a huge amount of native copper has been deposited.

Mining of the copper deposits is a multi-millenial business that started with the Old Copper Culture six thousand years ago and expanded with modern mining that began in the mid-1800s. Most of the native copper production came from mines on the Keweenaw Peninsula in Michigan, although smaller deposits and prospects are widespread in the Osler Group in Ontario

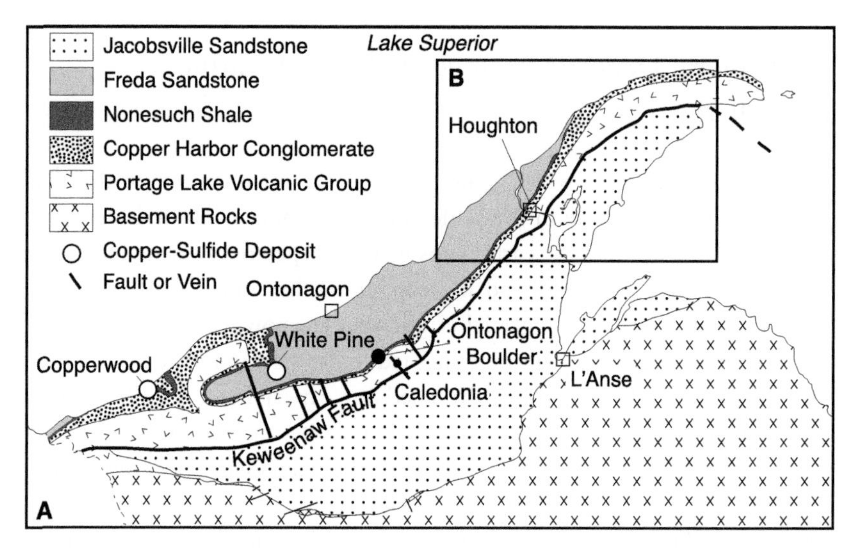

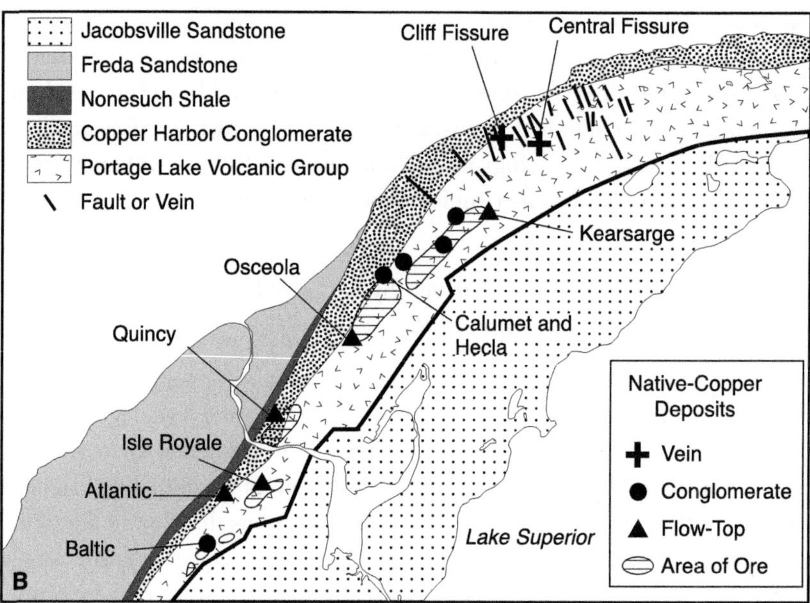

Figure 5.8. Copper deposits of northern Michigan: (**A**) location of native copper and copper sulfide deposits and the Ontonagon Boulder; (**B**) detail of the northern part of the Keweenaw Peninsula showing the location and names of important native copper mines. (Modified from Bornhorst and Barron 2011.)

and the Chengwatana (St. Croix) Group in Minnesota.[22] The only large copper sulfide deposit was the White Pine mine at the southwestern end of the Keweenaw Peninsula (fig. 5.8), although the nearby, smaller Copperwood deposit remains unmined and small copper sulfide veins are present in other places, especially Mamainse Point. According to Ted Bornhorst and Larry Lankton, native copper deposits in the Keweenaw produced about 5 million tons of copper and an unknown amount of silver between 1845 and 1968, and the copper sulfide deposits produced about 2 million tons of copper and 1,400 tons of silver between 1953 and 1996.[23] Although mining has stopped, large amounts of copper remain in deposits of both types, which continue to be explored and evaluated.

The native copper and copper sulfide ore have very different geologic features that conceal their common origin. The native copper ores of the Keweenaw Peninsula fill holes (porosity) in conglomerate layers (fig. 5.9A in color section) or vesicles and breccias in the tops of lava flows. [Many other minerals fill these vesicles, including thompsonite, a popular gemstone, especially around Grand Marais, Minnesota (fig. 5.9B in color section)]. In contrast, the copper sulfide ores at White Pine (fig. 5.9C in color section) consist of small grains of Cu and Cu-Fe sulfides that form layers in the Nonesuch Shale. The Cu and Cu-Fe sulfide layers follow layering of original early pyrite that formed during diagenetic reactions as the sediment was buried. Despite their very different appearances, both types of deposits formed by precipitation of copper from hot water, known as hydrothermal solutions. The solutions came from deep in the volcanic rocks of the Midcontinent Rift, where they were hot enough to leach copper from the rocks. As they moved upward toward the edges of the rift, the hydrothermal solutions precipitated their dissolved copper by cooling to form the native copper deposits and by reaction with sulfur in pyrite to form the White Pine deposit.[24]

We know that the White Pine (copper sulfide) deposits formed first because they are cut by veins containing native copper. Ted Bornhorst and others have reported an age of 1060 to 1047 Ma for the native copper deposits and suggested that they formed during late-stage compression and uplift of the rift, possibly around the time of the Jacobsville Sandstone. The exact age of the copper sulfide deposits is not known, but they must be younger than the 1062 Ma young age limit for the Oronto Group.

The main environmental legacies from mining of the copper deposits are piles of waste rock and emissions of mercury. The native copper ores were purified by a two-stage process that started with crushing the ores to separate the copper from the waste rock and then melting the copper to remove other impurities and cast ingots for shipment. Much of the crushing was done in stamp mills, which used a sharp blow to shatter the rock, releasing the more malleable metal. Waste from this process, known as stampsand, was disposed of in piles. The largest area of stampsand disposal, around Torch Lake in the Keweenaw, contains about 200 million tons of material. This area has been partially remediated, although sands remaining in the lake are still being moved by wave action. The copper ores contain small amounts of mercury around the Keweenaw that were released during processing and accumulated in lake sediment. Since the end of mining, mercury-bearing sediment layers in the lake have been covered and isolated by younger sediment in some areas, although problems remain in Torch Lake.[25]

5.4.2. Copper-Nickel (Magmatic) Deposits Formed in Dikes around the Midcontinent Rift

Although the historical value prize belongs to the copper-silver deposits, the future might belong to the copper-nickel deposits. Again we have two different types of deposits that have a common geologic origin, Eagle type and Duluth Complex type. Eagle-type deposits contain rich ores like those in the Eagle mine near Marquette, Michigan in dikes that formed early in the history of the rift. Duluth Complex type deposits include large, lower-grade (content of metal) deposits along the northern margin (lower contact) of the Duluth Complex in Minnesota (fig. 5.4B).

Both types of deposits formed from mafic to ultramafic magma, which is why they are referred to as magmatic deposits. The copper and nickel were concentrated when the magma underwent a process known as magmatic immiscibility in which an initially homogeneous mafic or ultramafic magma separates into two magmas that do not mix. The two magmas are a "silicate magma," enriched in elements such as silicon, aluminum, calcium, and magnesium, which make up normal rocks, and a "sulfide magma," enriched in sulfur, iron, copper, nickel, and platinum-group elements.[26] Sulfur is the key to the process, and the Eagle-type and Duluth Complex type deposits differ mainly in the source of sulfur that led to formation of the ore minerals. In the Eagle-type deposits, the magma contained its own sulfur and, at least in some deposits, formed large amounts of immiscible sulfides, which were heavier than the silicate magma and sank to the bottom of the magma chamber to form masses of very rich ore. In the Duluth Complex type deposits, ore formed when the magma dissolved sulfur-rich wallrocks, which caused precipitation of small grains of nickel-copper sulfides scattered (disseminated) through the rock. Both types of deposits contain significant amounts of platinum group elements, which adds to their value.

The rich ore in the Eagle deposit has allowed it to be mined underground with minimal surface disturbance (fig. 5.9D in color section). Other, similar deposits at Tamarack, Seagull, Current Lake, and BIC, which have been described by Ed Ripley and others, offer promise for the future (fig. 5.4B). The Duluth Complex type deposits, which would require larger open pit or underground mines, are the focus of continuing environmental concern because they are near the Boundary Waters Canoe Area Wilderness. However, they are also near the Mesabi Range iron mines (discussed in chapter 6), which would allow at least some of the deposits to be mined with minimal new disruption using infrastructure that is already in place.

5.4.3. The Nonesuch and Other Shales in the Midcontinent Rift Are Source Rocks for Oil and Gas

The Nonesuch Shale is clearly a source rock for hydrocarbons, and the porous sandstones that are beneath and above it could serve as good reservoirs for any oil or gas that it generated. Oil seeps are found here and there in the White Pine copper mine, and comparative analyses of organic matter in the oils and shales show that it originated in the Nonesuch Shale. Bill Kelly and Gail Nishioka found microscopic inclusions of oil in calcite from veins with an age of 1047 Ma, indicating that at least some oil had been generated from the Nonesuch by this time in the history of the Midcontinent Rift. At least five exploration holes have been drilled in the Midcontinent Rift to test its hydrocarbon potential. No economically interesting hydrocarbon reservoirs have been found so far, and the drilling has shown that maturity of the hydrocarbons in some parts of the rift is due to events that occurred during Paleozoic rather than Mesoproterozoic time.[27] One possible problem is the lack of one component of the hydrocarbon systems discussed in chapter 4: a cap rock that would have stopped oil or gas from leaking out of the porous sandstones. One way around this might be to produce oil or gas in situ by fracking the Nonesuch, as was discussed for the Antrim Shale in chapter 4.

5.5. The Evolution and Geochemistry of the Midcontinent Rift Are Consistent with a Plume Model

The geologic history reviewed above agrees with the plume model summarized in figure 5.3.[28] You can see in figure 5.5 that the igneous rocks that filled the Midcontinent Rift have been divided into five stages beginning at about 1115 Ma with the Logan Igneous Suite in the Nipigon area. This was followed by a short period between 1110 and 1105 Ma of voluminous basalt volcanism with very few felsic rocks, largely in the Osler, Powder Mill and North Shore groups. Between about 1105 and 1102 Ma, volcanism occurred only in the Mamainse Point area, and the rest of the rift was generally quiet. Then, from 1102 Ma to 1096 Ma, a second large volume of basalt was extruded in all the groups, including the Portage Lake Group. During a final (late) stage, beginning at 1094 Ma and ending at about 1083 Ma, volcanism in the Michipicoten Island and Mamainse Point areas was particularly active and a series of late basalts known as the Lake Shore Traps was extruded in the Copper Harbor Conglomerate.

The chemical composition of these igneous rocks provides insights into what was being melted to make this sequence. For a start, the compositions show that these rocks are part of the tholetic group (box 5.4), which forms by partial melting of the mantle. Studies by Suzanne Nicholson and others indicate that early magmas that filled the rift, including those that formed the copper-nickel sulfide deposits, formed by decompression melting as the plume rose from depths of 80 to 120 km. As the plume continued to rise, decompression melting took place at shallower depths of 50 to 100 km in the mantle. This shallow mantle, known as subcontinental lithospheric mantle (discussed in chapter 7), had a slightly different composition from deeper mantle and yielded basalt lavas with slightly different compositions.

Rhyolitic (felsic) magmas could not have been formed by melting of the mantle plume. Instead they were generated when basalt magmas ponded against the base of the crust, melting either the base of the basalt volcanic pile or the Precambrian crust. Their greater abundance in later volcanic rocks is ascribed to the thermal effects of magma that ponded beneath the continent during the short lull in magmatism that separated reverse and normal polarity volcanism.

BOX 5.4. PETROCHEMISTRY

The chemical and isotopic composition of igneous rocks tells us how magmas form, and this provides insights into the tectonic setting of the igneous activity. Figuring this out is the province of petrochemistry. Some magmas form by partial melting of the crust, but most come from the mantle, and one big question for these is whether they formed at a convergent tectonic margin where subduction was taking place or at a divergent margin where rifting was taking place. A first step in answering that question can be made by looking at the major element compositions of the igneous rocks, particularly in triangular diagrams showing their Na+K, Fe, and Mg content (usually expressed as oxides such as Na_2O). Two trends show up; a tholeiitic trend in which mafic rocks like basalt show strong enrichment in Mg because the first mineral to form as the magmas crystallize is olivine [Mg_2SiO_4] and a calc-alkaline trend in which mafic rocks are not enriched in Fe because magnetite is also an early mineral to crystallize.[29] This difference in trends happens because calc-alkaline magmas contain more water and are more oxidized. Rocks in the Midcontinent Rift follow the tholeiitic trend, confirming that they formed in a rift setting. Additional insights about the sources of magmas come from isotopic analyses.

As the plume collapsed and detached from its rising column, volcanism stopped and the rift began to subside, collecting sediment of the Oronto Group. Then everything changed; the crust was compressed, and the basin containing volcanic and sedimentary rocks was uplifted along the rift-margin faults. The big question, then, is why this would happen, and that takes us to the Grenville Orogen.

5.6. The Grenville Orogen Is a Continental Fragment That Collided with Laurentia

While the Midcontinent Rift was trying to tear Laurentia apart, plate tectonic collisions were building a huge mountain range known as the Grenville Orogen along the eastern side of the continent (fig. 5.10A). High peaks of the Grenville Orogen mountain range have long since been removed by erosion, but its roots form a well-defined belt that extends from Newfoundland to Mexico and beyond.[30] The belt consists mainly of highly metamorphosed and deformed rocks that were thrust westward onto Laurentia, as well as parts of the eastern margin of Laurentia itself that were deformed during the thrusting. As described in a series of studies by Toby Rivers, Tony Davidson, Fried Schwerdtner, and others, processes that formed the Grenville Orogen were active for so long that it has become the poster child for Precambrian "large, hot orogens."

Large, hot orogens make enormous mountain ranges (like the Himalayas) that do strange things, including collapsing on themselves and turning inside out. And, they take a long time to do it. The main part of the Grenville Orogen formed over a period of about 110 million years and involved two distinct phases that formed two belts of rock.

The first phase of the orogeny, which is known as the Ottawan phase,

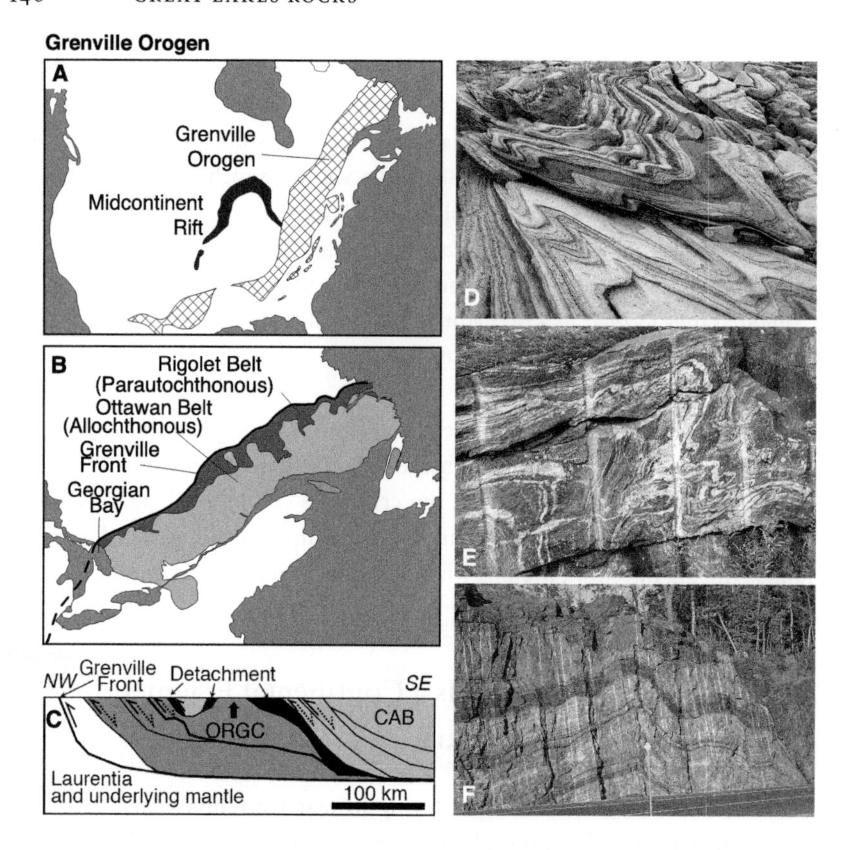

Figure 5.10. The Grenville Orogen. A) Grenville-age rocks along the eastern margin of North America (Laurentia); B) Allocthonous (Ottawan Phase) and Parautocthonous (Rigolet Phase) terranes in the Grenville Province; C) Cross section through the Grenville Province in southwest Ontario showing the results of gravitational collapse (dashed arrows) of the original thrust stack. The upper part of the thrust stack, which slid off to the southeast, is exposed today in the Composite Arc Belt (CAB). The lower part of the thrust stack was then uplifted along a detachment zone to form a metamorphic core complex known as the Ontario River Gneiss Complex (ORCG) (modified from McLelland et al., 2010; Rivers, 2015; Schwerdtner et al., 2016); D, E, F) extremely deformed rocks of the ORCG metamorphic core complex rocks east of Georgian Bay in Ontario. (Photos courtesy of W. M. Schwerdtner.)

lasted from 1090 to 1020 Ma. During this phase, island arcs and other rocks that had been forming in the ocean offshore, were pushed onto Laurentia creating a mountain belt known as the allochthonous terrane (fig. 5.10B).[31] This mountain belt consisted of a stack of thrust sheets so high that it collapsed from its own weight. During the collapse, the upper and lower parts of the thrust stack behaved very differently. The upper part, which we see today in the Composite Arc Belt (fig. 5.10C), flattened plastically and slid back down the former thrust faults, as you can see from the dashed arrows in the cross section. This process, which is known as gravitational collapse,[32] exposed the lower part of the thrust stack, which was undergoing metamorphism, causing it to flow upward. During upward flow, these rocks were pulled apart and plastically deformed creating rocks like those in figures 5.10D, 5.10E, and 5.10F, including an enormous dome of metamorphic rock known as a "metamorphic core complex" that is exposed today in the Ontario River Gneiss Complex (fig. 5.10C). The Ontario River Gneiss Complex and a similar domal uplift in the Adirondack Mountains are among the largest metamorphic core complexes recognized so far anywhere on Earth. Spectacular exposures of these rocks can be found in the Parry Sound area of Georgian Bay.

Don't miss an important point here. Metamorphic core complexes do not form when a region is compressed by plate tectonic collisions. Instead, they suggest that the region was stretched horizontally (otherwise, the top of the mountain range could not have slid off). If this is so, the compressive plate tectonic processes that caused Ottawan phase mountain building must have relaxed at an advanced stage, allowing extension, a point we will return to later when we talk about the supercontinent cycle.

Then, a second plate collision caused the Rigolet phase of the Grenville orogeny from 1005 to 980 Ma. This phase pushed the now-collapsed Ottawan-phase mountain range farther onto Laurentia and also deformed the Laurentian continent, forming the parautochthonous belt and the Grenville Front thrust fault that we first met in figure 5.2.

The processes that formed the Grenville Orogen were similar in some ways to the collision of the Indian subcontinent with the Asian plate, which formed the Himalaya Mountains. The collision deformed much of Asia north of the Himalayas, and it is reasonable to suspect that the Grenville event had a similar impact on Laurentia. It is this line of reasoning that leads to suggestions that continent-wide compression caused by the Grenville event stopped the opening of the Midcontinent Rift. The Midcontinent Rift and Grenville Orogen were not solitary features, however. Rather, they were part of the assembly of the Rodinian supercontinent, and they must be fit into that larger picture.

5.7. Debate Persists about How the Midcontinent Rift and Grenville Orogen Fit into the Supercontinent Cycle

Continental assembly is a peculiar time for a major rift to form, and this conundrum has been bothering geologists for a long time. Recognition that the Grenville Orogen was about the same age as the Midcontinent Rift made the matter more complex. As mentioned above, Bill Cannon, William Hinze and others, have suggested that the Midcontinent Rift was put out of business by the collision between the Grenville and Laurentia.

But how did things line up to start the Midcontinent Rift? To answer that question, we need more information on migration paths of the continents before the ~1100 Ma start of the Midcontinent Rift, and that involves paleomagnetism. Unfortunately, Laurentia is the only relevant continent fragment with abundant paleomagnetic poles for Mesoproterozoic time. That has resulted in a wide range of suggested reconstructions for the paths taken by continents as they assembled Rodinia, as you can see in figure 5.11.[33] From our Midcontinent Rift–Grenville perspective, the important requirement is that there is a continent abutting Laurentia during formation of the Grenville Orogen, and in most paleomagnetic studies the honor falls to Amazonia. Eric Tohver and his associates have provided strong geologic evidence to support a link between Laurentia and Amazonia, including suggested fragments of Amazonia in the Blue Ridge area of the Appalachians. David Chew and others have summarized a migration path for Amazonia as it collided with southern Laurentia (Mexico) at about 1200 Ma and scraped its way northward along the east side of Laurentia until it ran into Baltica at about 966 Ma.

This sequence of collisions correlates in general with the extended Grenville orogeny, and at first glance that might settle the matter: the Amazonia-Laurentia collision stopped the Midcontinent Rift. But it doesn't explain why the rift started. Carol and Seth Stein, working with a host of others, have reinvestigated this problem and come up with an interesting possibility. They suggest that there was a period of rifting during the long Grenville collision that moved the continents apart and that this rifting formed the Midcontinent Rift. They note also that the two-pronged rift suggested in figure 5.2 may have been the start of a microcontinent similar to the East African Rift today, rather than just a simple rift. According to their thinking, this rifting would have been competing with rifting to the east between Laurentia and Amazonia, and it might have stopped if an ocean had opened between the two continents. You can see how this could have happened in the paleomagnetic reconstruction in figure 5.11, where the positions of Amazonia relative to Laurentia between 1150 Ma and 980 Ma must have involved considerable rotation and possibly rifting. One final observation that supports this line of thinking is the gravitational collapse during the late Ottawan phase of the Grenville orogeny. Sliding of the top of the Grenville mountains downward and toward the east would obviously have been easier if Amazonia was moving away (rifting) from Laurentia at that time.

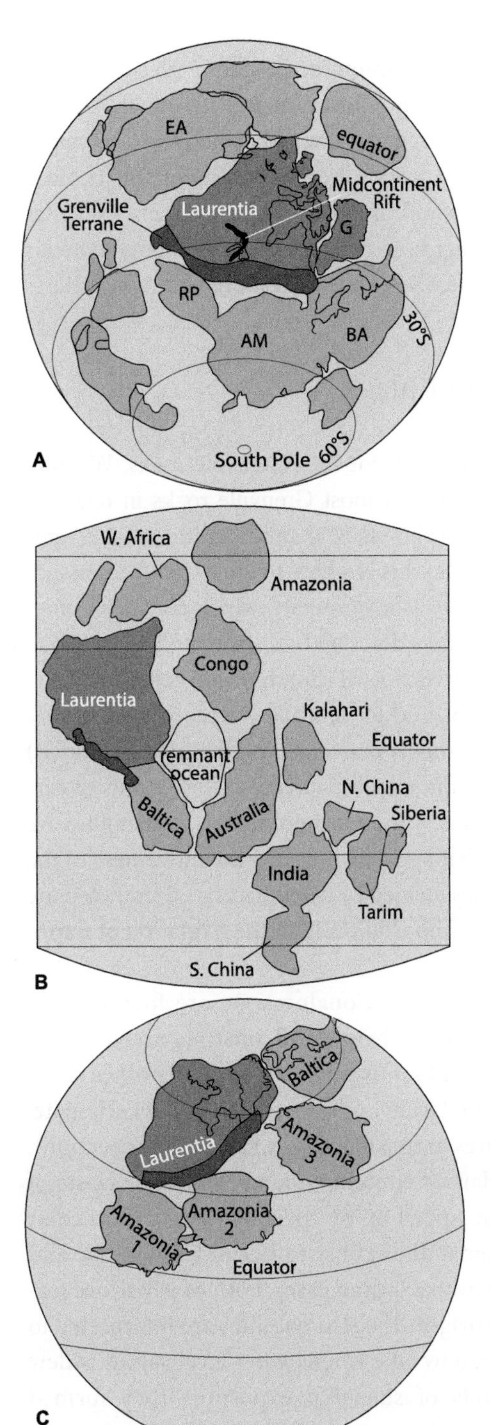

Figure 5.11. Alternate paleomagnetic reconstructions of Rodinia showing the location of the Midcontinent Rift and the Grenville terrane in Laurentia. Adjacent continent fragments include Amazonia (AM), Baltica (BA), Greenland (G), East Antarctica (EA), and Rio de la Plata (RP). Option C (from Chew et al. 2015) shows that Amazonia may have rotated along the eastern margin of Laurentia. (Modified from Weil et al. 1998; Chew et al. 2015; Evans 2017.)

Thus, we come to the end of our Mesoproterozoic story with several possible explanations for its main feature, the Midcontinent Rift. Maybe, despite its great importance to the geology, copper-trade archaeology, and even the economy of North America and the world, the Midcontinent Rift is but a footnote in the assembly of Rodinia and in Earth's supercontinent cycle. Or maybe it is the key to understanding how complex rifts form large igneous provinces and even microcontinents. Only time (and more work) will tell.

5.8. Life in the Midcontinent Rift

Evidence of life is scarce in the Midcontinent Rift–Grenville rocks. We have essentially no hope of finding anything in most Grenville rocks in our area because they are so intensely metamorphosed. Midcontinent Rift rocks resemble more closely their original form. So this is where to look, and the best evidence shows up in the Copper Harbor Conglomerate, which contains dome-shaped carbonate forms that were first described as stromatolites by Doug Elmore and studied in more detail by Nicholas Fedorchuk and colleagues.

The term *stromatolite* has been used to describe a wide variety of features in ancient rocks, but it has gradually converged on laminated (layered) carbonate forms that are fossilized microbial mats that collected bits of carbonate and other sedimentary grains.[34] The most common stromatolites are dome shaped or columnar forms that apparently reflect the movement of the microbes toward light. Stromatolites formed by cyanobacteria dominated the ancient fossil record and played an important role in the oxidation of Earth, as we will see in the next chapter.

Stromatolites in the Copper Harbor Conglomerate are found at the point of contact between conglomerate below and siltstone above or are entirely enclosed within siltstone and sandstone. Many stromatolites cover cobbles at the top of conglomerate layers and consist largely of carbonate, in which fine finger-like structures are visible (fig. 5.12A in color section). Adjacent parts of the Copper Harbor contain what appear to be clasts of stromatolites that may have been ripped up by currents (fig. 5.12B in color section). In some areas, the stromatolites consist of crinkled laminae and show a porosity thought to form by expanding gases, both of which are features characteristic of a biogenic origin. The stromatolites are interpreted to have grown at the margin of a shallow lake where water level varied widely and frequently, including periods of subaerial exposure. They formed microbial mats that covered the surface and probably contributed to the stability of the sedimentary environment.

A final life-related question is whether Midcontinent Rift time was associated with an extinction event. If we accept that Midcontinent Rift volcanism qualifies as a large igneous province, it may have generated enough CO_2 to change global climates. It turns out that some of our best information for evaluating this possibility comes from studies of paleosols (preserved ancient soils). Soils form from water that was in contact with the atmosphere, and where they have been preserved they are our best window into ancient atmospheric compositions. Nathan Sheldon has used analyses of paleosols in Midcontinent Rift volcanic and sedimentary rocks to show that atmospheric CO_2 levels actually declined during early Mesoproterozoic time, reaching a low about ten times less than modern preindustrial levels at about 1100 Ma. This is almost exactly the reverse of what we might expect if Midcontinent Rift volcanism contributed large amounts of CO_2 to the Mesoproterozoic atmosphere.

However, the Nonesuch Shale is somewhat similar to black shales, and it could still have been associated with an extinction event. Maybe the question is moot in view of the absence of an abundant fossil record. If stromatolites were all we had to exterminate, does that qualify? Midcontinent Rift time was definitely a time of change in stromatolite abundance, however, and the change paralleled the surprising change in atmospheric CO_2. According to Sheldon, the atmospheric CO_2 low during Midcontinent Rift time coincided with a peak in stromatolite abundance and diversity that had been identified earlier.[35] These results do not lend much support to Midcontinent Rift volcanism as a driver for Mesoproterozoic climate change, but they do leave the extinction door slightly ajar.

According to Robert Riding, stromatolites gave way after Midcontinent Rift time to other life forms, including the amazing Ediacaran biota. The Ediacaran biota showed up during a lull in the geologic history of the Great Lakes region, after the Midcontinent Rift closed down and before Paleozoic flooding of the continents. This was a time of huge changes, including massive continental glaciation between about 720 and 635 Ma, which may have led to a completely ice-covered planet known as snowball Earth, as suggested by Joseph Kirschvink and expounded by Paul Hoffman. The Ediacaran biota, which have been found in many parts of the world, are fossils of strange soft-bodied organisms that were among the first complex, macroscopic, multicelled organisms. They appeared at about 575 Ma near the end of a glacial period and spread around the world in what has been called the Avalon explosion. Almost immediately, geologically speaking, they gave way to other life forms in the Cambrian explosion, which was mentioned in chapter 4.[36]

Notes

1. The Old Copper Culture, also known as the Old Copper Complex, refers to indigenous groups that were involved in production, trade, and use of copper, most of which came from rock-hosted deposits in the Lake Superior region, as well as glacial sediments eroded from them. The copper was not smelted (it did not need to be because it was in the pure metal form), but it was shaped and annealed into complex forms. The oldest artifacts date to 6000 BCE. Early uses of the copper included a wide range of tools and weapons, even fishhooks and harpoons. Later use was largely for ornaments, which has been interpreted to indicate a stratification of society. Copper Culture State Park in Wisconsin is a burial ground used during this period (Martin 1999; Pleger 2002; Pompeani et al. 2015).

2. For more on Douglass Houghton and the fascinating copper history of Lake Superior, especially after European settlement, see Pantell (1971); Krause (1992); Rosemeyer (1999); Bornhorst and Lankton (2009).

3. Geophysical methods include measurements of density (gravity), magnetic and electrical properties, and radioactivity, all of which can be done remotely (without actually touching the rock). Seismic methods, which determine the structure of buried rock by recording the movement of shock waves, commonly require application of a shock and placement of recording devices. The choice of a method depends on the composition of rocks that are being studied. Basalt (see below), which makes up the Midcontinent Rift, has a density of about 2.8 to 3.0. This is much higher than the densities of 2 to 2.6 in the average sedimentary rocks and granites that surround and overlie the rift. Additional early geophysical studies over Lake Superior are summarized in Hinze et al. (1982); Cannon et al. (1989); Thomas and Teskey (1994); and Allen et al. (1997).

4. Magmas and the igneous rocks, which form when the magmas cool and crystallize, consist largely of silicon and oxygen, with smaller amounts of aluminum, iron, magnesium, calcium, sodium, and potassium. Ultramafic rocks are enriched in magnesium, mafic rocks are enriched in iron, magnesium, and calcium; and felsic rocks are enriched in sodium and potassium. Magmas can reach the surface to form extrusive (volcanic) rocks or remain at depth to form intrusive (plutonic) rocks. The most common ultramafic volcanic rocks are picrite and komatiite (discussed in chapter 7). The most common mafic volcanic rock is basalt, which is the main rock in the Midcontinent Rift. Basalt magma that does not reach the surface forms the intrusive rock known as gabbro or diabase. The most common felsic rocks are granite and granodiorite (slightly more mafic than granite and sometimes referred to as intermediate), both of which crystallize at depth. Their extrusive equivalents are rhyolite and andesite.

5. Sometimes magma rising from the mantle stops (ponds) at the bottom of the crust because compressive forces limit faults through which it can rise or because it is too dense to continue rising. The heat in this ponded or underplated magma can melt overlying crust, forming felsic magmas that reach the surface to form rhyolite volcanic rocks.

6. Although the Midcontinent Rift is generally accepted as a member of the LIP club, there are some concerns. Its ~24 million-year span of volcanism is longer than the 1 to 5 million-year life of most other LIPs. For volcanic rocks of the main Midcontinent Rift, early estimates of magma production rates were 0.15 to 0.2 km^3/year, which are within the range estimated for other LIPs. However, the data of Miguel Merino and others show magma production rates for the entire Midcontinent Rift (including the Lake Superior

region and buried continuations) of only 0.05 to 0.08 km^3/year. Regardless of this, the Midcontinent Rift is clearly an unusual volcanic feature that must be to fit into global supercontinent history.

7. The 2 million km^3 estimate (Cannon and Hinze, 1992) is favored over a 1.3 million km^3 estimate (Hutchinson et al. 1990) in the most recent summary of Midcontinent Rift geology by Miller and Nicholson (2013). The basalt magma that underplates the crust in the geologic cross section (fig. 5.2B) is considered to contain a similar amount of magma, bringing the total magma of the Lake Superior region to an enormous 4 million km^3. Additional reviews of the volcanic history of the Midcontinent Rift are provided in Green (1982); Green and Fitz (1993); Nicholson et al. (1997); Miller (2007); Vervoort et al. (2007); Hollings et al. (2007).

8. Pyroclastic volcanic rocks consist of fragments of volcanic rocks in a matrix of volcanic rock; they can form by explosive volcanic eruptions or by breaking up of lava flows. Ignimbrites are a special type of pyroclastic rock that are deposited from a cloud of hot gas, felsic magma, and rocks; some ignimbrites are so hot when they are deposited that they become welded into massive rock also called welded tuff. The presence of these felsic rocks in an igneous province like the Midcontinent Rift, which is dominantly mafic, is probably due to melting of the lower crust, although some could have formed by differentiation (progressive crystallization) of the mafic magmas.

9. Thick lava flows can form when lava expands the central part of a flow that has already cooled and crystallized on the outside, inflating it like a balloon. A more likely origin for the giant Greenstone Flow is some sort of ponding, as proposed by Longo (1984), making the flow a small version of a magma ocean.

10. The few pillows that have been found in Midcontinent Rift basalts are thought to have formed when the magma was emplaced in freshwater lakes rather than the ocean (Nicholson et al., 1997).

11. "Reversed paleomagnetic poles" refers to the fact that Earth's magnetic poles were reversed, with the south pole at the north pole (for further information, see https://www.nasa.gov/topics/earth/features/2012-poleReversal.html).

12. Fairchild et al. (2017) reports ages of 1083 Ma for volcanic rocks on Michipicoten Island, which extends the age range of Midcontinent Rift volcanism from the 1110–1094 Ma that is usually quoted.

13. The magnetic field in which most basalts were emplaced is preserved when the rock cools below the Curie temperature of 585°C for magnetite, the most common magnetic mineral in basalts.

14. Dikes are generally vertical planar intrusions; sills are generally horizontal planar intrusions. See Weiblen (1992) for a review of Midcontinent Rift intrusive rocks; Miller and Chandler (1997) for information on the Beaver Bay Complex, which intrudes the Duluth Complex; Hollings et al. (2010) for information on dikes and sills; and Walker et al. (1993) for information on the Coldwell Complex.

15. Anorthosite is an igneous rock that consists almost entirely of calcium-rich plagioclase (anorthite). The origin of anorthosite magmas is controversial, but most proposed origins start with a basalt magma and invoke some process to separate the feldspars from the mafic minerals, with or without melting of lower crustal rocks to add feldspar components (Arndt 2013). Alkalic magmas contain enough sodium and potassium to combine with all available silicon, producing a rock that lacks quartz (SiO_2).

Carbonatites are magmas that consist largely of calcite. Some alkalic and carbonatitic magmas form during early stages of continental rifting (Weidendorfer et al. 2017).

16. As described in Rogala et al. (2007), Sibley Group sediments were deposited in lakes that evolved into playas starting at about 1450 Ma. The lake sediments were covered with dune deposits at about 1110 Ma. Ages of the smaller sandstone and quartzite exposures in other parts of the Midcontinent Rift are not known, but they are assumed to be the same as those of the Sibley dune sands, about 1110 Ma.

17. Information on the sedimentary history of the Midcontinent Rift comes from Ojakangas and Morey (1982a,b); Kalliokoski (1982); Daniels (1982); Anderson (1997); Suszek (1997); Cullers and Berendsen (1997); Ojakangas et al. (2001a,b); Ojakangas and Dickas (2002); Stewart and Mauk (2017).

18. The debate about a marine versus lacustrine origin for the Nonesuch Shale pits geologic against geochemical evidence. Geology argues for a freshwater origin mainly because the Nonesuch Shale overlies the Copper Harbor, which was almost certainly formed in a nonmarine system. Geology also notes that any arm of the sea would have extended 800 km from the east side of Laurentia, a pathway for which no evidence exists. Geochemistry points out that a source of sulfur is needed to form the abundant pyrite (FeS_2) in the Nonesuch. The grains of sediment in the Nonesuch contain iron but no sulfur. So the sulfur must have come from the water in which the Nonesuch was deposited (because the pyrite formed too soon after the sediment was deposited for other fluids to have invaded the rock). Seawater contains lots of dissolved sulfur and is a common source of sulfur for pyrite in sediments. At this point, the controversy continues (see Mauk et al. 2015).

19. See Ojakangas and Dickas (2002) for a review of these sedimentary rocks and Houlihan et al. (2015) for paleomagnetic measurements on the Hinkley and Fond du Lac sediments.

20. For the age of the faulting, see Bornhorst and Williams (2013). For zircon studies, see Stein et al. (2015); Brown and Hampton (2016); Wirth (2016) and Malone et al. (2016). The main point of contention regarding the age of the Jacobsville is whether some of it was deposited after Grenville-age compression, which probably ended at about 980 Ma.

21. There were quite a few other big ones, including what might be the largest, a 28.2-ton boulder originally on loan to Presque Isle Park in Marquette, Michigan.

22. See figure 9 in Miller and Nicholson (2013) for the location of native copper deposits and prospects outside the Keweenaw Peninsula.

23. In 2017 US dollars, the native copper production was worth about $35 billion and copper sulfide production about $15 billion. Silver from the copper sulfide deposits was worth about $1.5 billion.

24. As summarized by Bornhorst and Barron (2011) and Bornhorst and Mathur (2017), copper-bearing solutions flowed out of the rift twice. Early, low-temperature solutions (~100°C) flowed out as the lavas were being compacted beneath the accumulating sediments. These solutions flowed upward through the volcanic rocks and the Copper Harbor Conglomerate where they encountered the overlying Nonesuch Shale. There copper in the solutions reacted with pyrite (FeS_2) in the shale, exchanging dissolved copper for iron in the pyrite to form Cu- and Cu-Fe sulfides. Later solutions (~225°C) flowed up through the Midcontinent Rift volcanic and interflow conglomerates and deposited native copper as they cooled, mostly in vesicles in the volcanic rocks and holes (pores)

in the interflow conglomerates. For additional information, see studies by White and Wright (1954), White (1968); Brown (1971); Bornhorst et al. (1988); Mauk et al. (1992); Swenson et al. (2004); and the summary paper by Nicholson et al. (1992).

25. For more information on ancient and modern environmental issues, see Jaebong et al. (1999); Pompeani et al. (2015); Kerfoot et al. (2016).

26. Magmatic immiscibility is a counterintuitive process; why would a perfectly happy magma split into two separate, coexisting magmas? The answer relates to the fact that the concentration of elements in the magma changes as it cools and crystallizes. As crystallization proceeds, the concentration of iron and sulfur increases, causing them to bond to form iron sulfides. Normally, the iron sulfides would crystallize from the magma as a sulfide mineral such as pyrrhotite ($Fe_{1-x}S$), but the mafic and ultramafic magmas are too hot, and so the iron sulfide remains liquid, an iron-sulfide magma. Copper and nickel in the original magma also go into the new iron sulfide magma. For more on magmatic immiscibility as an ore-forming process, see Kesler and Simon (2015). For more on these deposits, see Ripley (1986, 2014); and Miller and Nicholson (2013).

27. Gallagher et al. (2017) have shown that the thermal history of the Midcontinent Rift was relatively mild. Analysis of organic matter in the Nonesuch shows that it has undergone sufficient burial to generate oil. Similar analyses in sedimentary units that correlate with the Nonesuch in buried parts of the Midcontinent Rift in Minnesota (Solor Church Formation), Iowa, and Kansas show the organic matter to be mature or overmature (Hieshima et al. 1989; Imbus et al. 1990; Palacas 1995; Hegarty et al. 2007).

28. In addition to a review by Miller and Nicholson (2013), this summary is based on information in Van Schmus (1982); Cannon (1992); Klewin and Shirey (1992); Nicholson et al. (1997); Vervoort et al. (2007) and Keays and Lightfoot (2015).

29. Compositions of related igneous rocks can be shown in AFM diagrams (fig. 5.13) that show relative abundances of alkalis (sodium and potassium), iron and magnesium. Part A shows the two most common compositional trends. Tholeiitic rock groups plot along a curved trend starting with mafic rocks (basalts), continuing toward the iron apex and then turning downward toward felsic rocks (rhyolite) at the alkali apex. Calc-alkaline rocks have lower relative iron contents and plot in a linear trend that extends from basalts toward the alkali apex. Part B shows that rocks of the the Midcontinent Rift follow a tholeiitic trend.

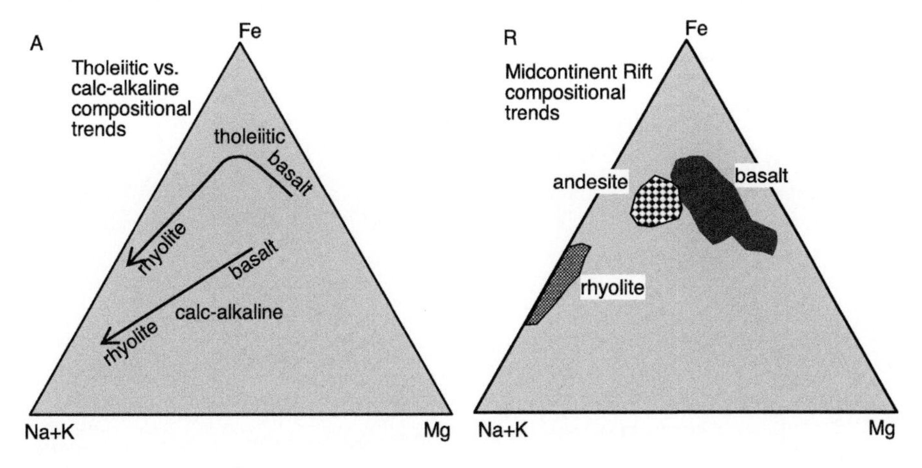

30. Grenville Orogen rocks exposed in the Canadian Shield are known as the Grenville Province, and this term is sometimes extended to include the Adirondack Mountains. The entire belt, which extends into Mexico and parts of other continents, is commonly referred to as the Grenville Orogen.

31. Autochthonous rocks are found where they formed; allochthonous rocks have been displaced a long distance from their place of origin; parautochthonous rocks are not in place but have not moved very far.

32. In simplest terms, gravitational collapse happens when the strength of rocks in a mountain belt is not sufficient to hold up their weight. The mountain usually collapses by flattening of former thrust sheets and sliding backward along the same faults along which it was originally stacked (Liu and Yang, 2003; Rutte et al., 2017). To make room for the mountain top to slide downward, the region must be in "extension," which requires either that plates be moving apart or laterally with respect to each other (rather than converging).

33. See also Cawood et al. (2016); and Fairchild et al. (2017).

34. For more on Precambrian stromatolites, see Grotzinger and Knoll (1999); and Allwood et al. (2006).

35. Sheldon (2013). See Awramik and Sprinkle (1999) for diversity data and Walter and Heys (1985) for abundance data.

36. For background on the Ediacaran biota and various views on its characteristics and paleontological significance, see Shen et al. (2008); MacGabhann (2014); Droser and Gehling (2015).

CHAPTER 6

Building the Continent

Paleoproterozoic Basins, Mountains, and Meteorites

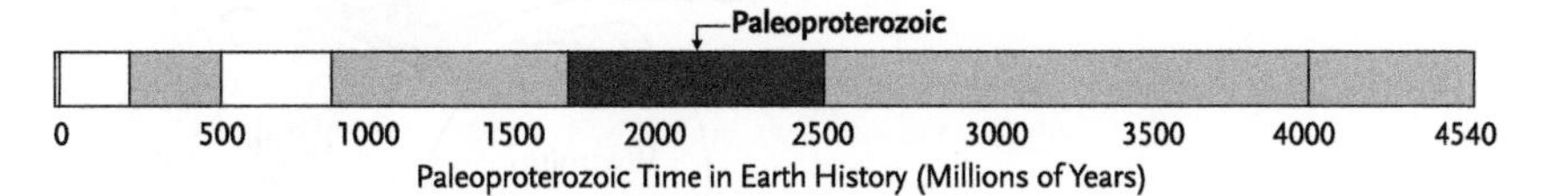

6.1. The Superior Craton Broke Away from the Superia Supercraton at the Start of Paleoproterozoic Time

Moving back to the start of the Paleoproterozoic about 2500 million years ago, our time traveler would have had a new problem. There was a scarcity of continents. There was a global ocean, of course, but it contained far fewer land masses than during Midcontinent Rift time. There were only a few small crustal fragments called cratons and even fewer larger clusters called supercratons—nothing that you would call a continent. These cratons and supercratons, which had formed during Archean time, were in the earliest stages of coming together in collisions called orogenies to form the larger land masses that we know today as continents.

One of the largest supercratons in the early Paleoproterozoic sea, and probably the best one to settle down on, was Superia, which makes up the northern part of the Great Lakes region. Unfortunately, almost as soon as our time traveler settled down on Superia, it began to break apart. You can see in figure 6.1 that it broke into five smaller cratons, Wyoming, Hearne, Karelia, Kola, and Superior.[1] Rifts that cut through the Superia supercraton as it broke apart were filled with volcanic and sedimentary rocks, much like the Midcontinent Rift. Some of the rifts widened to form oceans, which flooded the margins of the cratons, depositing more sediment and forming some of Earth's first passive margin sedimentary basins (box 6.1). In figure 6.1, these sedimentary and volcanic rocks are shown as Paleoproterozoic Cover Sequence, indicating that they were deposited on (and cover) the older Archean-age cratons that are discussed in the next chapter.

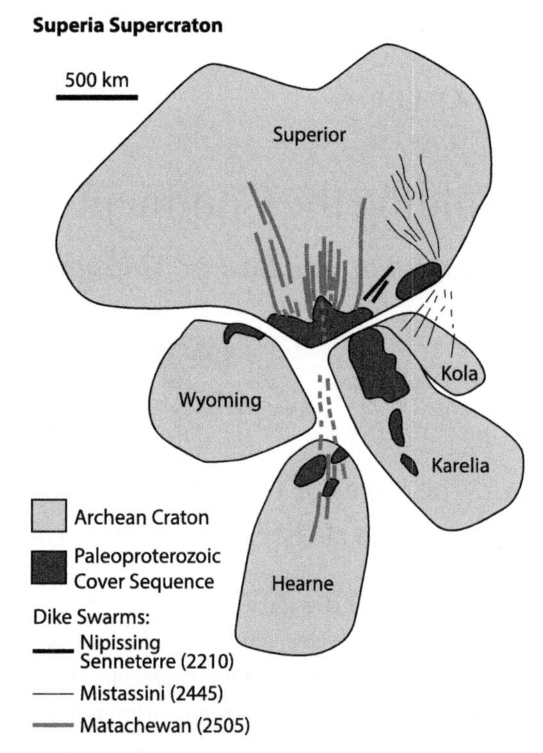

Figure 6.1. The early Paleoproterozoic supercraton known as Superia (including the Superior craton) showing the cratons, dike swarms, and cover sequence rocks that formed as it broke apart. (Modified from Ernst and Bleeker 2010.)

BOX 6.1. RIFT AND DRIFT: PASSIVE MARGIN BASINS

Rifts in cratons usually start as relatively narrow valleys flooded by volcanic rock that is gradually covered with sediment eroded from the rift walls. If rifting stops, as was the case with the Midcontinent Rift, we are left with a "rift basin" partially to totally filled with volcanic and sedimentary rocks. If this rift-and-drift process continues far enough, however, continent margins on both sides sink below sea level, forming shallow, flooded shelves that become passive margin or craton margin sedimentary basins containing limestone, evaporites and other chemical sediments. Passive margin basins became common only in Paleoproterozoic time, when cratons had combined to make widespread continents that could be flooded. This dramatic change in the pattern of Earth's sedimentation is reflected in the abundance of biogenic and chemical sediments discussed in this chapter. These sediments are also important because the carbonates provided a sink for CO_2 in the atmosphere and therefore an important new part of the carbon cycle that is so important to global climate and sea level.

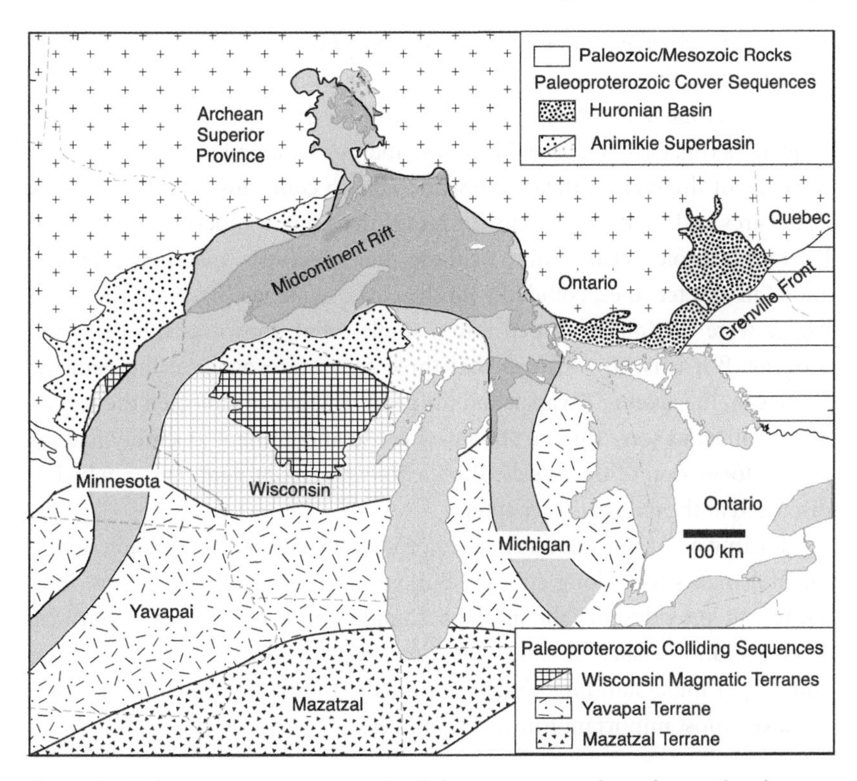

Figure 6.2. Paleoproterozoic cover and colliding sequences along the south side of the Superior craton. (Parts of the Wisconsin magmatic terranes and Animikie Superbasin [shaded] and all of the Yavapai and Mazatzal terranes are concealed beneath younger sedimentary rock). Modified from Cannon et al. (2007) and NICE Working Group (2007).

As the rifted cratons migrated farther apart, they took on lives and alliances of their own. By the end of Paleoproterozoic time, Karelia and Kola had joined the Baltic supercraton in Europe. Hearne and Wyoming, after long migrations, actually rejoined Superior to form Laurentia, the continent that is at the core of North America. It is the assembly of Laurentia that we are interested in here, and that story is best told by the geologic events that took place on the south side of the Superior craton.

The story can be broken into two parts. The first part concerns the Paleoproterozoic Cover Sequence rocks that are exposed in the Huronian Basin on the east and the Animikie Superbasin[2] on the west (fig. 6.2). These sedimentary basins record the rift-and-drift process that accompanied the break-up of Superia as well as several other important events in Earth history Among them are Earth's earliest glaciation events, one of its largest meterorite impacts, the change to an oxygen-rich atmosphere, and deposition of peculiar sediments that are society's main source of iron and uranium.

The second part of the story concerns the Paleoproterozoic Colliding Sequences (fig. 6.2). Because the passive margin sedimentary basins were on the edge of the craton, they were vulnerable to collisions with the volcanic island arcs and small continents (terranes) that were wandering about in the Paleoproterozoic ocean. Three of these, the Wisconsin magmatic, Yavapai and Mazatzal terranes, collided with the cover sequence (fig. 6.2). These collisions were originally lumped into a single event known as the Penokean orogeny. But it is now known that three separate orogenies, the geon 18[3] Penokean, the geon 17 Yavapai, and the geon 16 Mazatzal, affected the region. These collisions were the start of a new phase of continent assembly that led toward formation of the middle Proterozoic supercontinent, Nuna, which is discussed at the end of this chapter.

Sorting all of this out has taken a prodigious amount of geologic sleuthing and makes for a long chapter. But, reading it is worth the effort, both because it provides insight into the complex processes of supercontinent assembly, and because it shows how long-term research reveals new ideas and deeper understanding. Finally, it provides the geologic context for some of Earth's most important mineral resources.

6.2. The Eastern Cover Sequence in the Huronian Basin Includes the Huronian Supergroup, the Sudbury Igneous Complex, and the Whitewater Group

The Huronian Basin, which is exposed along the north shore of Lake Huron (fig. 6.3), includes sediments of the Huronian Supergroup, as well as the Sudbury Igneous Complex, which resulted from an enormous meteorite impact, and the Whitewater Group, a curious sequence of sediments that is exposed only within the Sudbury Igneous Complex. To the north and west, the Huronian sediments lap up over the Archean craton. To the east, they terminate against the Grenville front,[4] discussed in chapter 5, and to the south they are covered with Paleozoic sediments.

6.2.1. The Huronian Supergroup Includes an Early Rift Phase and Late Passive Margin Phase

Original mapping of the Huronian Supergroup by geologists from the Geological Survey of Canada and the Ontario Geological Survey divided it into

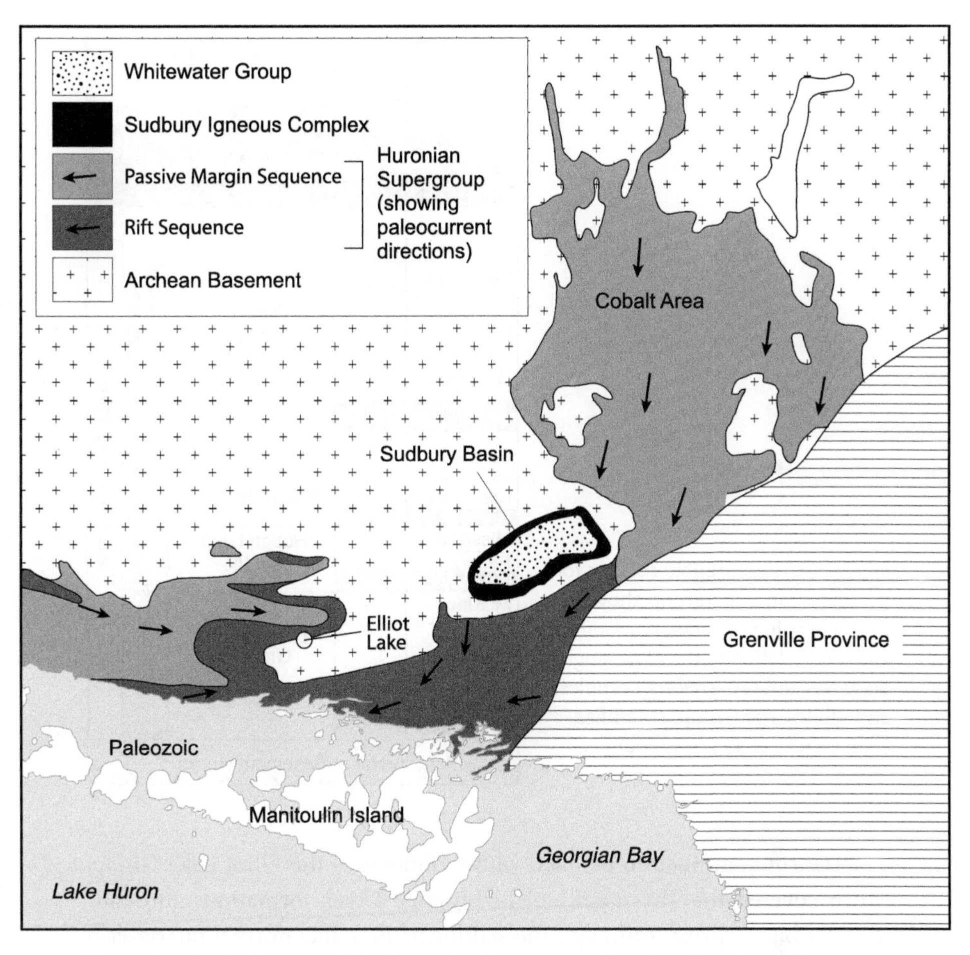

Figure 6.3. Geologic map of the Huronian Supergroup showing directions of sediment transport (paleocurrent) and the location of the Sudbury Igneous Complex, Whitewater Group, and Elliot Lake uranium deposits. (Modified from Roscoe and Card 1992.)

four groups, which record the transition from rift to passive margin basin (fig. 6.4).[5] Rocks at the bottom of the sequence, including the Livingstone Creek and Thessalon formations, were deposited during rifting. The Thessalon Formation consists of basaltic and rhyolitic volcanic rocks that were erupted into the rift basin, and the Livingstone Creek Formation consists of sandstones and conglomerates containing debris that was eroded from the Archean craton to the north. Thessalon lavas lack pillows, indicating that the Superior continental margin was not yet submerged, and their age of about 2450 Ma marks the start of rifting.[6]

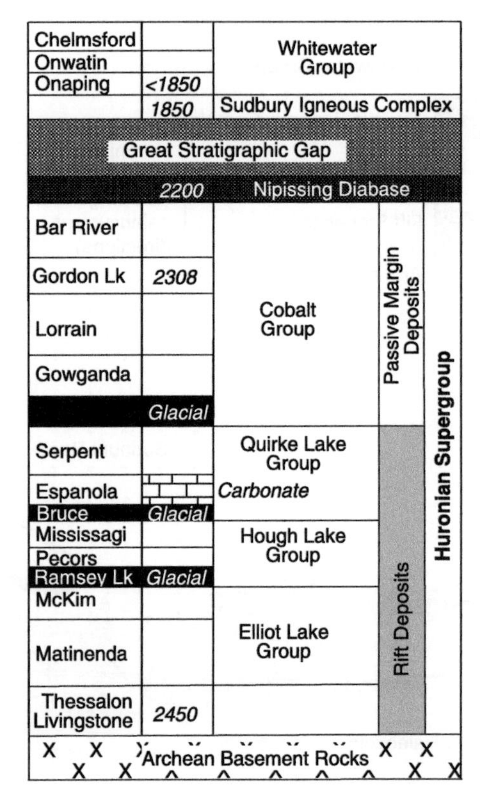

Figure 6.4. Stratigraphic column for the Huronian Supergroup showing the four main sedimentary groups and their relation to the Nipissing diabase, Sudbury Igneous Complex, and Whitewater Group. (Modified from Piercey et al. 2007; Young 2015.)

The rift continued to be filled with sediments of the Elliot Lake Group, which covered the Thessalon and Livingstone Creek formations and contained a big surprise. Although the sediments look like normal quartz-rich conglomerates, sandstones, and siltstones, the conglomerates contain small detrital grains that simply should not be there (figs. 6.5A, B, C in color section). This is especially true of the oldest unit, the Matinenda Formation,[7] where the small grains include pyrite and uraninite (UO_2). These two minerals are extremely rare as sedimentary grains because they decompose in the presence of oxygen-rich water. These small sedimentary grains made the Matinenda Formation simultaneously a major uranium resource and an important milestone in the evolution of Earth's atmosphere, as discussed in the section on uranium deposits.

The rest of the Huronian sedimentary sequence, including the Hough Lake, Quirke Lake, and Cobalt groups (fig. 6.4), contains some of Earth's oldest glacial deposits, including both tillites and varved sediments. Tillites look just like the glacial tills that we saw in chapter 2, but they are called tillites because they have been converted by time and burial into massive rock (fig. 6.5D, E in color section). The Cobalt Group glacial deposits in the Gowganda Formation, with their distinctive pink cobbles in a gray matrix, are probably the most famous ancient glaciogenic sediments in the world (figs. 6.5D, E in color section). The varved sediments consist of thinly laminated, alternating fine- and coarse-grained sediment pairs (known as varves) with a few large cobbles and boulders that appear to have been dropped into the sediment from above (fig. 6.5F in color section). These are weird rocks. How could a large boulder land in the middle of a sequence of very fine-grained sediment? The most likely answer is that the boulders were dropped into open water by melting ice rafts. If so, the entire sediment package probably formed along the margins of a glacier, with the till deposited from ice and the varved sediment deposited in lakes that cycled annually between ice-free and ice-covered conditions.

The glacial sediments are overlain by sandstones and conglomerates and, in the Quirke Lake Group, by carbonate sediments of the Espanola Formation (fig. 6.5G in color section), one of Earth's oldest carbonate sediments. This sequence from glacial to clastic and even carbonate sediments was repeated three times, and it reflects major transgressions of the sea related to episodic, probably global, glacial periods (box 6.2). As these transgressive sequences were being deposited, the craton margin was transitioning from a rift to a passive-margin (or craton-margin) basin. The Elliot Lake, Hough Lake, and Quirke Lake groups were deposited in the rift, but the last transgressive cycle, the Cobalt Group, was deposited more widely over the flooded craton margin.

The upper part of the Cobalt Group, which consists of the Lorrain, Gordon Lake, and Bar River Formations contains limonite and hematite (fig. 6.5H in color section) rather than pyrite that is abundant in the lower part of the Huronian sequence. This change from reduced to oxidized iron in the sediments reflects the Great Oxidation Event, a pivotal point in Earth's Paleoproterozoic history as discussed further below. Lorraine Formation quartzite, known as puddingstone, has distinctive bright red, hematite-rich clasts (fig. 3.2A).[8] Puddingstone glacial erratics, including one shown in figure 3.2A, are widespread in the southern part of the Great Lakes region.

BOX 6.2. GLOBAL PALEOPROTEROZOIC GLACIATION: WHERE AND HOW?

The Huronian glacial sediments are not alone. Grant Young, Dick Ojakangas, and others have described similar deposits in Paleoproterozoic rocks around the world and raised the possibility that they record a global glacial event. In the United States, Paleoproterozoic glacial deposits are recognized in the Marquette Range Supergroup (Animikie Superbasin) of Michigan and Wisconsin and the Medicine Bow Mountains of Wyoming. In Canada, there are deposits in the Hurwitz Basin in Nunavut and the Chibougamau area of Quebec. Elsewhere deposits have been found in the Kola Peninsula of Russia, adjacent parts of Finland, the Transvaal Basin of South Africa, and western Australia. Because it is difficult to determine exact isotopic ages for glacial deposits, correlations among these deposits are uncertain. The issue is complicated by the presence of multiple glacial deposits in several areas, and by different basin conditions. In the Huronian, for instance, the Ramsay Lake and Bruce glacial deposits formed in restricted rift basins where glaciation may have been caused by local effects. The Gowganda glacial deposits formed over a larger, passive margin area, however, making them more likely to reflect a global glacial event.

Global correlation of Paleoproterozoic glacial deposits raises the question of what could have caused and mediated such a widespread event. Having the continents in a polar position would have been a good start, but help would have probably been needed from global climate. The lack of land plants would have made weathering and erosion of mountains formed by any orogeny much faster and more intense. This might have caused rapid draw-down of CO_2 and consequent cooling, and subsequent warming could have happened if cooling slowed weathering, allowing CO_2 to build up again. Other elements of the carbon cycle that we used for Paleozoic and Pleistocene glaciation are not likely to have been much help. Organic deposits to store reduced carbon (plants) were missing, and carbonate sediments were less common and have a strange isotopic composition that is the subject of continuing research.[9] So maybe something else affected global glaciation, which raises a possibility that is unique to Paleoproterozoic time, the faint young sun paradox (box 6.3).

Huronian Supergroup deposition probably started at about 2450 Ma, the age of the Thessalon Formation. John Craddock and his team have found detrital zircons in the Matinenda Formation as young as 2497 Ma, suggesting that sediments began to pour in shortly after that.[10] The younger age limit for the Huronian sediments is at least 2308 Ma, the age of detrital zircon in the Gordon Lake Formation reported by Birger Rasmussen and associates. The Nipissing diabase, which was intruded into the sediments at 2217 Ma (fig. 6.4),[11] marked the end of Huronian sedimentation; nothing more happened.

But, something is wrong here. In most rift-and-drift sequences of this type, the end of sedimentation marks the start of an orogeny as a continent arrives from somewhere to smash into the basin. This did not happen to the Huronian, however, and instead it waited for about 350 million years before a collision took place. Grant Young has called this long period of time the Great Stratigraphic Gap (fig. 6.4), and it will figure in the supercontinent story that we are working our way toward. But first we have a meteorite impact to deal with.

BOX 6.3. THE FAINT YOUNG SUN PARADOX AND GLOBAL GLACIATION

One of the great paradoxes of the early Earth is why it did not freeze. Carl Sagan and George Mullen pointed out in 1972 that solar radiation during Archean time was only about 70 percent as intense as it is today and that this was probably not enough to keep the oceans from freezing. So the early Earth should have been mantled in ice, much like the so-called End Proterozoic snowball Earth.[12] But, water-deposited sediments and pillowed basalt flows provide strong evidence that water was present far back into early Archean time. How the water could have existed is part of the "faint young sun paradox."

Although several explanations have been suggested for this paradox, the best candidate is the greenhouse effect, which can raise the surface temperature of a planet significantly. Estimates of early atmospheric CO_2 abundances range from 10 to almost 50 times present-day preindustrial levels, but it is not clear that this would be enough, stimulating the search for another greenhouse gas that might help. Sagan and Mullen emphasized ammonia (NH_3), but more recent attention has focused on methane (CH_4). Methane is particularly attractive because it is a common gas in primitive planetary atmospheres and it is a more powerful greenhouse gas than CO_2.[13]

The methane story is also interesting because it will react with oxygen to form water vapor and carbon dioxide, which are much weaker greenhouse gases. Thus, an increase in the oxygen content of the atmosphere would cause a loss of methane and global cooling. Alexander Pavlov and others have suggested that just such a cooling during the Great Oxidation Event caused Paleoproterozoic global glaciation, although the timing is somewhat problematic.[14] If it did happen, we are fortunate that oxygen appeared at a point late in Earth history when solar radiation had become strong enough to prevent formation of a long-term snowball Earth.

6.2.2. The Sudbury Igneous Complex Formed When a Giant Meteorite Struck the Huronian Supergroup

After deposition of the Huronian sediments and intrusion of the Nipissing diabase, nothing happened in the region for about 350 million years. And then the Sudbury meteorite impacted near Sudbury, Ontario. That's a long wait, so you have time to imagine what it was like for Robert Dietz in the 1960s as he began to think that the Sudbury Igneous Complex was formed by a meteorite. Today, when we are familiar with the idea of giant impact features on Earth, this is not a difficult concept. Back then it was definitely at or beyond the margin of reasonable hypotheses, and it was greeted with incredulity in some circles. Dietz persisted, however, and now we accept that the Sudbury Igneous Complex was formed by a meteorite impact. Isotopic measurements, recently refined by Donald Davis, indicate that it happened at 1849 Ma.

Model simulations have provided insight into the remarkable series of events associated with the Sudbury impact. In one such study, Michael Zieg and Bruce Marsh (2005) calculated that the meteorite created an instantaneous cavity and melted zone that reached at least 30 km to the base of the crust (fig. 6.6A).[15] This cavity could have relaxed within about two minutes to form a large shallow crater containing a superheated (1700°C) sheet of melted rock about 3 km thick and an overlying layer of broken rock (breccia layer) about 2 km thick. The Sudbury Igneous Complex, which was originally thought to be a typical intrusive rock, is actually the part of that melt sheet that escaped later erosion; it is a roughly elliptical body up to 3 km thick and 60 by 27 km in area (fig. 6.6B).

The Sudbury Igneous Complex that we see today consists of a thick mafic basal zone and a smaller upper felsic zone known as granophyre, a form of granite. Chemical and isotopic analyses summarized in separate studies by Ann Therriault, Stephen Prevec and others confirm that it is indeed melted crust (rather than a magma that rose from the mantle), making the Sudbury Igneous Complex by far the largest known impact melt sheet in the world and the only one that cooled slowly enough to separate (differentiate) into two zones of different composition.[16]

Before we leave the Sudbury Igneous Complex, it is worth asking just how someone could find proof for its impact origin. After all, generations of geologists had studied it without invoking an extraterrestrial impactor origin. The answer has a lot to do with timing. The impactor concept became widespread when the geologic community was intimately involved with the

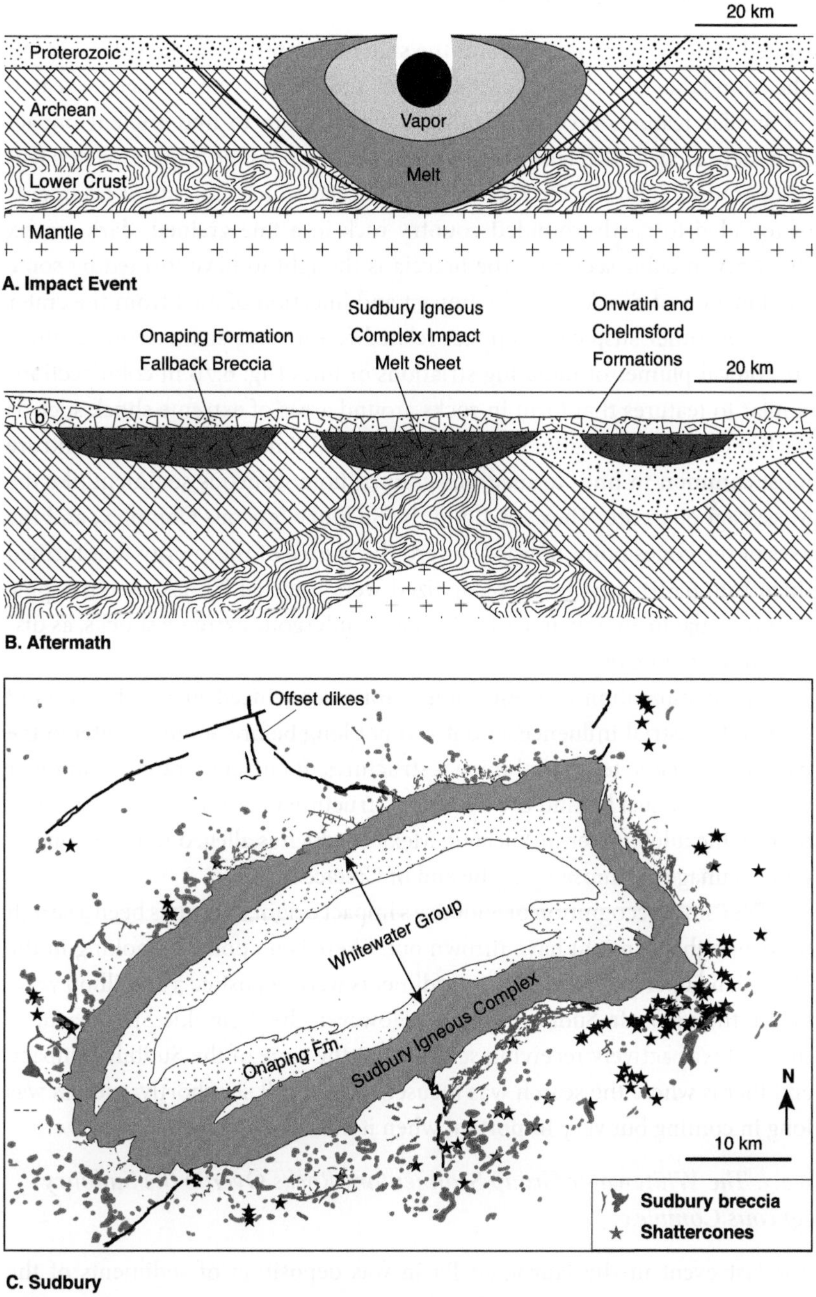

20 km
Proterozoic
Archean
Vapor
Melt
Lower Crust
Mantle

A. Impact Event

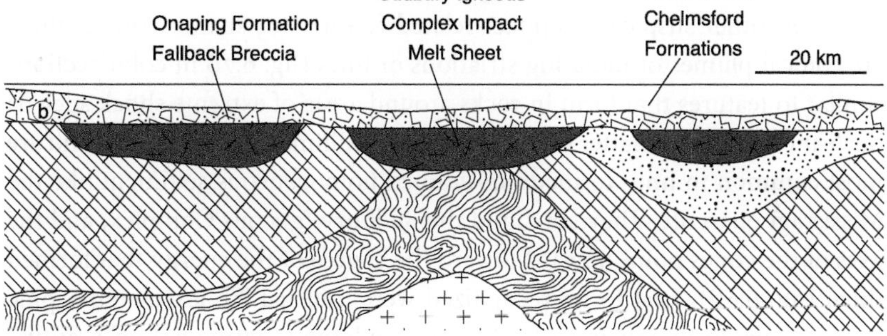

B. Aftermath

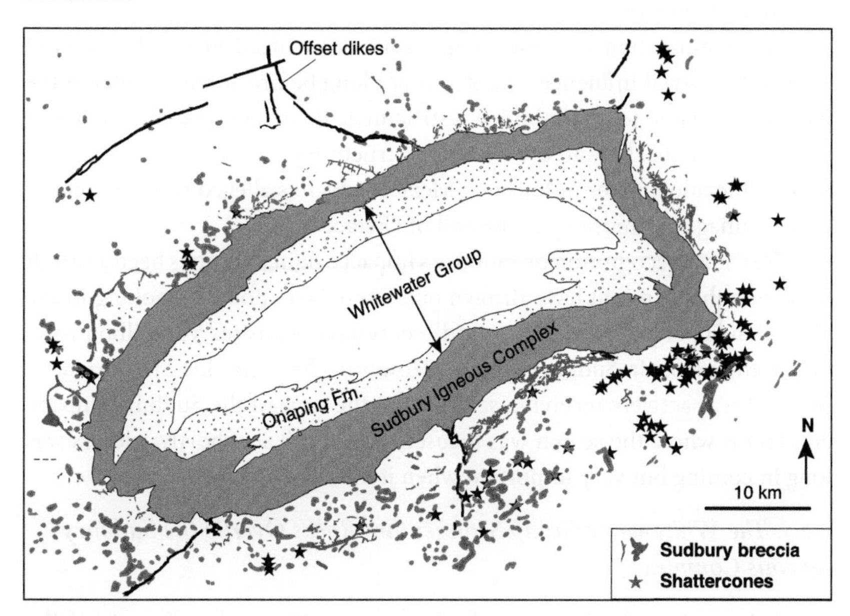

C. Sudbury

Figure 6.6. (**A, B**) Diagrams showing the sequence of events during and after the Sudbury impact; (**C**) map of the Sudbury Igneous Complex and Whitewater Group showing distribution of Sudbury breccia and shatter cones. (Modified from Zieg and Marsh 2005; Ames et al. 2008).

United States' lunar missions and the study of cratering in the solar system. The abundance of cratering on the Moon and Mercury encouraged a search for similar features on Earth, and geologists began to find ancient impact sites there as well. This led to recognition of three possible impact-related features around Sudbury. One is Sudbury breccia, which consists of fragments of rounded to partly rounded country rock in a fine-grained, dark matrix (fig. 6.7A in color section). The breccia is thought to have formed by some combination of shock from the impact and injection of melt from the crater floor.[17] Another suspicious impact-related feature was shatter cones, which are conical plumes of radiating striations or lines (fig. 6.7B in color section) similar to features that form in rocks around areas of extreme shock such as nuclear explosions. Shatter cones range from millimeter to meter size and are most common in fine- to medium-grained homogeneous sediment like the Mississagi quartzite. They surround the Sudbury Igneous Complex, and the axes of the cones point upward toward the source of the energy that formed them (fig. 6.6B). Last to be recognized were unusual "shock metamorphic features" that form only in rocks that have undergone extreme shocks, as discussed further below.

The combination of these three features convinced many observers of an extraterrestrial influence, and it was not long before Sudbury entered the pantheon of large terrestrial impact structures. It currently ranks as number two in the list, just behind the Vredefort structure in South Africa and ahead of the Chicxulub crater in Yucatan, Mexico, which is alleged to have done so much damage to dinosaurs at the end of Cretaceous time.

Ever since recognition of Sudbury's impact origin, there has been a search for debris that it might have thrown out. There is no point in looking in the Huronian Supergroup because its sediments were deposited 350 million years before the 1849 Ma Sudbury impact. However, the Animikie Superbasin to the west was actively receiving sediment at the time of the Sudbury impact, and that is where the search was focused. As we will see shortly, success was long in coming but very important when it did.

6.2.3. The Whitewater Group Sedimentary Rocks Overlie the Sudbury Igneous Complex

The last event in the Huronian Basin was deposition of sediments of the Whitewater Group, which overlies the Sudbury Igneous Complex. It consists of three formations that are found only inside the oval outcrop of the complex (fig. 6.3 and fig. 6.6B), where they were fortuitously preserved from erosion. Undoubtedly, they covered a much wider area originally. These formations record a much-delayed final gasp in the history of the Huronian Basin. They are obviously younger than the 1849 Ma Sudbury Igneous Complex (because they overlie it) and they mark the end of the Great Stratigraphic Gap, which began at the close of Huronian sedimentation at about 2200 Ma.

The lowermost of the Whitewater sedimentary rocks is the Onaping Formation (fig. 6.4), which is a thick layer of breccia and glass shards made up of the rocks and debris from the Sudbury impact; it is essentially a "fallback breccia" consisting of rocks that were thrown into the air by the impact. The other two formations are more typical water-deposited sediments that were eroded from an Archean continent to the east. This easterly source is our first hint that some kind of collision might have taken place shortly after the Sudbury impact. To get a better idea of what was happening, we need to look at the western cover sequence in the Animikie Superbasin.

6.3. The Western Cover Sequence in the Animikie Superbasin Contains the Marquette Range Supergroup

Deciphering the geologic history of the Animikie Superbasin is complicated by the fact that it was deformed by the Penokean, Yavapai and Mazatzal orogenies and then split into two parts by the Midcontinent Rift. Rocks in the superbasin are exposed today in thrust sheets and curious dome-and-keel structures, that make it hard to correlate rocks from one place to another. To top it off, most early studies of the rocks focused on separate areas of iron mining known as iron ranges[18] without much attention to correlation across the entire area (fig. 6.8 and fig. 6.9). As a result, this work did not lead to widely used formation or group terms for all of the sediments and volcanic rocks in the superbasin (or for the superbasin itself, as discussed in note 2). Marquette Range Supergroup has been used for most of the rocks on the east side of the rift in Michigan and Wisconsin, and is extended here to rocks on the west side in Minnesota and Ontario. The debris layer from the Sudbury meteorite impact turns out to provide a time-line that helps correlate among isolated areas and across the Midcontinent Rift.

The Sudbury impact debris layer is the Holy Grail of Great Lakes geology and its discovery has allowed a quantum leap in correlating rock units in the Animikie Superbasin. In 2005, in the scoop of the decade, William Addison and Gregory Brumpton found the debris layer near Thunder Bay, Ontario (fig. 6.8). Shortly afterward, Bill Cannon, Klaus Schulz and co-workers found it in deformed parts of the superbasin in Michigan (fig. 6.8). Some of these layers

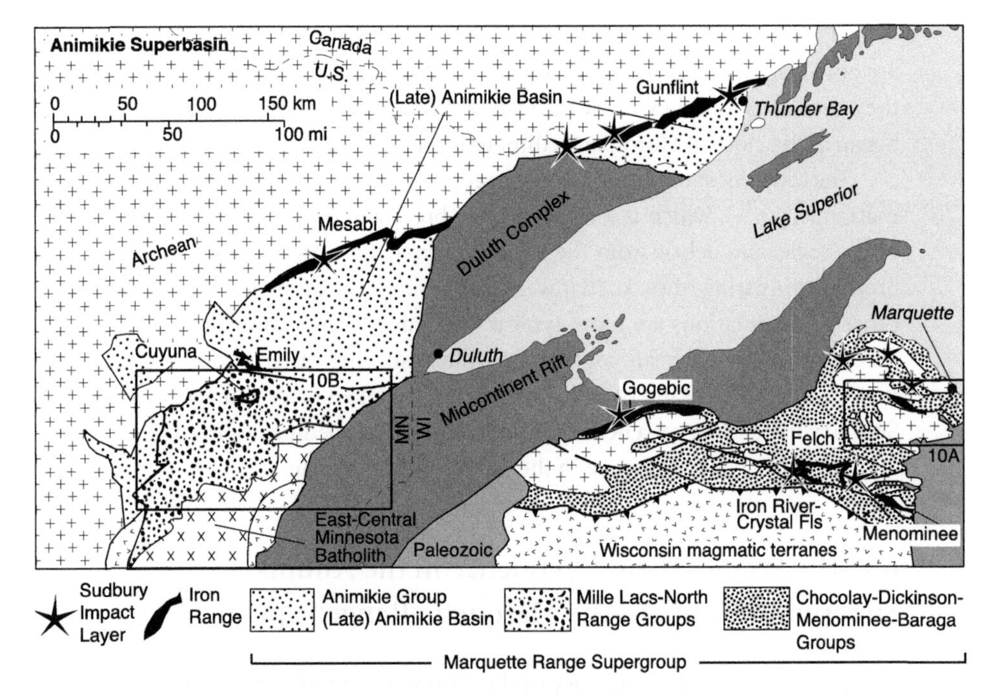

Figure 6.8. The Animikie Superbasin (including the [Late] Animikie Basin, containing rocks of the Animikie Group). Also shown are locations of iron ranges and the more detailed maps in fig 6.10A and B. (Modified from Schulz and Cannon 2007; Ojakangas et al. 2011.)

were in plain sight, and geologists had puzzled over them for years.[19] They were amazingly chaotic (fig. 6.7 C, D in color section) and clearly required a special explanation. The key observation that showed their impact origin was recognition of shock metamorphic features and ejecta particles.[20] The Sudbury debris layer gives us a 1849 Ma layer that splits the superbasin into two parts, pre-Sudbury and post-Sudbury. Using this debris layer and ages of detrital zircons, rocks of the entire superbasin can be included in the Marquette Range Supergroup and this, in turn, helps unravel its geologic history.

6.3.1. Pre-Sudbury Rocks of the Marquette Range Supergroup Include the Chocolay, Menominee, and Dickinson Groups in the East and Mille Lacs and North Range Groups in the West

Pre-Sudbury rocks of the Marquette Range Supergroup record two distinct transgressions of the sea while Superia was breaking apart. This transgression is well displayed in formations on the east side of the rift, where formations

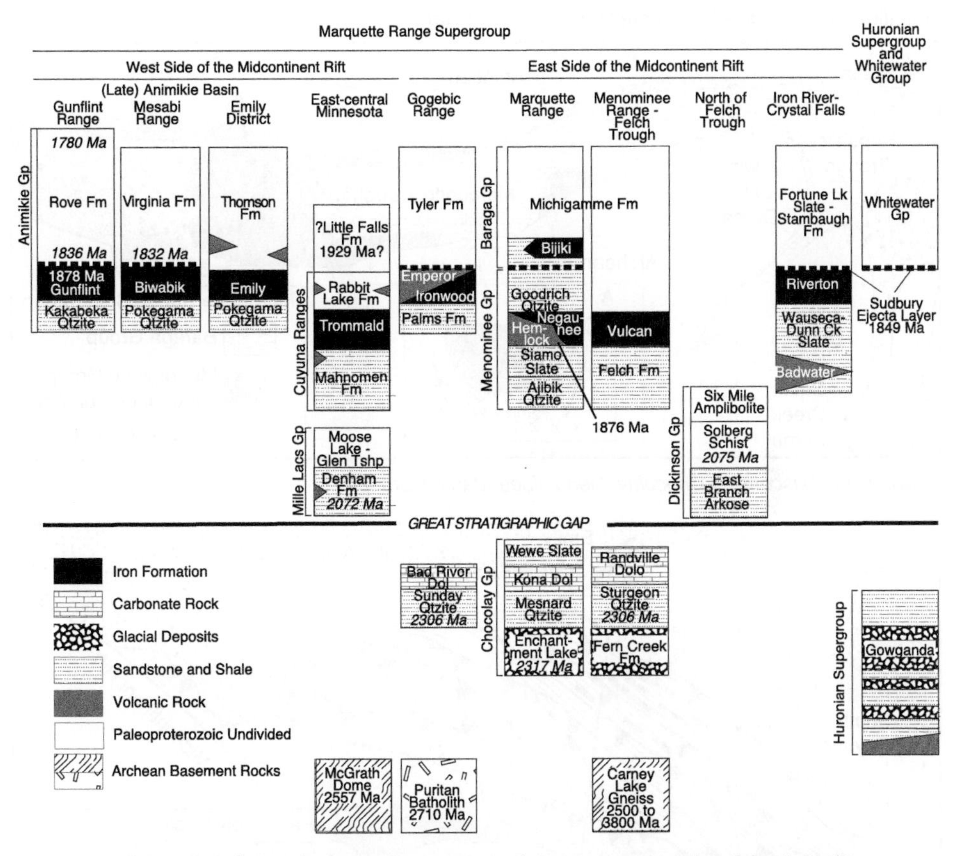

Figure 6.9. Correlation diagram showing age relations among rocks of the Animikie Superbasin and its relation to the Huronian Basin and Sudbury impact debris layer. Numbers show isotopic ages discussed in the text. (Modified from Vallini et al. 2006; Schulz and Cannon 2007; Ojakangas et al. 2011; Cannon et al. 2010; Craddock et al. 2013; Pietrzak-Renaud and Davis 2014; Cannon et al. 2018.)

are exposed in tight synclines called keels or troughs (fig. 6.10A). It is more difficult to reconstruct on the west side where rocks have been thrust into a stack of sheets (fig. 6.10B). The history of the east side, which has been summarized by Klaus Schulz and Bill Cannon, includes two transgressions that are represented by the Chocolay and Menominee groups (fig. 6.9). A third group, the Dickinson Group, is found in only one area and has not yet been fully integrated into the story.[21]

In the *Chocolay Group*, transgression of the sea was preceded by glaciation. The Enchantment Lake and Fern Creek formations contain diamictites (fig. 6.11A in color section) and varved sediments with dropstones that were

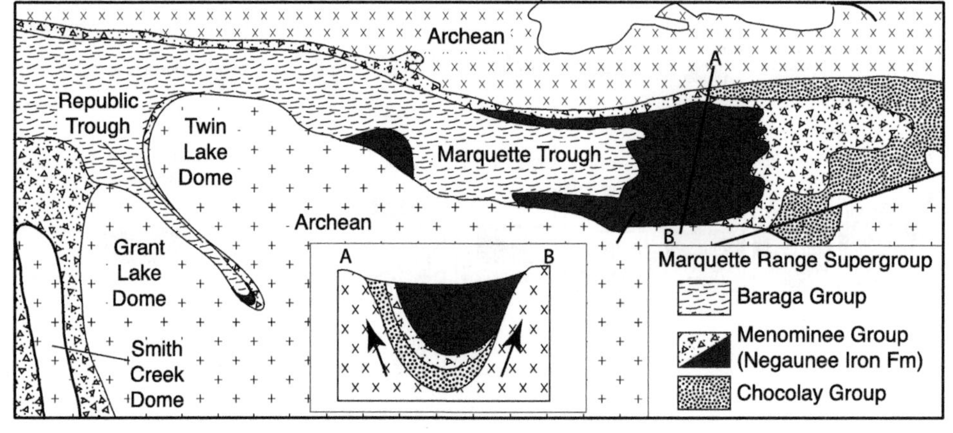

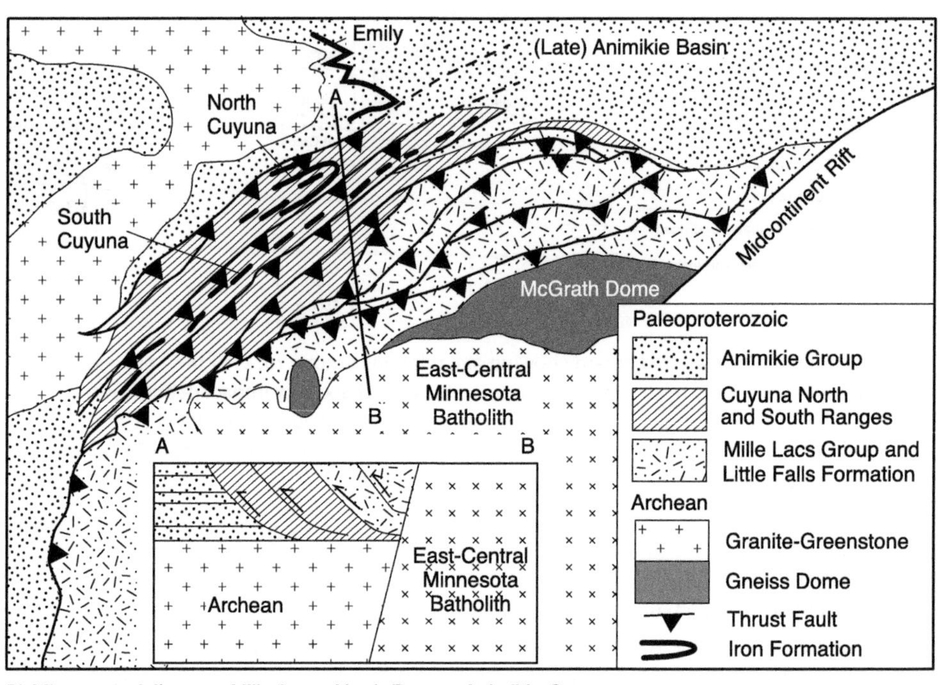

Figure 6.10. Geologic maps and cross section showing rocks of the Marquette Range Supergroup in: (**A**) Marquette and Republic troughs (keels) in Michigan (Felch trough is just south of the area of this map); (**B**) central Minnesota. (Modified from Sims and Carter 1996; Boerboom 2011.)

first called glacial deposits by Francis Pettijohn. These glacial deposits are correlated with the youngest of the three glacial deposits in the Huronian Supergroup (the Gowganda), as shown in figure 6.9, providing the best link that we have between the Huronian and Animikie sequences. The Mesnard and Sturgeon quartzites (fig. 6.11B in color section), which overlie the glacial deposits in the Marquette Range Supergroup, were deposited in the beaches that formed as the sea transgressed onto land after the ice left, and the Kona and Randville dolomites and the Wewe Slate were deposited as transgression brought deeper water (fig. 6.11C in color section).[22] These rocks have a maximum age of about 2300 Ma, as indicated by the youngest detrital zircons analyzed by Daniele Vallini and John Craddock and coworkers (fig. 6.9), which agrees with ages of the Huronian rocks.

After a pause of at least 100 million years, the *Menominee Group* was deposited over the Chocolay during a second transgression of the sea that started with near-shore deposits of the Ajibik Quartzite followed by deeper water deposits of the Siamo Slate. Instead of continuing with deposition of carbonate deposits, however, the basin filled with mafic volcanic rocks (basalt) of the Hemlock Formation and a curious new sedimentary rock called iron formation. The basalts show that rifting was producing faults along which magmas could rise from the mantle, and the iron formations show that the ocean had a strange composition. The iron formations consist of alternating layers of silica and iron carbonates or oxides that were deposited largely in shallow water. They were given different names in each mining region, including Negaunee in the Marquette area, Vulcan in the Menominee area, and Riverton in the Iron River-Crystal Falls area (fig 6.9). These iron formations are chemical sediments that were precipitated from iron-bearing seawater, as discussed later in this chapter. In addition to being a spectacularly unusual aspect of Earth's chemical evolution, they are the main source of iron for steel that supports our civilization.

The age of the Menominee Group is fixed by a 1876 Ma age for the Hemlock volcanic rocks reported by David Schneider and others, and this makes it just a little older than the Sudbury impact.

The search for pre-Sudbury rocks on the west side of the Midcontinent Rift is difficult because the rocks are strongly deformed and the area is almost completely covered by Pleistocene glacial deposits. The Paleoproterozoic rocks have been divided into thrust sheets known as panels, which have been detected in geophysical surveys (fig. 6.10B). In recent summaries by Terry Boerboom, Val Chandler and others, two main rock packages have been recognized, the Mille Lacs Group and the North Range Group.[23] The *Mille Lacs Group* includes the Denham Formation, which consists of siltstone, conglomerate, dolomite, and pillowed to fragmental basalt. Detrital zircons as young as 2072 Ma have been reported from the Denham Formation by John Craddock, suggesting that it (and maybe the Mille Lacs Group) correlates with the Menominee and Dickinson groups to the east (fig. 6.9).[24] The main thrust panels north of the Mille Lacs Group contain rocks of the Cuyuna iron ranges (fig. 6.10B). The *North Range Group* includes the Mahnomen and Rabbit Lake formations and the intervening Trommald iron formation, which are exposed in the North Cuyuna iron range (fig. 6.9). Its age is not known, although the presence of iron formation suggests a correlation with the Menominee Group in the Marquette Range Supergroup as shown in figure 6.9.[25] Stratigraphic relations between these two groups are not clear. If they are both related to the Menominee Group, however, Chocolay-equivalent rocks are missing west of the Midcontinent Rift.

Regardless of the uncertainties about specific groups, it is clear that the Great Stratigraphic Gap between the Huronian and Whitewater was also present between the Chocolay and Menominee Groups, although it was probably shorter.[26]

6.3.2. Post-Sudbury Rocks of the Marquette Range Supergroup Include the Baraga and Animikie Groups

As you can see in fig. 6.9, the Sudbury debris layer is just above iron formation layers in most parts of the Animikie Superbasin. In most areas, the post-Sudbury sedimentary rocks that overlie the impact layer start with near-shore sandstones and continue with a thick succession of deeper water shales. On the east side of the rift these are the Michigamme, Tyler, and Fortune Lake Formations (fig. 6.11E in color section), which make up the *Baraga Group*.[27] On the west side, post-Sudbury sediments are called the Rove, Virginia, and Thomson Formations, all of which are included in the *Animikie Group*, which is found in the (Late) Animikie Basin. (Read carefully here; the Animikie Group is not the Animikie Supergroup and the Animkie Basin is the not the Animikie Superbasin.)[28] Ages of zircons in ash beds and sedimentary rocks in shales of the (Late) Animikie Basin range from about 1878 Ma to as young as 1780 Ma in the northernmost part.

Animikie Group rocks in the (Late) Animikie Basin on the west side of the rift are generally flat lying, whereas much of the Baraga Group on the east side is deformed.[29] This difference might distract you from the important point that all of these rocks were originally part of a single sedimentary basin that extended across the entire region during post-Sudbury time. Dick Ojakangas and others have called this basin the Animikie Basin, but it is referred to here as the Animikie-Baraga Basin to avoid confusion with the larger Animikie Superbasin. This Animikie-Baraga Basin received sediment from the Archean rocks to the north, but also from a southern source. The southern source was made up of rocks that had been thrust northward by a collision, and this brings us to the colliding sequences.

6.4. The Colliding Sequence Includes the Wisconsin Magmatic, Yavapai, and Mazatzal Terranes

During its long existence, the Animikie Superbasin was a sitting duck along the south side of the Superior supercraton. It turns out that three different terranes collided with the Animikie Superbasin. First came the Wisconsin magmatic terranes during geon 18, then the Yavapai terrane in geon 17, and finally the Mazatzal terrane in geon 16 (fig. 6.2).

6.4.1. The Wisconsin Magmatic Terranes Include the Pembine-Wausau and Marshfield Terranes

The Wisconsin magmatic terranes are exposed largely in Wisconsin, although a small part was split apart and moved into Minnesota by the Midcontinent Rift (fig. 6.8). In Wisconsin, the Wisconsin magmatic terranes are separated from the Animikie Superbasin by the Niagara fault zone. Elsewhere the fault zone is hard to find; in Minnesota it was obliterated by the East-Central Minnesota batholith and in Ontario to the east the colliding sequence is covered by Paleozoic sediments on the south.

As you can see in figure 6.12, the Wisconsin magmatic terranes have two main parts, as outlined by the work of Paul Sims, Klaus Schulz, Randy Van Schmus, and others. The largest part is the *Pembine-Wausau terrane*, a group of volcanic island arcs that formed at subduction zones in the Paleoproterozoic ocean. The earliest of the arcs includes a dismembered ophiolite complex[30] (the seafloor on which the arc formed) overlain locally by pillow lavas, felsic flows, and fragmental rocks (figs. 6.13A, B, C in color section) that were part of the volcanic island arc. The petrochemistry[31] of these volcanic rocks confirms that they are largely calc-alkaline and formed at a subducting convergent margin. Volcanic rocks along the north side of the terrane have ages of 1889 to 1860 Ma, whereas those along the south side have ages of 1845 to 1835 Ma. This is interpreted to mean that subduction formed the older volca-

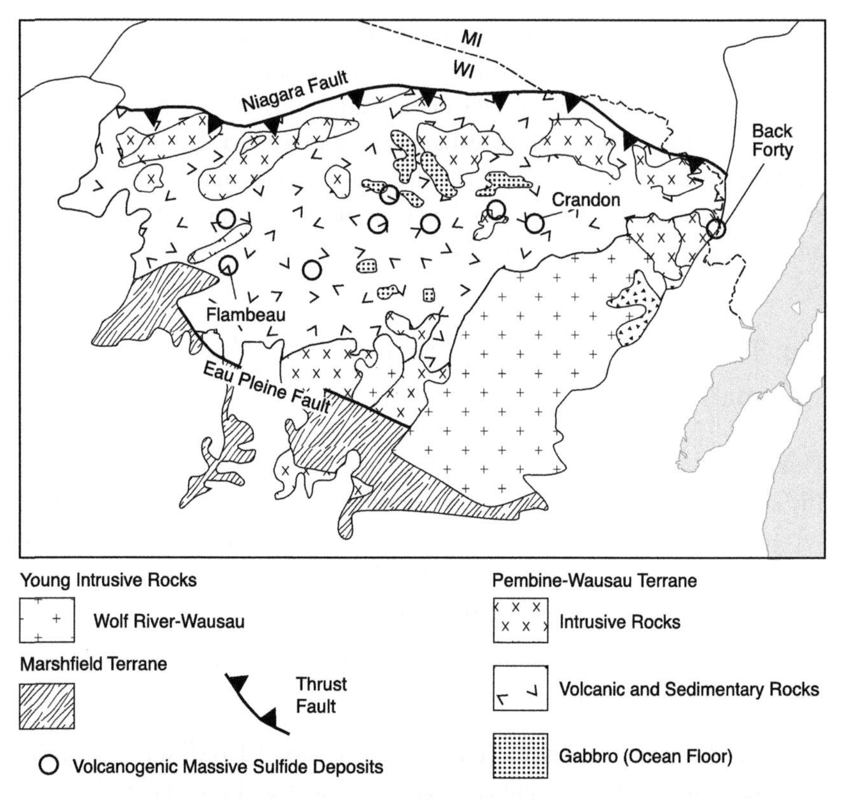

Figure 6.12. Geologic map of the exposed part of the Wisconsin magmatic terranes in Wisconsin showing location and major rock types in the Pembine-Wausau and Marshfield terranes. See figure 6.2 for the location of the Wisconsin magmatic terranes and figure 6.8 for their extension into Minnesota. (Modified from Sims et al. 1989; Schultz and Cannon 2007.)

nic arc along the north side of the terrane first, and then switched to the south side forming the younger volcanic arc.

The other part of the Wisconsin magmatic terranes is the *Marshfield terrane* (fig. 6.12), a continent fragment consisting of Archean gneiss (fig. 6.13D in color section) that is at least 2800 Ma old according to Randy Van Schmus and coworkers. These Archean rocks are included in the Wisconsin magmatic terrane because they are overlain by a Paleoproterozoic volcanic arc similar to the ones to the north. The only difference is that this arc formed on Archean continental crust rather than seafloor.[32] Volcanic and intrusive rocks that make up the Paleoproterozoic Marshfield arc have an age of 1870 to 1860 Ma, generally similar to the older arc in the Pembine-Wausau terrane. The volcanic rocks are also calc-alkaline and formed at a subduction zone. As we will see shortly, it was collisions from these combined terranes that caused the Penokean orogeny.

6.4.2. *The Yavapai and Mazatzal Terranes Consist Largely of Juvenile Volcanic and Intrusive Rocks*

The *Yavapai and Mazatzal terranes* that we see in the Great Lakes region are the far eastern extension of terranes that extend northeastward from Arizona (fig. 6.14). Both are juvenile terranes, which means that they consist largely of volcanic and intrusive rock like the Pembine-Wausau terrane, rather than older Archean or early Paleoproterozoic cratons.[33] Yavapai juvenile rocks are largely geon 17 whereas Mazatzal rocks are geon 16. Yavapai intrusive and volcanic rocks in the Great Lakes region include the Montello batholith with an age of about 1760 Ma (fig. 6.14) and large sheets of rhyolite welded tuff of the same age that formed when still-liquid rhyolite magma was extruded explosively to form thick, solid sheets. Other Yavapai-age intrusions, such as the 1787–1772 Ma East-Central Minnesota batholith (fig. 6.10B), were emplaced into rocks of the Animikie Superbasin north of the actual Yavapai terrane. Mazatzal and younger intrusions and volcanic rocks are also scattered through the southern part of the region. The largest of these is the geon 15 Wausau syenite-Wolf River granite batholith complex in central Wisconsin.[34]

6.4.3. *Baraboo Interval Quartzites Covered the Amalgamated Terranes*

After the Animikie Superbasin, Wisconsin magmatic and Yavapai terranes collided, they were covered by a layer of sandstone known as the Baraboo Interval Quartzites. Remnants of this layer are found in outcrops scattered through the southern part of the Great Lakes region, including the Baraboo Hills in Wisconsin that was mentioned in chapter 4 as an island in the Cambrian sea (fig. 6.14). The quartzites consist almost entirely of grains of quartz and minor kaolinite, and they were deposited in shallow water (fig. 6.13E in color section) on a surface of low relief. Gordon Medaris, who has worked extensively on these rocks with numerous colleagues, reported that paleosols on Precambrian basement rocks underlying the Baraboo quartzites were so strongly weathered that they consist largely of quartz, hematite, and kaolinite; all other minerals have been removed by weathering.

The Baraboo Interval Quartzites are younger than the 1760 Ma Montello batholith, which they overly unconformably, and older than the 1468 Ma Wolf River batholith, which intrudes the Waterloo and McCaslin Quartzites. Most

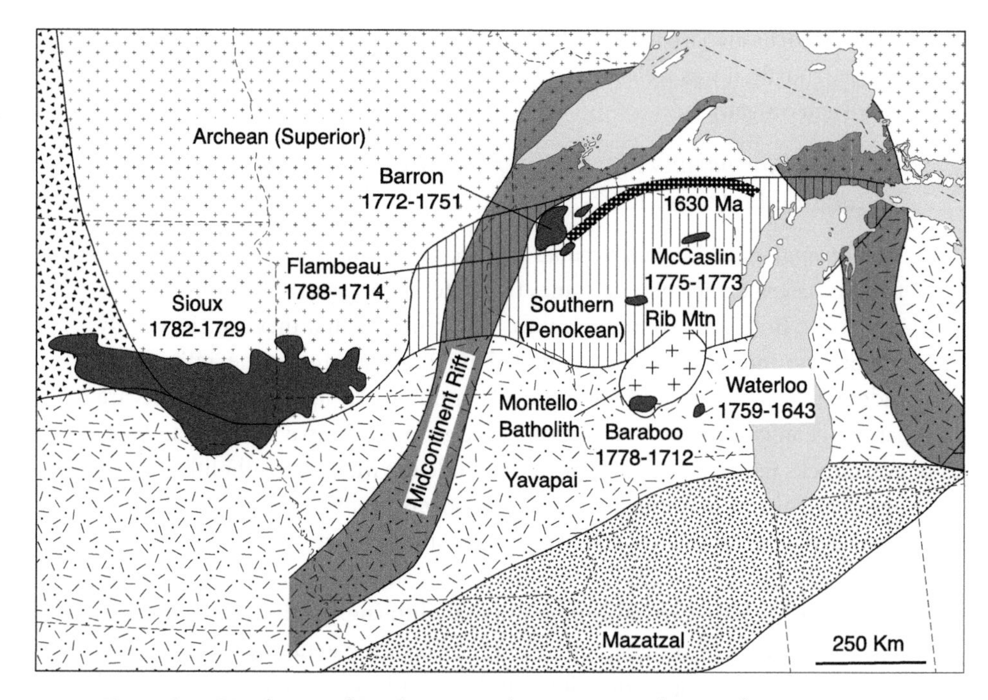

Figure 6.14. Distribution of Baraboo Interval Quartzites (with ages of youngest detrital zircons). Heavy line labeled 1630 Ma separates quartzites that have been deformed on the south from those that are flat lying on the north. Deformed rocks of the Animikie Superbasin and Wisconsin magmatic terranes make up the Southern Province. (Modified from Anderson 1980; Medaris et al. 2003; Czeck and Ormand 2007; NICE Working Group 2007; Schwartz et al. 2018.)

detrital zircons in the quartzite have ages of 1788 to 1711 Ma (fig. 6.14), which was originally thought to indicate that the quartzite was deposited shortly after the Yavapai orogeny. But, Joshua Schwartz and others have reported zircons as young as 1643 Ma in one Baraboo outcrop area (the Waterloo Quartzite) (fig. 6.14) and indicated that this probably represents the maximum age of all Baraboo Interval Quartzites. Most of the Baraboo outcrops in central and southern Wisconsin are folded and deformed, whereas outcrops in the much larger Barron and Sioux areas to the north and west are flat-lying, a point we will return to shortly. As we will see in a moment, this adds an interesting complication to our effort to unravel the deformation history of the Penokean, Yavapai, and Mazatzal orogenies.

6.5. The Cover Sequence Underwent the Penokean, Yavapai, and Mazatzal Orogenies

The Animikie Superbasin is variably deformed and metamorphosed, and clearly underwent a major orogenic event—or was it several? It was originally thought that all of the deformation took place during the geon 18 Penokean orogeny. But, it is now thought that the Wisconsin magmatic, Yavapai and Mazatzal colliding terranes docked sequentially against the south side of the continent and caused three separate orogenies, the geon 18 Penokean, geon 17 Yavapai, and geon 16 Mazatzal.

6.5.1. The Penokean Orogeny Involved Thin-skinned Deformation That Was Difficult to Detect with Isotopic Age Measurements

Research to sort out the ages of these different collisions and orogenies was getting underway at about the same time that isotopic age measurements were being developed. So early ages gave tantalizing clues that had to be followed up repeatedly as methods improved. Efforts to measure the age of the "Penokean orogeny" started in 1961 with geologist Samuel Goldich and physicist A. E. Nier, who pioneered isotopic age analyses. They assigned an age of 1800 to 1600 Ma to metamorphic rocks thought to be from the Penokean orogeny. Shortly afterward they changed the age to 1800 to 2000 Ma. Then, in 1965, Thomas Aldrich and G. L. Davis teamed up with Harold James, who had mapped the geology of much of the Marquette Range Supergroup, right in the middle of the old, now eroded Penokean mountain range, to report more than 100 isotopic ages that ranged from 2700 Ma 1100 Ma. These ages were bedeviled by the problem of closure temperatures (box 6.4) but clearly signaled that the so-called Penokean orogeny was more complicated than originally thought.

More recent studies have reinforced this view. For instance, although Daniel Holm and others reported geon 18 ages for some metamorphic rocks in the Marquette Range Supergroup, many other rocks yielded geon 17 and 16 ages. Farther to the east in the Huronian Supergroup, Patricia Piercey, working with David Schneider and Daniel Holm, reported similarly young geon 17 and even geon 14 ages. These ages were a surprise. If you go back and look at the ages of volcanic rocks in the Wisconsin magmatic terranes, you can see that they are almost entirely geon 18. On that basis, it was thought that ages of metamorphic rocks should be largely in the geon 18 range in the deformed Huronian and Animikie cover sequence rocks. Their scarcity raised the obvious question of whether the Penokean orogeny happened at all and, if it did, whether its effects could be distinguished from those of the later orogenies.

BOX 6.4. ISOTOPIC AGES AND CLOSURE TEMPERATURES

Most isotopic ages are based on measurements of relative abundances of parent and daughter isotopes. For instance, ^{40}K decays in part to ^{40}Ar and the ratio between these two isotopes can be used along with the half-life of the ^{40}K decay reaction to determine the age of a sample containing potassium. However, if the abundances of either ^{40}K or ^{40}Ar in the sample are changed, the ratio will not represent the real age of the sample. The most common change is loss of the daughter isotope from the sample, which results in an age that is too young. This is a particularly serious problem for rocks that cool slowly or that are reheated by a later event (heat allows ^{40}Ar to escape from most samples). Reheating is common in complex metamorphic terranes that have undergone multiple orogenies, making it hard to find samples that retain their original metamorphic age.

As a rock cools, the rate of loss of daughter isotopes like ^{40}Ar gradually diminishes and the temperature at which it ceases is known as the closure temperature. Closure temperatures are different for different isotopic systems and minerals, and this can be used to sort out geologic events. To measure the age of an intrusion or a high-temperature metamorphic event, choose a mineral with a high closure temperature like zircon and use the U-Pb system. On the other hand, choose a mineral like muscovite, which has a lower closure temperature, and use the K-Ar system if you want to measure the age of a later, low-temperature metamorphic event. Early isotopic age measurements were not always interpreted in terms of closure temperatures, which made the ages difficult to understand.

It turns out that the best evidence for the Penokean orogeny comes from geologic relations. The three main features that we see in the deformed rocks, and that resulted from the three orogenies, are thrust faults (fig. 6.15A), curious dome-and-keel structures (fig. 6.15B), and an even more curious concentric zonation of minerals that indicate metamorphic intensity (fig. 6.15C). Klaus Schulz and Bill Cannon have shown how these three features formed during a two-stage process related to the Penokean and Yavapai orogenies.[35]

According to Schulz and Cannon, the Penokean orogeny involved *thin-skinned deformation*, during which collision with the Wisconsin magmatic

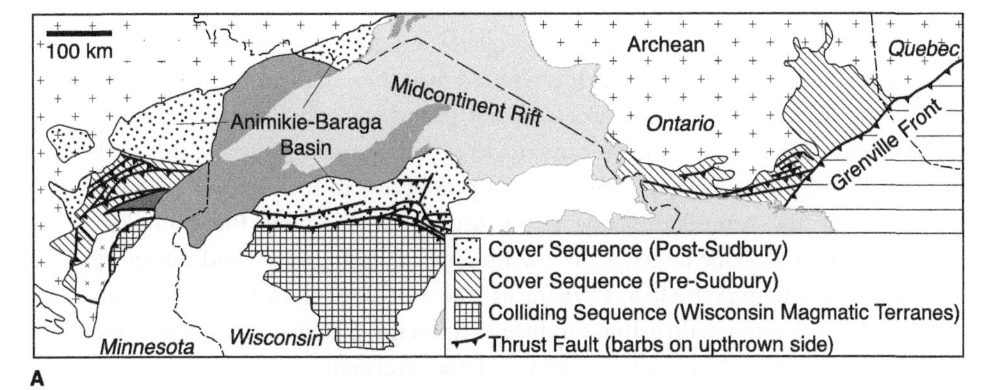

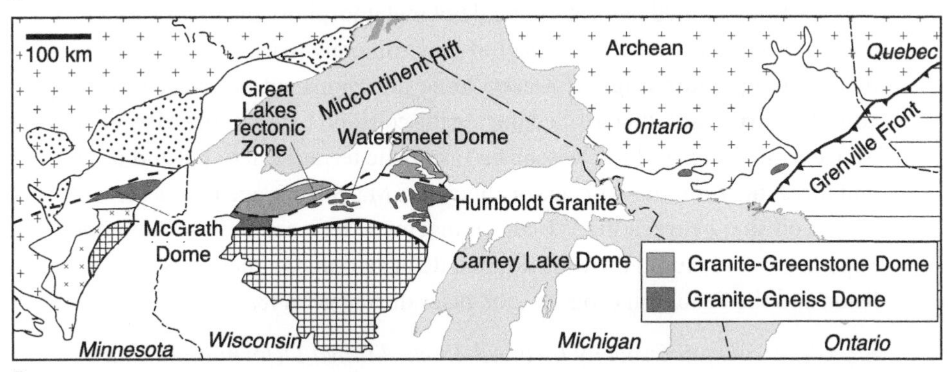

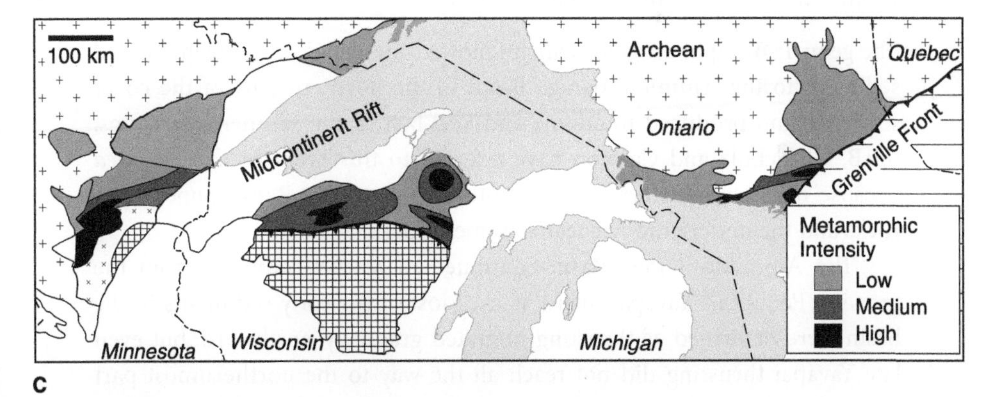

Figure 6.15. Maps of the Huronian-Animikie sequence showing (**A**) approximate locations of major thrust faults; (**B**) distribution of dome-and-keel structures (and Great Lakes Tectonic Zone); and (**C**) concentric zones of metamorphic intensity. (Modified from Sims and Carter 1996; James 1955, 1958; Southwick and Morey 1991; Tinkham and Marshak 2004; Chandler et al. 2007; Schluz and Cannon 2007; Boerboom 2011; Ojakangas et al. 2011; Cannon et al. 2018.)

terranes caused large sheets of Marquette Range Supergroup rocks to be thrust northward (fig. 6.15A). It is referred to as thin-skinned because underlying Archean basement rocks were not disturbed; only the overlying thin layer of sedimentary and volcanic rocks was deformed. However, the stack of thrust sheets was big enough to warp the crust to the north downward, forming the Animikie-Baraga Basin. Sediments that filled the Animikie-Baraga Basin came largely from the thrust stack to the south and ranged in age from about 1878 Ma to as young as 1780 Ma, as noted earlier. This means that Penokean-age thrusting produced sediments that were then vulnerable to deformation during later Yavapai and Mazatzal collisions.

The thin-skinned deformation did not metamorphose the rocks strongly, and so it left very little in the way of new minerals that could be used to determine its isotopic age. However, some intrusions cut the Niagara fault zone, the thrust fault that formed during the collision between the Marquette Range Supergroup and the Wisconsin magmatic terranes (figs. 6.8 and 6.12), and these intrusions have an age of about 1835 Ma. So, this age marks the end of the collision between the Wisconsin magmatic terranes and the Animikie Superbasin that caused the thin-skinned deformation. On that basis, the collision and the Penokean orogeny took place during geon 18.

6.5.2. The Yavapai Orogeny Involved Thick-skinned Deformation That Produced the Dome-and-Keel Structures

The geon-17 Yavapai orogeny, which followed the Penokean orogeny, added sediment to the Animikie-Baraga Basin in the northern part of the cover sequence and produced the dome-and-keel belt in the southern part (figs. 6.15B, C). Schulz and Cannon have referred to this event as *thick-skinned* because it involved deformation of both the Marquette Range Supergroup rocks and the underlying Archean basement rocks.

The Animikie-Baraga Basin continued to receive sediment from the growing Penokean-Yavapai thrust stack. Most of the early sediments in the basin were deformed as thrusting migrated gradually northward, but even late Yavapai thrusting did not reach all the way to the northernmost part of the (Late) Animikie Basin. Here we find largely undeformed sedimentary rocks of the Animikie Group (figs. 6.8 and 6.10B) that have 1780 Ma detrital zircons. In fact, these zircons are the same age as the East-Central Minnesota batholith (ECMB) just to the south (figs. 6.8, 6.10B, 6.13F [in color section]). So, the youngest sediments that filled the Animikie-Baraga Basin came from erosion of intrusive rocks that were emplaced during the Yavapai orogeny. Daniel Holm and Terry Boerboom have reviewed the history of this basin and showed how its southern part was deformed as Yavapai deformation progressed northward through time.

The final stage of Yavapai deformation involved development of the dome-and-keel structures. These curious structures occupy a belt that extends across the region from the Carney Lake dome in Michigan to the McGrath dome in Minnesota (fig. 6.15B). They formed when the Archean basement rocks rose in domes and the overlying Paleoproterozoic cover rocks sank downward in tightly folded domes and keels. You can see what it looked like in the cross section in figure 6.10A, which shows that the Paleoproterozoic Marquette Range Supergroup rocks were folded into a syncline (keel, or trough as it is often called) surrounded by Archean basement rocks (dome). Some Archean basement rock domes consist of deformed and metamorphosed granite whereas others consist of less deformed and younger volcanic and intrusive rock.[36]

The curious concentric metamorphic mineral zonation formed at this time (fig. 6.15C), probably by heating from the underlying domes and intrusions. This is shown by age measurements on metamorphic minerals by Eric Tohver and others that range from as young as 1730 Ma in the center of the domes to near 1800 Ma on the margins, as well as by the presence of Yavapai-age intrusions in the domes. The best example is the Humboldt granite in the central part of the large concentric metamorphic zones just south of Marquette.

Exactly how the Archean basement rose to form domes is a matter of controversy. The most likely possibility is that the Penokean-Yavapai thrust stack grew so thick (and high) that its upper part slid away, removing overlying rocks that allowed hot Archean basement to flow upward, as described by David Schneider, Douglas Tinkham, and Stephen Marshak. This is essentially the same extensional collapse process that we reviewed in chapter 5 for formation of the metamorphic core complexes in the Grenville Province, although the domes are smaller here and more of the overlying sedimentary rocks have been preserved.

6.5.3. Evidence for the Mazatzal and a Possible Younger Orogeny Is Found in the Baraboo Interval Quartzites

Could the Mazatzal orogeny have had an impact on the Great Lakes region as well? The most striking support for this possibility came when Daniel Holm, Denise Romano, and others drew a 1630 Ma line through the Wisconsin magmatic terranes (fig. 6.14). South of this line, minerals with low closure temperatures showed signs of a heating event at 1630 Ma; north of the line, they did not. This was interpreted to represent effects of the geon 16 Mazatzal orogeny in the colliding sequence.

Just what happened to the Baraboo-age quartzites during geon 16 is less clear. The fact that they are folded south of the 1630 Ma line but flat-lying north of the line was originally thought to indicate that the quartzites were folded during the geon 16 Mazatzal event. Maybe this was a widespread event; Joshua Bailey and others have even reported folding of this age in the Sudbury structure. But the presence of geon 16 zircons in the quartzites, as mentioned above, requires that they were deposited after geon 16 uplift and folding. If so, they must have been folded by an even younger event.

Evidence for a younger event is seen in the mineralogy of the quartzites. In many places, the original detrital quartz and kaolinite in the quartzite has reacted to form pyrophyllite, an unusually soft mineral that can be carved (box 6.5). Reaction of quartz and kaolinite to form pyrophyllite requires temperatures of more than 300°C, far too high to have been caused by burial beneath the thin cover of younger rocks. Muscovite is also present in the Sioux Quartzite and some other areas, and its formation requires addition of potassium to the quartzite. Gordon Medaris has suggested that both pyrophyllite and muscovite were formed by a regional fluid migration event related to uplift during emplacement of the Wolf River batholith, and has raised the possibility that this event was associated with folding of the quartzites. If so, there was an even younger, geon 14 Wolf River, orogenic event in the colliding terrane.

BOX 6.5. THE SIOUX QUARTZITE AND CATLINITE

The Sioux Quartzite is home to catlinite, a very fine grained layer in the Sioux Quartzite at Pipestone National Monument, Minnesota (fig. 6.16A in color section). Catlinite is made up largely of pyrophyllite, a soft mineral that can be carved into various forms, including pipes (fig. 6.16B in color section), along with muscovite, diaspore, hematite, and rutile. Catlinite was named for George Catlin, an American painter and historian who wrote about the area in the 1830s. The catlinite layer, which is about 0.3 to 0.45 m thick, is sandwiched between layers of quartzite and can be accessed in shallow pits. According to park policy, it can be mined only by registered American Indians. Mining involves removal of the overlying quartzite and then the pipestone layer, all using hand tools. Catlinite objects were items of trade before Europeans showed up, and ceremonial catlinite pipes were among the first art objects they saw as they entered the Great Lakes region. Catlinite pipes were even used in the far north during annual meetings between the Cree and Hudson's Bay Company fur traders.

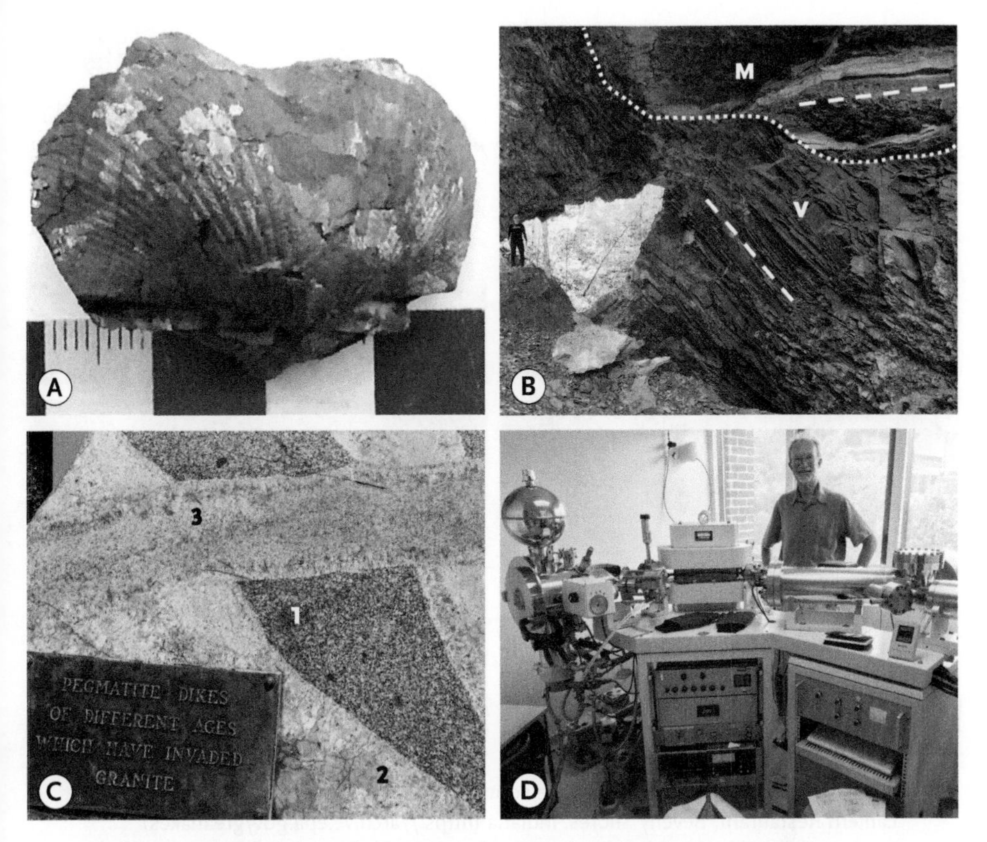

Figure 1.4. Determining geologic ages and time with (**A**) fossils—brachiopod (replaced by pyrite) (Devonian, Ohio, scale in cm); (**B**) superposition –Late Cambrian Munising Sandstone (M) overlying Paleoproterozoic Vulcan Iron Formation (V), Quinnesec Mine, Michigan (small person at left side of photo gives scale; long dashed lines show orientation of layering in the two rock units and short dashed line shows the unconformity between them); (**C**) cross-cutting relations – three ages of igneous intrusion (1 oldest, 3 youngest), North University Building, University of Michigan; (**D**) James Gleason at one of the mass spectrometers used to measure isotopic ratios, University of Michigan.

Figure 2.5. (**A**) Shoreline erosion at the south end of Lake Michigan at the Red Lantern Restaurant, Beverly Shores, Indiana (https://archive.epa.gov/greatlakes/image/web/html/viz_iss1.html); (**B**) actively eroding Lake Erie shoreline in Pennsylvania (Hapke et al., 2009); (**C**) shoreline stabilization project at Concordia University, near Milwaukee, Wisconsin, which lost 5 acres of lakefront land to erosion between 1982 and 2000 (http://greatlakesresilience.org/stories/wisconsin/stabilizing-concordia-university's-bluff-0); (**D**) homes such as this one, near Manistique, Michigan, along the north shore of Lake Michigan, are vulnerable to the storm waves that built the dunes on which they rest.

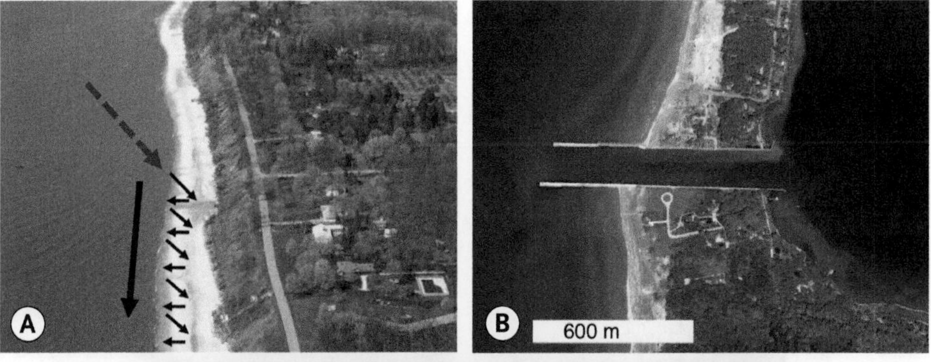

Figure 2.8. (**A**) Longshore drift of beach sand. The large dashed arrow shows the direction of waves, smaller arrows show the movement of sand along the beach, and the large solid arrow shows the net movement of sand. (**B**) Jetties protecting the entrance to Portage Lake at Onekama, Michigan. Lake currents have piled sand up on the south side of the southern jetty. (**C**) Air view of the St. Clair Delta extending into the north end of Lake St. Clair.

Figure 2.9. Lake Michigan dunes: (**A**) Warren dunes showing tree roots exposed by wind erosion; (**B**) Grand Sable dunes (above the white line) perched on glacial sediments (below white line), which supplied sand to the wind that made the dunes.

Figure 2.12. Waterfalls and potholes: (**A**) Tahquamenon Falls, Michigan (courtesy of Dan Troyka); (**B**) Minnehaha Falls, Minnesota; (**C**) Kakabeka Falls, Ontario; (**D**) Giant pothole (arrow) about 30 m deep at Taylors Falls, Minnesota.

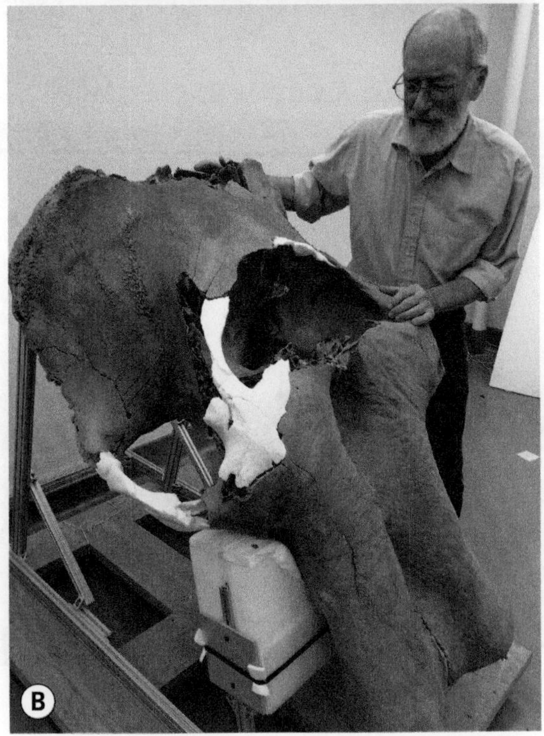

Figure 2.14. The Bristle Mammoth: (**A**) skull and tusks from the excavation being loaded onto a truck; (**B**) Dan Fisher examining the Bristle Mammoth skull prior to display at the University of Michigan Natural History Museum.

A
B
C
D
80
70
60
50
40
30
20
mm

Figure 3.4. Glacial drift: (**A**) fine-grained till of the Des Moines lobe at Morton, Minnesota; (**B**) coarse-grained till of the Rainy lobe south of Kabetogama Lake, Minnesota; (**C**) till with some rounded cobbles; (**D**) layered water-sorted drift of the Mille Lacs Drift, Superior lobe, Minnesota.

Figure 3.2. (*facing page*) (**A**) This boulder erratic, which is known as puddingstone because of its peculiar red clasts, is from the Lorrain Formation discussed in chapter 6. It was picked up by the glacier near Sault Ste. Marie, Ontario, and deposited about 300 miles to the south in Ann Arbor, Michigan. Dale Austin provides scale; (**B**) giant glacial grooves in Devonian age limestone at Kelleys Island in Lake Erie; (**C**) the Three Maidens at Pipestone National Monument in southwestern Minnesota, fragments of a single unusually large erratic that was broken apart by frost action (there is another unusually large erratic, the Bleasdell Boulder, near Glen Miller, Ontario); (**D**) faceted cobbles showing flat surfaces formed by grinding along the base of the glacier.

Figure 3.10. (**A** and **B**) Tree trunk and roots from buried forest, Gribben locality, Michigan, described by Pregitzer et al. (2000) (photo courtesy of Patrick Martin and T. J. Bornhorst, Michigan Technological University); (**C**) divers examining boulders at the bottom of Lake Huron that served as caribou drive lanes for prehistoric hunters (photo by Tane Casserley, National Oceanographic and Atmospheric Administration, courtesy of John O'Shea, University of Michigan Museum of Anthropological Archaeology).

Figure 4.2. (**A**) Cambrian Mt. Simon Sandstone overlying weathered Precambrian intrusive rock north of Chippewa Falls, Wisconsin; (**B**) Cambrian conglomerate overlying Precambrian Baraboo Quartzite that made up an island in the Cambrian sea at Devils Lake, Wisconsin. The arm rests along the contact, with layered Baraboo Quartzite of the island below and rounded boulders of Baraboo Quartzite, which make up the basal Cambrian sediment, above.

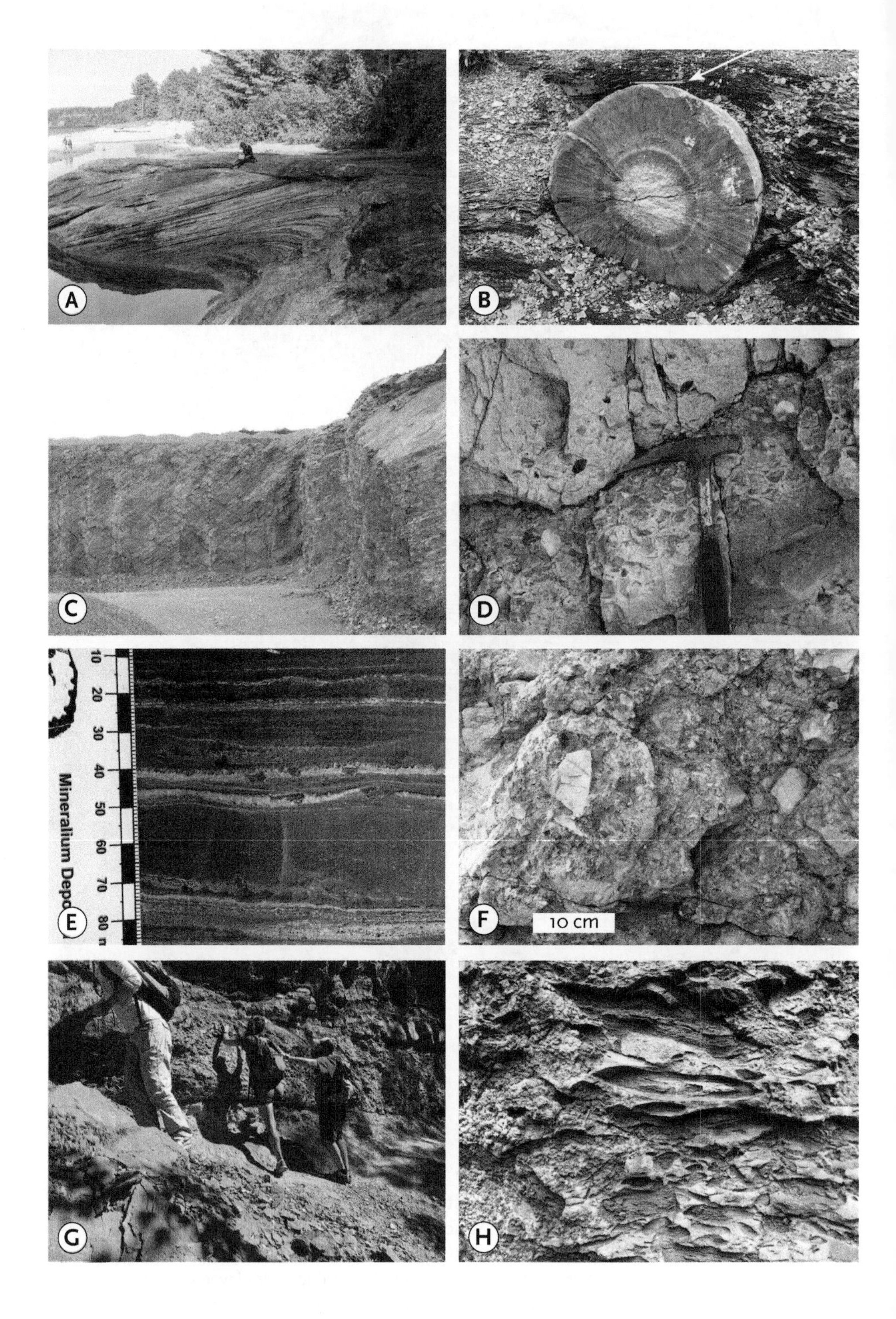
A
B
C
D
E
F
10 cm
G
H

Figure 4.10. Paleozoic sediments of the Hollandale Embayment: (**A**) Upper Cambrian Tunnel City (Franconia) Formation near Taylors Falls, Minnesota; (**B**) Middle Ordovician St. Peter Sandstone overlain by the Platteville Formation, St. Paul, Minnesota; (**C**) Upper Ordovician Decorah Shale, St. Paul, Minnesota; (**D**) Upper Ordovician Galena Limestone, Mt. Horeb, Wisconsin.

Figure 4.9. (*facing page*) Paleozoic sediments of the Michigan Basin. (**A**) Inclined possible dune deposits overlain by flat-lying beach deposits in the Cambrian Munising Formation, Miners Beach, Pictured Rocks National Lakeshore; (**B**) concretion with concentric growth zones from the Devonian Kettle Point Shale, Kettle Point, Ontario (the arrow points to pencil that provides scale); (**C**) flank deposits of the Silurian Pipestone Junior Reef, Pipestone Junior Quarry, Indiana; (**D**) detail of reef margin deposit in C showing talus consisting largely of brachiopods; (**E**) drill core showing dark and light layers of salt from the Silurian Salina Group, Michigan Basin; (**F**) Mackinac breccia near St. Ignace, Michigan; (**G**) Pennsylvanian-age stream deposits, Lincoln Brick Park, Grand Ledge, Michigan, showing coarse basal lag deposits (directly in front of students) disconformably overlying finer-grained flood plain deposits; (**H**) close up of the lag deposit in G showing clasts of clay (dark) that were eroded at the cutbank of the stream enclosed in a sandstone (light) matrix (field of view is 0.5 m wide). (Photos A, B, C, D, G, H, and captions courtesy of Kyger Lohmann, University of Michigan.)

Figure 4.15. Paleozoic life: (A) Cambrian seafloor with sponges (S) and trilobites (T); (B) Devonian seafloor with trilobites (T), crinoids (C), and corals (CL); (C) trilobite from the Middle Ordovician Trenton Group, Ontario; (D) eurypterid (sea scorpion) from the Upper Silurian Salina Group, Ontario; (E) crinoids from the Middle Devonian Traverse Group (Silica Shale), Ohio. (All images from the University of Michigan Museum of Natural History.)

Figure 5.6. Volcanic rocks of the Midcontinent Rift: (**A**) unconformity between lava flow (T) and Archean-age basement rock (R) north of Sault Ste. Marie, Ontario; (**B**) southeast-dipping lava flows, Mamainse Point, Ontario; (**C**) northwest-dipping Portage Lake lava flows and Copper Harbor conglomerate, Keweenaw Peninsula, Michigan; (**D**) aerial view of Portage Lake lava flows at the north end of the Keweenaw Peninsula; (**E**) vesicles (V) in Mamainse Point lava; (**F-G**) basalt flows with oxidized (red) tops (T) and ropy textured flow tops, North Shore Group, Minnesota; (**H**) columnar structures in basalt of the North Shore Group, Minnesota.

Figure 5.7. Intrusive and sedimentary rocks of the Midcontinent Rift: (**A**) layering in the Duluth Complex, Eldes Corner, MN; (**B**) Beaver Bay Complex (dark) with large inclusions of anorthosite (white) at Silver Bay, Minnesota; (**C**) interflow sedimentary rock covered with columnar basalt flow, North Shore Group, Minnesota; (**D**) Copper Harbor Conglomerate with clasts of volcanic rocks; (**E**) Copper Harbor Sandstone showing cross-bedding from Copper Harbor, Michigan; (**F**) Nonesuch Shale near White Pine, Michigan.

Figure 5.9. Midcontinent Rift ores: (**A**) native copper filling porosity in conglomerate; (**B**) thompsonite (pink) filling vesicles in lava; (**C**) chalcocite and pyrite along a layer in the Nonesuch Shale; (**D**) massive iron-nickel-copper sulfide ore (right) in contact with silicate wallrock (left) at the Eagle mine (6-in wire mesh gives scale). In A and C, the green color resulted from weathering of the copper minerals to form copper carbonates such as malachite.

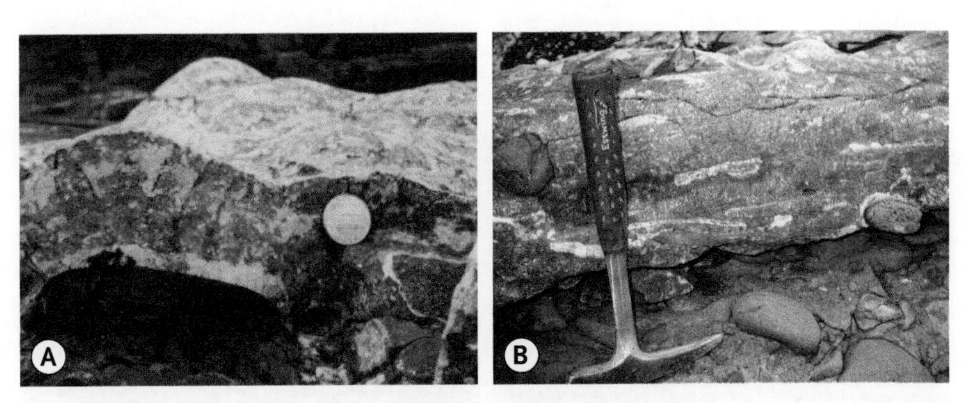

Figure 5.12. Stromatolites from the Copper Harbor Conglomerate: (**A**) dome-shaped form; (**B**) broken clasts of stromatolite.

A
M
A
B
C
D
E
F
G
H

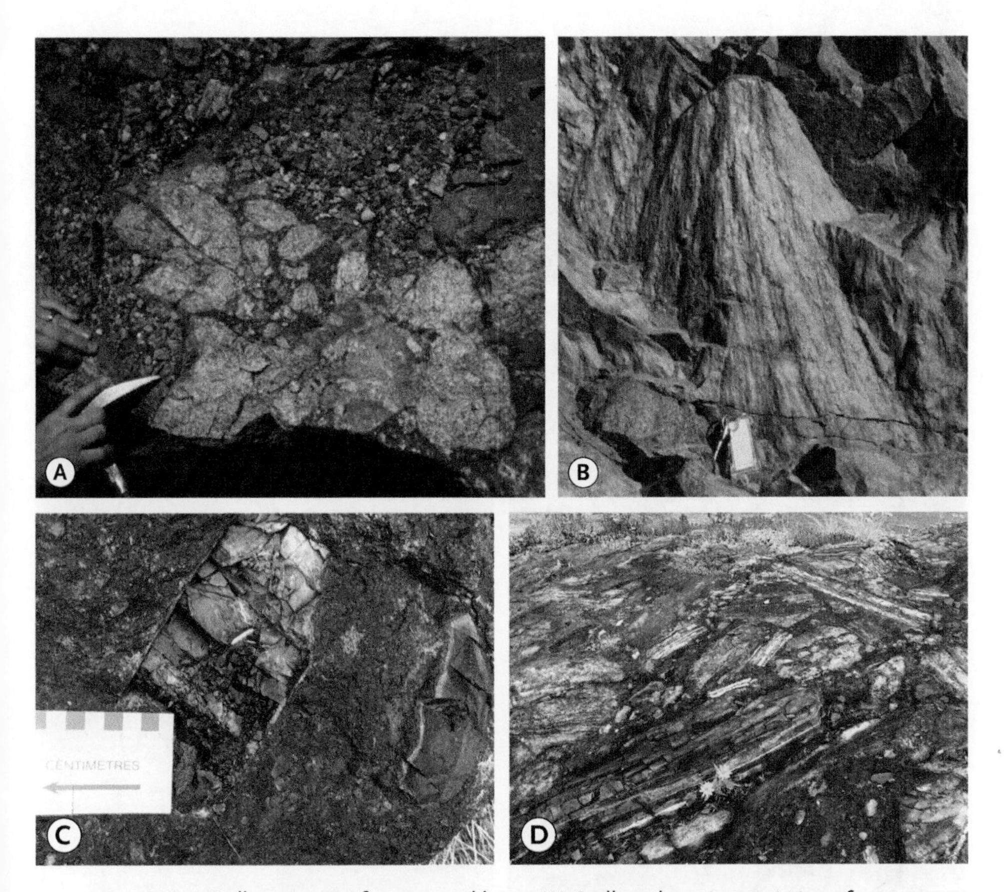

Figure 6.7. Sudbury impact features and layer: (**A**) Sudbury breccia consisting of disoriented fragments of granite (point of hammer gives scale), Sudbury, Ontario; (**B**) shatter cone (keys give scale), Sudbury, Ontario; (**C, D**) Sudbury impact layer consisting of jumbled fragments of silicified Gunflint Formation sediments, Thunder Bay, Ontario (scale in C is 9 cm wide; view in D is 1 meter wide)

Figure 6.5. (*facing page*) Huronian Supergroup rocks: (**A**) unconformity between Archean rocks (A) and Matinenda Formation (M); (**B**) Matinenda Formation sandstone with cross-bedding and detrital pyrite; (**C**) Mississagi Formation with ripple marks; (**D**) Ramsey Lake Tillite; (**E**) Gowganda Tillite with characteristic pink granite clasts; (**F**) dropstone of pink granite in finely layered (varved) Gowganda siltstone; (**G**) Bruce limestone member of the Espanola Formation; (**H**) Gordon Lake Formation redbeds with small white reduction spots. The yellow global positioning system (GPS) unit used for scale is about 10 cm in length.

Figure 6.11. Animikie Superbasin rocks: (**A**) Fern Creek Tillite; (**B**) Sturgeon River Quartzite with ripple marks; (**C**) Randville Dolomite with stromatolites, all in the Menominee district, Dickinson County, Michigan; (**D**) deformed Negaunee banded iron formation, Marquette trough, Michigan; (**E**) Michigamme Formation, Iron Mountain, Michigan; (**F**) undeformed Biwabik iron formation, Hibbing, Minnesota; (**G**) Archean basement rock overlain by the Gunflint iron formation, Kakabeka Falls, Ontario; (**H**) Rove Formation, Grand Portage, Minnesota.

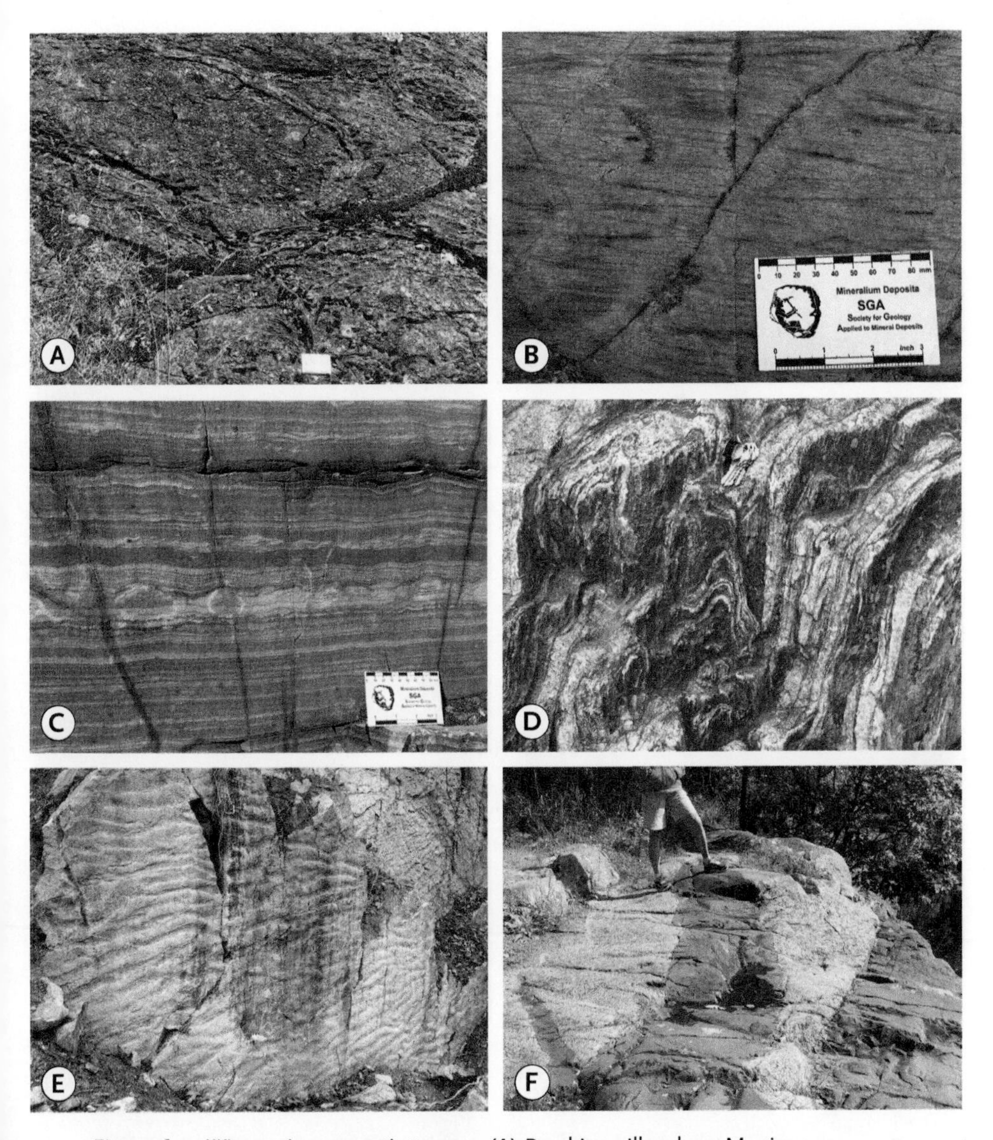

Figure 6.13. Wisconsin magmatic terrane: (**A**) Pembine pillow lava, Monico, Wisconsin; (**B**) felsic volcanic rock, Dells of the Eau Claire River, Wisconsin; and (**C**) layers of fine-grained volcanic ash, Wausau, Wisconsin; (**D**) Hatfield gneiss in the Archean-age Marshfield terrane, Lake Arbutus, Wisconsin. Other rock units: (**E**) Baraboo Interval Quartzite with ripple marks, Rock Springs, Wisconsin; (**F**) granite (light) of the East Central Minnesota batholith intruded by dark diabase dikes, St. Cloud, Minnesota.

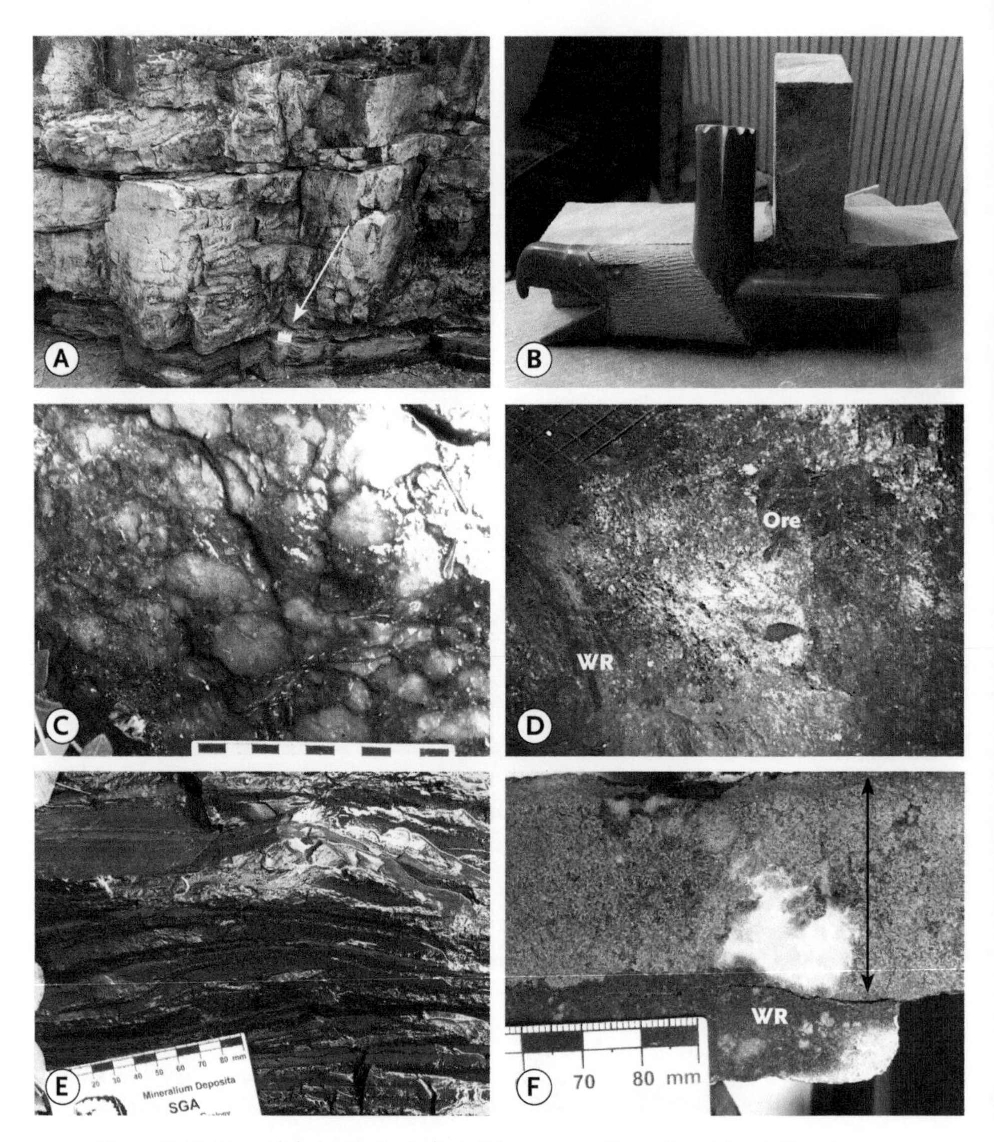

Figure 6.16. Mineral deposits, including: (**A**) quarry in Sioux Quartzite at Pipestone National Monument showing the thin catlinite layer at the base of the pit (arrow; the small white rectangle is 9 cm long); (**B**) rough-cut and carved catlinite pipes; (**C**) uranium ore, Matinenda conglomerate with quartz pebbles and small grains of pyrite and uraninite, Elliot Lake, Ontario; (**D**) nickel ore, contact between rich nickel-sulfide ore (upper right) and barren wallrock (WR) in the Creighton mine, Sudbury, Ontario (width of photo 2 m); (**E**) banded iron formation with black layers of iron minerals and red layers of silica, Marquette trough, Michigan; (**F**) vein containing quartz (white) and native silver (dark) from Cobalt, Ontario (arrow shows width of the narrow vein; WR = wallrock).

Figure 7.2. (**A**) Pillow lava (top left), North Caribou terrane, Red Lake, Ontario; (**B**) pillow lava (top), Wawa terrane, Gilbert, Minnesota; (**C**) spinifex texture in komatiite lava flow (top right), LaMotte-Vassan Formation, Abitibi terrane, Val d'Or, Quebec (courtesy of François Robert); (**D**) intermediate (andesitic) volcaniclastic rock, Val d'Or Formation, Abitibi terrane, Val d'Or, Quebec; (**E**) reworked volcanic tuff, Wawa terrane, Vermillion district, Minnesota; (**F**) metasedimentary rock, Quetico terrane, Seine River Village, Minnesota; (**G**) Temiskaming Conglomerate, Pamour mine, Abitibi terrane, Timmins, Ontario; (**H**) vermillion granite with inclusions of wallrock, Quetico terrane, Kabetogama Lake, Minnesota.

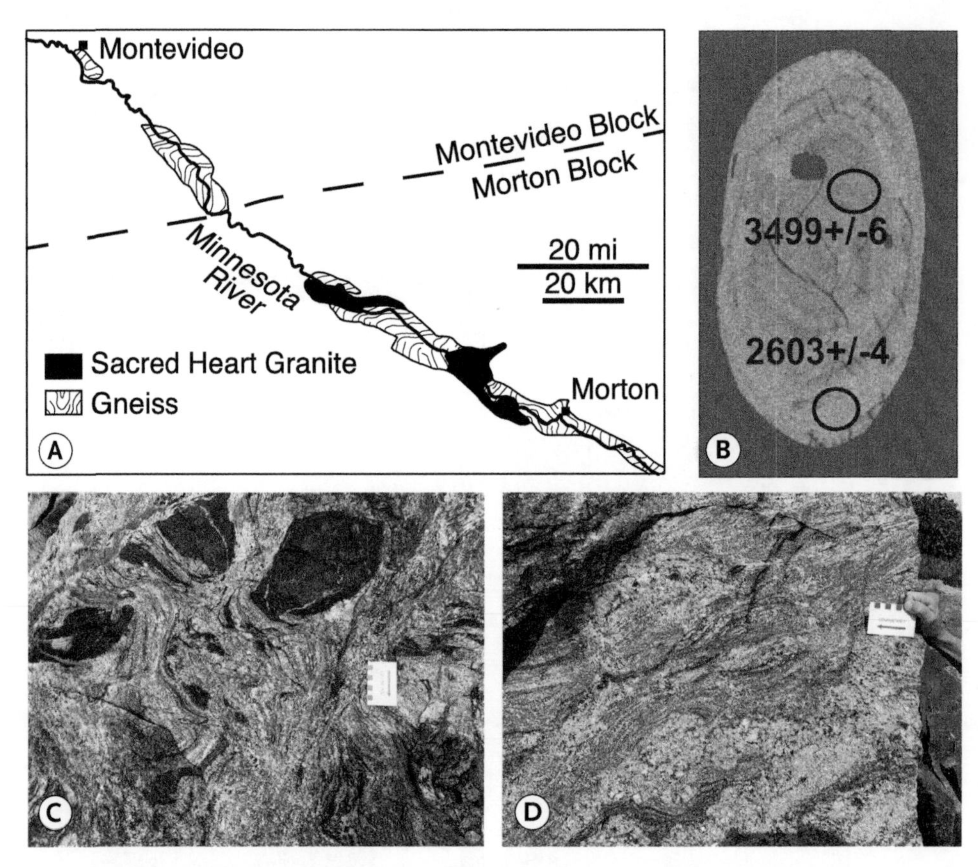

Figure 7.3. Minnesota River Valley terrane: (**A**) geologic map of the exposed part of the Minnesota River Valley terrane; (**B**) image of a zircon from the Montevideo gneiss showing age measurements on the old core and young rim (from Bickford et al. 2006); (**C, D**) Morton gneiss. The two photos in C and D, from the same large outcrop, show the wide lithologic variation in the Morton gneiss; large mafic inclusions of amphibolite (probably metamorphosed supracrustal basalt) in gray gneiss (left) and pink granite gneiss with small lenses of gray gneiss (right). The scale bar is 10 cm in length.

Figure 7.11. Archean ore deposits: (**A**) massive sulfide ore (2 m wide), Millenbach VMS deposit near Rouyn-Noranda, Quebec; (**B**) orogenic gold vein, Pamour mine, Timmins, Ontario, showing layered (sheeted, ribbon) vein; (**C**) orogenic gold vein, Pamour mine, showing near-horizontal veins extending from the side of the main vein (indicating fluid pressures greater than rock pressures); (**D**) high-grade vein segment, Pamour mine, showing jigsawlike wallrock (dark) cemented by a quartz vein (white) with abundant visible gold (arrows) (courtesy of E.H.P. van Hees); (**E**) overview of the Victor mine showing circular form of the kimberlite pipe outcrop (**F**) rough diamonds from the Victor mine.

Figure 7.12. (**A**) Giant stromatolite mounds; (**B**) overview of stromatolites from the Mosher carbonate, Steep Rock Lake, Atikokan, Ontario. (Photos courtesy of Howard Poulsen.)

6.6. Plate Tectonic Migration during the Penokean, Yavapai, and Mazatzal Collisions Formed the Continent of Laurentia and the Supercontinent Known as Columbia or Nuna

By now it is obvious that there was no such thing as "the collision." Instead Paleoproterozoic rocks on the southern side of Superior experienced an episodic terrane wreck of collisions between at least 1880 and 1630 Ma. This terrane wreck was part of the assembly of the continent that became Laurentia, as well as its joining with others to form a new late Paleoproterozoic supercontinent, known as Columbia or Nuna. The belt of deformed rocks that resulted from this terrane wreck is the Southern Province, and it is similar to other deformed belts that surround Superior on the west and north (fig. 6.17).

The plate tectonic process that joined Superior with other cratons in the Great Lakes area to form Laurentia has been summarized by Klaus Schulz and Bill Cannon. It began when the Pembine-Wausau terrane advanced on the Superior craton from the south, with south-directed subduction of ocean crust beneath the Pembine-Wausau terrane forming its volcanic arc. At about 1875 Ma, the Pembine-Wausau terrane collided with the Superior craton along the Niagara fault zone, which separates Pembine-Wausau rocks from Marquette Range Supergroup rocks. Then subduction of ocean crust flipped to north directed along the south side of the Pembine-Wausau terrane, forming the younger volcanic rocks of the Pembine-Wausau terrane, as well as those in the Menominee Group in the Marquette Range Supergroup. By 1850 Ma, the Marshfield terrane had collided along the south side of the Pembine-Wausau terrane, forming a single craton and causing the first-stage, thin-skinned folding and metamorphism in the Animikie Supergroup sediments that is considered to be the geon 18 Penokean orogeny.

Plate tectonic events after the Penokean orogeny, as depicted by Daniel Holm, Randy Van Schmus and others, involved continued subduction from 1800–1750 Ma of a Yavapai oceanic plate that generated the East-Central Minnesota, Montello, and other intrusions, followed by collision of the Yavapai terrane with the amalgamated cover-collision sequence terranes around 1750 Ma. This long period of convergence and subduction must have included at least one period of extension in order to account for the orogenic collapse of the Penokean-Yavapai mountains that formed the geon 17 dome-and-keel structures and concentric metamorphic zoning. The Mazatzal terrane then collided from the south with all of these rocks at about 1630 Ma and shortly afterward they were leveled by erosion, deeply weathered and covered by the

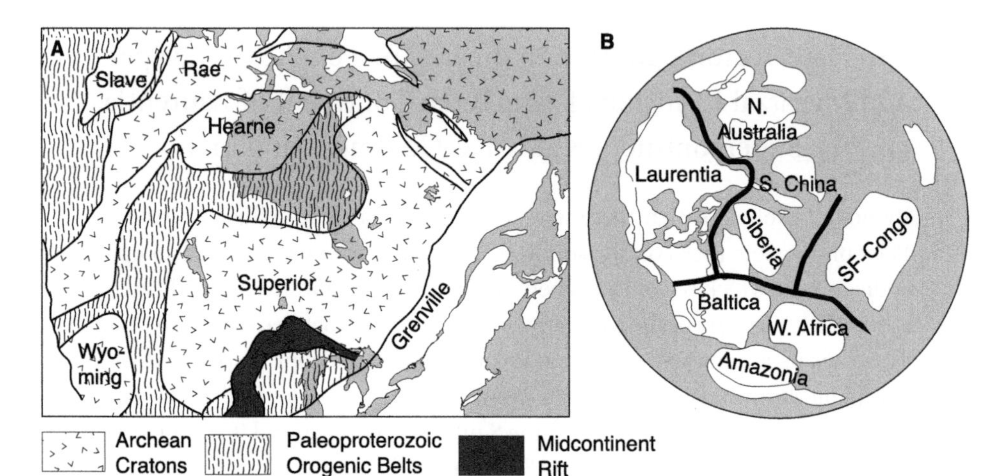

Figure 6.17. Supercontinents and Superia: (**A**) the continent of Laurentia, which consisted of Superior and other cratons surrounded by orogenic belts formed during their Paleoproterozoic amalgamation; (**B**) position of Laurentia in the supercontinent Nuna at about 1400 Ma. (Modified from Davidson 2008; Zhao et al. 2011; Evans et al. 2016.)

Baraboo Interval Quartzites. The final event was emplacement of the 1468 Ma Wolf River batholith and formation of the regional fluid flow system that probably formed the pyrophyllite and muscovite-bearing catlinites.

By the end of Paleoproterozoic time, Superior was surrounded by other orogenic belts that are geologically similar to the Southern Province, which we have spent so much time dissecting (fig. 6.17A). The largest of these is the Trans-Hudson Belt, which borders Superior on the west and north. When it was all over, Superior had combined with the Hearne, Rae, and Slave Archean cratons, as well as the Trans-Hudson, Southern, Yavapai, Mazatzal, and other Proterozoic orogenic belts, to form Laurentia, which is the heart of North America (fig. 6.17B). John Rogers and M. Santosh suggested that Laurentia went on to combine with Siberia, Baltica, and parts of Australia to form a new supercontinent that they named Columbia. More recent work suggests that this supercontinent, which has been renamed Nuna, persisted from about 1600 to 1350 Ma, after which it began to break apart, possibly at about the time of deposition of the Sibley Group sediments that we saw in chapter 5. According to David Evans and others, Laurentia then returned to its status as an independent continent and waited for the Midcontinent Rift to start.

6.7. The Cover and Colliding Sequences Host Important Deposits of Iron, Uranium, Copper, Nickel, Zinc, Silver, and Cobalt

The Paleoproterozoic cover and colliding sequences have an unusually large and varied mineral endowment, including Earth's most important iron deposits and the largest suite of nickel-copper deposits on the planet. In addition to their economic importance and their role in settlement of the region, the deposits provide important insights into the evolution of Earth and its oceans and atmosphere.

6.7.1. Uranium Deposits Formed as Placers in Ancient Streams of the Huronian Basin

The uranium story is centered on the Elliot Lake, or Blind River, district in the southern part of the Huronian Basin (fig. 6.3). Uranium was discovered in the area by prospectors in 1948, just at the time that world governments were trying to stock up on this amazing element, which could be made into weapons and, just possibly, might yield cheap energy. By the late 1950s, mining was under way, and Elliot Lake was supplying a quarter of world uranium production.[37] The mines finally closed in the 1990s, in part because the good ore was exhausted but also because uranium prices decreased in response to cessation of government purchases and discovery of richer deposits in the Athabasca area of Canada.

The Elliot Lake deposits were the focal point of an extended argument about their origin and its bearing on Earth's atmosphere. Recall that the ores consist of small grains of uraninite (UO_2) in conglomerates that were deposited by ancient braided streams (fig. 6.16C in color section). (Individual stream deposits form buried layers that range from 2 to 13 m thick and were followed by miners for as much as 10 km underground.) Stuart Roscoe, who carried out the first detailed studies of these deposits, concluded that the ores were paleoplacer deposits, in which small grains of heavy uraninite had collected at bends, rough spots, and other irregularities in the braided streams, much like placer gold does in today's streams. (The term *paleoplacer* indicates that the deposits were not active placers in a modern stream.) His early reports were greeted with a chorus of disbelief and published discussions, mostly in the journal *Economic Geology*, which make for interesting reading today.

Many geologists were reluctant to accept Roscoe's paleoplacer conclusion because uraninite is not stable under oxidizing conditions of the sort that prevail today. To make matters worse, the conglomerates also contain grains of pyrite, which is even less stable in an oxidizing environment. The nail in the skeptics' coffin, however, came from a study by Nicholas Theis, who showed that sizes of large quartz cobbles and small pyrite grains were perfectly correlated in the conglomerates (in other words, sediment with lots of pyrite grains also had lots of large quartz cobbles). This happened because flowing water could move large cobbles of quartz but only small grains of the much heavier pyrite. (Similar tests for uraninite were complicated by the fact that it had changed composition through time, partly due to damage from radioactive decay of uranium.) These results meant that oxygen must have been missing in the streams, and it was the start of efforts to quantify the compositional evolution of Earth's atmosphere. Acceptance of a paleoplacer origin for the uraninite deposits made the Elliot Lake deposits an important data point in calculations of the history of Earth's atmosphere as discussed further below. Unfortunately, although everyone realized that uraninite would oxidize in modern environments, they did not see the pollution that would be caused by mining.[38]

6.7.2. Nickel-Copper Deposits Associated with the Sudbury Igneous Complex Formed by Magmatic Processes

The Sudbury Igneous Complex is one of the largest mining regions in the world. Production plus remaining reserves total about 1650 million tons of ore containing 1.2 percent nickel and 1.1 percent copper worth about $300 billion in 2017 prices, and that doesn't even count the silver, gold, platinum, palladium, osmium, iridium, ruthenium, selenium, and tellurium that are by-products recovered during smelting of the ores. The ores consist of massive nickel-copper-iron sulfide minerals that accumulated in deposits scattered along the basal zone at the bottom of the Sudbury Igneous Complex (fig. 6.16D in color section). These are magmatic deposits that formed when the metals combined with sulfur in the Sudbury magma to form droplets of metal sulfide (just like the Eagle deposit discussed in chapter 5). Because the Sudbury Igneous Complex was superheated (remember that it was a melt sheet), the metal sulfide droplets had time to sink all the way to the bottom of the magma chamber, where they accumulated in the basal zone of the Sudbury Igneous Complex or in extensions that were intruded into the surrounding country rocks. Early deposits were found at the surface around the perimeter of the intrusion (where they had been exposed by erosion), and exploration has found deposits down to depths of about 3000 m. Exactly how far the deposits extend downward remains to be determined.

Early mining at Sudbury was carried out at the surface, but mining today is underground, and really deep. The Creighton mine has workings more than 2100 m deep, and plans are being made to mine other deposits at depths of 2700 m. In the early days of mining, smelting of ore mined at Sudbury released large amounts of SO_2 (from the sulfide minerals) into the atmosphere. This created locally acid rain that left an area of about 100 km^2 around Sudbury with essentially no vegetation and a larger area with only bushes and grass. In fact, the landscape was so barren that it was used by the National Atmospheric and Space Administration (NASA) to train US lunar astronauts to search for evidence of meteorite impacts. Installation of SO_2 recovery systems have cut emissions from about 2 million tons annually in 1970 to only 20,000 in 2019, and ongoing reclamation efforts have already brought vegetation back to much of the affected area.

6.7.3. Banded Iron Formations Are Sediments That Formed in an Iron-rich Ocean

Banded iron formations consist of large amounts of iron that was precipitated from the Paleoproterozoic ocean. As mentioned above, this is problematical because iron does not dissolve in water in the presence of oxygen (oxidizing conditions). Instead, it forms ferric (Fe^{3+}) ions that react with water to form goethite [(FeO)OH)] and other minerals found in what we call rust. However, iron in the ferrous oxidation state (Fe^{2+}) is soluble in water if the water is in a reducing environment that is isolated from abundant oxygen. This made iron formations another player in the debate about the oxygen content of the early atmosphere and ocean.

Most Precambrian iron formations consist of alternating layers of iron minerals such as hematite (Fe_2O_3), magnetite (Fe_3O_4), or siderite ($FeCO_3$) and layers of silica (fig. 6.16E in color section), and this has led to the common appellation banded iron formation (BIF).[39] These rocks present problems at both ends of their life cycle. Where did the iron come from, and why was it deposited as a chemical sediment? Mafic igneous rock and associated hot springs are by far the most likely source of iron, and the Hemlock volcanic rocks, which are interlayered with the iron formation in Michigan, are unusually rich in iron. Why the iron was deposited is not so simple. It was originally thought that the iron was deposited when the oxygen content of the atmosphere (and ocean) increased during the Great Oxidation Event, which happened at about 2400 Ma as discussed below. However, we now know that Marquette Range Supergroup iron formations have ages of about 1850 to 1890 Ma, far younger than the Great Oxidation Event, and other iron formations around the world are considerably older than this. More complex theories for deposition of the iron involve microbial processes. Sediment layers associated with some iron formations contain fossils that might be from these organisms, including *Grypania spiralis*, discussed below, and Kurt Konhauser and others have noted that the estimated density of microorganisms is sufficient to deposit all the iron found in the sediment.

For iron formation to be deposited after the Great Oxidation Event, however, the ocean must have contained dissolved iron. But the oxygen content of the atmosphere had increased long before deposition of the Negaunee, Biwabik, and other iron formations, and this should have stripped iron from the ocean. One work-around is for oxygenation of the ocean to have lagged behind the atmosphere, as described recently by Noah Planavsky and others. During at least part of its history, this probably involved stratification, with a lower reducing layer that was isolated from the oxidizing atmosphere. Currents upwelling from the deeper ocean could have brought up iron-rich waters, perhaps seasonally, to be deposited in oxygen-bearing shallow water.[40] But then, why didn't the ocean remain stratified and continue to deposit iron formations forever? Several possibilities can be suggested. Maybe the oxygen content of the atmosphere became high enough to affect the entire ocean. Or maybe circulation of deep water eventually moved all the iron-rich water to the surface. One variant of this was suggested by John Slack and Bill Cannon, who pointed out that the Sudbury meteorite could have mixed up the stratified water in the Animikie Superbasin ocean causing it to become oxidized from top to bottom. Another possibility, suggested by Simon Poulton and others, is that the content of dissolved sulfur in the ocean increased enough to deposit pyrite, thus stripping dissolved iron from the deep ocean. Regardless of the exact cause, after about 1850 Ma, Earth's oceans largely abandoned the iron formation business. In fact, iron formations of this type only reappeared one more time in Earth history, and then only briefly. That was at the end of Proterozoic time between about 800 and 600 million years ago, when Earth could have undergone a snowball condition that allowed parts of the world ocean to become reducing.

Mining has taken place in all of the iron ranges in the Great Lakes region, although most production has come from the Mesabi and Marquette ranges (fig. 6.9). Early mining concentrated on special, "enriched" ores with iron content of 60 percent or more, which could be shipped directly to blast furnaces. In the enriched ores, silica has been dissolved and removed, and remaining iron minerals have been recrystallized. When and how the silica-removing enrichment occurred remains unclear, with possibilities ranging from modern or ancient flow of groundwater to Penokean or later metamorphism, and isotopic age measurements and geologic relations provide support for all of these.[41] As enriched ores were exhausted in the Great Lakes deposits, mining turned to unenriched banded iron formations, which contain only about 20 to 30 percent iron. This ore, which is known as taconite, has to be pulverized so that the iron mineral grains can be separated from the silica. In most mines, the iron minerals were formed into small balls known as pellets, which are transported to blast furnaces, most of which are at the south and east ends of the Great Lakes. Although this makes a much cleaner blast furnace feed, it is more expensive to mine and is the main reason that North American iron ores are not able to compete on world markets with enriched ores from Australia and Brazil.

6.7.4. Cobalt-Silver-Arsenide Veins Formed in Huronian Sediments around Nipissing Intrusions

The great silver mining district in the Huronian Basin at Cobalt, Ontario (fig. 6.3), produced almost 500 million ounces of silver, which would have been worth about 8 billion dollars in 2017. The silver is found as native silver in very narrow veins (fig. 6.16F in color section), usually just a few centimeters wide, along with cobalt-nickel arsenides. The veins surround Nipissing diabase that intruded Cobalt Group sedimentary rocks, and elements in the veins were probably leached from the diabase by water circulating through the sedimentary rocks and precipitated when the water cooled as it flowed away from the diabase. As the Cobalt mines were exhausted, miners fanned out throughout northern Ontario looking for more ore deposits, giving Cobalt credit for starting the era of metal mining in Canada but also for leaving arsenic-rich mine wastes.[42]

6.7.5. Mineral Deposits in the Colliding Sequence Are Largely Volcanogenic Massive Sulfide (VMS) Deposits Containing Copper, Zinc, and Lead

Volcanic arcs like the Pembine-Wausau terrane commonly form volcanogenic massive sulfide (VMS) deposits, which contain iron, copper, lead, and zinc sulfides. These deposits are particularly abundant in Archean volcanic arcs, and their origin is discussed further in the next chapter. Their discovery in Paleoproterozoic volcanic rocks in Wisconsin was based in part on geologic similarities between the Pembine-Wausau rocks and Archean rocks to the north in Ontario. As described by Gene LaBerge, exploration for these deposits began in Wisconsin in the 1960s when geologic studies showed that the terrane was an ancient volcanic arc. At least eight VMS deposits were discovered, including the Crandon deposit, one of the largest VMS deposits in the world (fig. 6.12).[43]

6.8. Paleoproterozoic Life Was a Cause and Consequence of the Great Oxidation Event

Although you wouldn't think of fossil hunting in rocks this old, fossils and evidence of life are actually surprisingly abundant in Paleoproterozoic rocks. By far the best-known examples are stromatolites, which are widespread in the Marquette Range Supergroup rocks (fig. 6.18A) and have been described by Hans Hoffman and others in the Huronian. One very curious fossil known as Grypania (fig. 6.18B) was found in sedimentary rocks interlayered with banded iron formation near Marquette, Michigan. Tsu-Ming Han and Bruce Runnegar described it as eukaryotic algae. Eric Hiatt and others have also reported fossil bacteria from the Michigamme Formation, and Carolyn Hill, Patricia Corcoran, and others have ascribed sedimentary structures in the Huronian sequence to fossil bacteria. Most of the rest are microscopic fossils found in the silica layers in iron formations, as described by Elso Barghoorn, Andrew Knoll, and others. Most images of these fossils show them to be segmented filaments, but they lack the third dimension. Three-dimensional images of these filaments obtained by David Wacey and others provide better information on their actual morphology (fig. 6.18C).

The real life-related story of the Paleoproterozoic, however, is the increase in oxygen in the atmosphere and ocean. As we have seen, Earth's early atmosphere contained CH_4 and other reducing gases and was very different from our present-day atmosphere, which has almost 21 percent oxygen. The Paleoproterozoic sedimentary rocks have been at the heart of the long-running debate about when Earth's atmosphere became enriched in oxygen. Geological help with this debate comes from soils and sedimentary rocks that form at Earth's surface and should reflect the composition of the atmosphere and water. Elements with more than one oxidation state can provide a sensitive indication of the oxygen content of the water and the coexisting atmosphere.

The first element to attract attention was uranium in conglomerates of the Matinenda Formation at the base of the Huronian sequence. Uranium has two oxidation states: U^{6+}, which forms soluble compounds in (oxidizing) water containing oxygen; and U^{4+}, which forms insoluble compounds such as the mineral uraninite (UO_2) in (reducing) water that lacks oxygen. Preservation of detrital uraninite grains in the Matinenda strongly suggests that

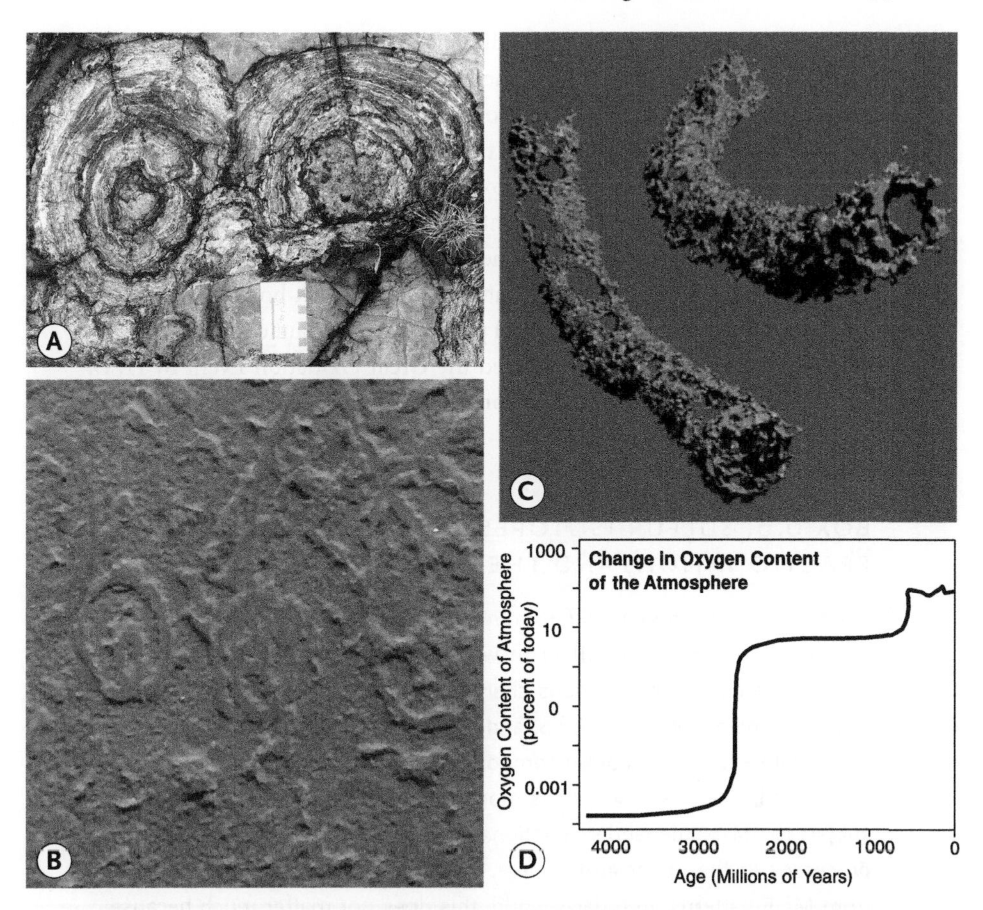

Figure 6.18. Paleoproterozoic life: (**A**) Stromatolites from the Gunflint Formation underlying the Sudbury impact debris layer, Thunder Bay, Ontario; (**B**) *Grypania spiralis* from the Empire mine, Marquette County, Michigan; (**C**) 3-D reconstruction of two Gunflintia filaments (courtesy of David Wacey); (**D**) graph showing changes in the oxygen content of Earth's atmosphere through time. (Modified from Kump 2008; Lyons et al. 2014.)

Earth's early streams and atmosphere were still reducing at the start of Paleoproterozoic time. Another element that attracted attention was iron, which is relatively soluble in reducing environments but almost completely immobile in oxidizing environments, as explained above. In the lower part of the Huronian sequence, sedimentary rocks and paleosols contain reduced iron, but iron in sedimentary rocks near the top of the sequence is oxidized (fig. 6.6H). Thus, it appeared that the Huronian succession was deposited during the transition from a reducing to an oxidizing atmosphere.

Studies in other paleosols and sedimentary rocks around the world confirmed that Earth's atmosphere became more oxidizing during Paleoproterozoic time. But questions remained about just how much oxygen entered the atmosphere and how it changed with time. Was the Paleoproterozoic transition abrupt or gradual and when did oxygen in the atmosphere reach present levels? Discovery of mass-independent fractionation in sulfur isotopes (box 6.6) provided new answers to these questions and resulted in renewed efforts to quantify the evolution of Earth's atmosphere. In fact both evidence and interest were so strong that the critical period of significant change in oxygen content at about 2400 Ma, was termed the Great Oxidation Event by H. D. Holland, who had studied the evolution of Earth's atmosphere and oceans for decades.[44]

BOX 6.6. SULFUR ISOTOPES, MASS-INDEPENDENT FRACTIONATION, AND THE GREAT OXIDATION EVENT

Isotopes of an element differ in mass. This difference causes mass-dependent fractionation (MDF) in which isotopes of different mass respond differently during various chemical processes. For instance, sulfur has three main isotopes, ^{32}S, ^{33}S, and ^{34}S, and MDF causes ^{34}S to concentrate relative to ^{32}S in oxidized compounds like sulfate rather than reduced compounds like H_2S, where ^{32}S is more abundant. Sulfur isotopes are also subject to mass-independent fractionation (MIF) caused by photochemical processes in the upper atmosphere, and MIF patterns are very different from MDF patterns. In today's world, this does not matter much because the amount of sulfur undergoing MIF in the upper atmosphere is so much smaller than the amount of sulfur undergoing MDF in waters and soils at the surface. But, if oxygen is not present in the atmosphere, MIF dominates, giving sulfur very different isotopic compositions from those that we see today. This effect is most obvious in isotope ratios involving ^{33}S. In 2000, James Farquhar and his colleagues reported the first observation of this effect in sulfur from natural materials, and measurements gradually grew to show that fractionation patterns involving ^{33}S for sulfur-bearing minerals changed abruptly between 2500 and 2000 Ma. This has since been accepted as important proof of the timing of the Great Oxidation Event.[45]

The use of several different geological and chemical constraints has allowed relatively firm estimates of the change in oxygen content of Earth's atmosphere through time. In the early Proterozoic, it rose in a two-step process, first to somewhere between 1 and 10 percent of the present level at about 2400 Ma and then to near present levels at about 600 Ma (fig. 6.18D). Oxygen in the early Earth was probably produced by cyanobacteria, whereas today it is produced by a wide array of photosynthetic processes. Oxygen is consumed by all sorts of reactions, including oxidation of reduced gases, such as methane to water vapor and carbon dioxide, and weathering of rocks containing reduced sulfur, iron, and carbon. In the early Earth, oxygen simply was not abundant enough to carry out all these duties. The transition to an oxygen-rich atmosphere began only when production of oxygen outpaced its consumption. This could have been caused by a decrease in the production of reducing gases by volcanoes, perhaps related to the transition to larger continents during Paleoproterozoic time, or by an increase in the production of oxygen, possibly caused by an evolutionary change in cyanobacteria or other organisms. The relationship between oxygen and reducing gases was probably dynamic and could have swung back and forth through time. In fact, evidence is accumulating that oxygen-rich oases formed at special times and places during Archean time, which is our next stop.

Notes

1. Superia was highly vulnerable to destruction at the beginning of Proterozoic time. Plumes of hot mantle rose from depth, splitting the supercraton and filling the fractures with dikes of diabase, an intrusive equivalent of basalt. Each fracturing event formed distinctive radiating or parallel dike systems called swarms, which are preserved in the remnants of the cratons. Along the southern margin of the Superior craton are found the remains of the large 2505 Ma Mistassini and 2445 Ma Matachewan dike swarms. Richard Ernst and Wouter Bleeker (2010) have used these dike swarms like bar codes to show how some of the ancient cratons were assembled to form the Superia supercraton shown in figure 6.1. Supercratons were also attacked from above. The lack of land plants made it easy for rivers to erode, and sediment was shed into the growing oceans as the cratons rifted apart.

2. The term Animikie Superbasin, as used here, includes all basins and parts of basins that host Paleoproterozoic sedimentary rocks in Michigan, Wisconsin, Minnesota, and adjacent western Ontario north of the Niagara fault zone. This "Superbasin" was originally called the "Animikie Basin" by James (1958) and Morey (1983) and that term has been used recently by Young (2015) and Planavsky et al. (2018). However, the term Animikie Basin has also been used by Ojakangas et al. (2001) to refer to the late stage basin that hosts the Animikie and Baraga Groups, and it has even been used by some to refer to the basin that hosts only the Animkie Group. In this book, the term "(Late) Animikie Basin" is used for this late stage basin. The Animikie Superbasin as defined here and the Wisconsin magmatic terranes are sometimes referred to as the Penokean Province, but this term gives the erroneous impression that deformation was entirely of Penokean age, which is not the case as explained in this chapter. Southern Province is a more correct and widely used term (fig. 6.14 shows the Southern Province).

3. The term *geon* refers to a 100 Ma time interval. Geon 17, for instance, refers to ages in the range of 1800 to 1700 Ma.

4. A classic report called "The Disappearance of the Huronian" written in 1930 by T. T. Quirke and W. H. Collins, delved into the difficulty of recognizing deformed Huronian rocks at the Grenville front.

5. Among the leaders in the early studies were Terence Quirke, William Collins, and J. A. Robertson. Later researchers included Ken Card and Stuart Roscoe, whose work was stimulated by discovery of the large uranium deposits in the lower part of the sequence. The description here is based on summaries by Roscoe (1969, 1973), Bennett et al. (1991), Ojakangas (1988), Ojakangas et al. (2001), Young (2013, 2014, 2015), and Young et al. (2001).

6. The ages of Thessalon and related rocks are from Krogh et al. (1984) and Ketchum et al. (2013). The Thessalon age overlaps ages of the Matachewan dike swarm (see note 1) and was likely associated with rifting.

7. The Matinenda Formation sandstones and conglomerates have cross bedding, ripple marks and other sedimentary structures indicating deposition in shallow, south-flowing braided streams. The overlying McKim Formation consists of fine-grained sandstones and mudstones that were deposited in deepening water as the rift continued to open.

8. Bleeker et al. (2018) has suggested that the red clasts were eroded from as yet undiscovered Paleoproterozoic-age deposits of jasper (red silica sediment with iron oxides) in the Animikie Superbasin to the west.

9. The Paleoproterozoic carbon isotope story is a long and ongoing one. Early analyses noted that the carbonate sediments were strongly enriched in ^{13}C, a geochemical deviation known as the Lomagundi Event. This anomaly was originally interpreted to indicate that a large amount of organic carbon was buried when the sediments were deposited. But large deposits of organic matter have not been found. An alternative explanation suggests that carbon in the carbonates was produced by reactions involving methane in sediments (Sekine et al. 2010; Wallmann and Aloisi 2012).

10. Isotopic age measurements can be made on the mineral zircon ($ZrSiO_4$) because it contains small amounts of uranium, which undergoes radioactive decay to form lead. Zircons that grew in an igneous rock provide an age for the host rock. Detrital zircon grains in sedimentary rocks are used to estimate the maximum age of sedimentation, which must be less than the youngest detrital zircon grain that it contains. See Craddock et al. (2013) for information on zircons in the cover sequence sedimentary rocks.

11. Curiously, there are no obvious faults and rifts in the Huronian sequence that might have guided the Nipissing magmas upward from the mantle. Grant Young (2015) suggested that the Nipissing magmas are the distal end of the 2218 Ma Senneterre dike system to the northeast in Quebec (fig. 6.1).

12. The term *snowball Earth* was suggested by Joseph Kirschvink (1992) to describe widespread glacial deposits that formed at the end of Proterozoic time sometime between about 800 and 550 million years ago. The cause of this event, which has been documented by Paul Hoffman and his coworkers (1998), might also have been related to changes in atmospheric greenhouse gas concentrations. See Pavlov et al. (2000) for information on the possible role of methane and Kasting and Ono (2006) for a review of paleoclimates during the first two billion years of Earth history. For more information on the faint young sun paradox, see Gaidos et al. (2000); Rosing et al. (2010).

13. For discussions of Proterozoic atmospheric CO_2 levels, see Kaufman and Xiao (2003); Wallman and Aloisi (2012); Sheldon (2013). Recent calculations of Stephanie Olson and others (2016) suggesting that methane may have been limited to levels below about 10 ppm in the early atmosphere reopened the faint young sun paradox. Mingyu Zhao and others (2017) tried to close it again with calculations showing that microbial mats covering only 8 to 10 percent of Earth's surface would have been able to maintain sufficiently high atmospheric methane levels.

14. Glacial deposits in both the cover sequence sedimentary rocks underlie sedimentary rocks that show evidence (oxidized iron) for increased oxygen in the atmosphere.

15. The impact crater depth proposed by Zieg and Marshak (2005) is at the deep end of a spectrum of suggested depths. See Wang et al. (2018) for a review and recent evidence.

16. The Sudbury Igneous Complex is a layered igneous complex. It looks somewhat like the Duluth Complex described in chapter 5, although its magma originated as a melt sheet formed by a meteorite impact rather than partial melting of the mantle (which supplied the Duluth Complex magma). As the Sudbury magma crystallized, mafic minerals settled downward to form layers that make up the lower part of the complex, leaving a residue of felsic material to make the granophyre that forms the upper part of the complex. The Sudbury Igneous Complex lacks layers rich in chromite or magnetite such as those found in the Duluth Complex, but the base of the complex (below the mafic layer) contains enormous magmatic nickel-copper ores. For descriptions of the Sudbury Igneous Complex and its impact melt sheet, see Riller (2005); Roussell and Brown (2009); Lightfoot (2017).

17. O'Callaghan et al. (2016) have shown that the Sudbury breccia consists largely of melted country rock with little or no contribution from the melt sheet. Whether the breccia formed by melting of the rock matrix or by fragmentation (cataclasis) is still being debated.

18. Early mapping of Marquette Range Supergroup rocks by the US Geological Survey (James et al. 1956, 1958, 1968; Bayley et al. 1966; Gair and Thaden 1968; Cannon and Gair 1970; Gair 1975) focused on the iron ranges and recognized most of the units described here. Additional information on Animikie Supergroup sedimentary and volcanic rocks, including their regional relations, is in Schmidt (1963); Morey (1978, 1983); Sims and Carter (1996); Fralick et al. (2002); LaBerge et al. (2003); Cannon et al. (2007); Schulz and Cannon (2007); Ojakangas et al. (2002, 2011). The term Marquette Range Supergroup is used for all of the rocks in the Animikie Superbasin in an effort to simplify a very complex geologic story; it has not been formally proposed or approved.

19. The impact sediment and ejecta layers are well exposed in the back yards of several houses that border Hillcrest Park in Thunder Bay, Ontario, and they form half of the front yard of a nearby house (Addison and Brumpton 2012). So, the original problem was not finding the layers; it was interpreting their origin.

20. Meteorite impacts generate extremely high, instantaneous pressures. In addition to melting some of the impacted crust, as happened at Sudbury, these pressures generate shock metamorphic features. One common feature of this type consists of planes in the mineral quartz that look like closely spaced breaks or cleavage. But, quartz has no natural cleavage and the only known way to form these features is through an intense shock. The crystal structure of quartz can also be changed to high-pressure forms, especially stishovite. Melted rock that is thrown into the air by an impact commonly takes the form

of accretionary lapilli, small spherical pellets consisting of fine-grained rock or volcanic ash that clustered around gas bubbles. Recognition of these features in sediment layers helps identify them as ejecta from a meteorite impact. For a review of shock metamorphic features see French (1998).

21. The *Dickinson Group* is an isolated sequence of metamorphosed rocks that is found only in the Felch trough as described by Bill Cannon and Klaus Schulz (2007). It consists of clastic sediments and mafic volcanic rocks that were originally thought to be Archean in age. But, John Craddock found detrital zircons as young at 2100 Ma in the sedimentary rocks, showing that they are about the same age as the Menominee Group (Figure 6.9).

22. In the Gogebic iron range, glaciogenic rocks are missing, and the Sunday quartzite rests directly on Archean basement. Note that the Marquette Range Supergroup lacks basal rift-type volcanic rocks and coarse clastic sediments like the Thessalon and Livingstone Creek Formations in the Huronian Supergroup. Grant Young has suggested that early rift-related rocks in the Animikie Superbasin, if they were ever present, were carried away on the continent that formed the southern side of the rift (and, of course, we don't know where that continent is today).

23. Early studies in the western part of the Animikie Superbasin by Schmidt (1963), Southwick et al. (1988), Southwick and Morey (1991), Morey and Southwick (1993), and Morey (1978, 1983, 1999) provided the basis for later studies by Chandler et al. (2007), Boerboom (2011), Craddock et al. (2013), and Boerboom et al. (2014) on which this summary is based.

24. Parts of the Mille Lacs Group might be older or possibly even younger. An Sm-Nd isochron age of 2197 Ma has been reported for basalt from the Denham Formation, and an age of 1883 Ma has been reported for magmatic zircon in a mafic intrusion in the Moose Lake-Glen Township area. Furthermore, the Little Falls Formation, which is reported to overlie the Denham Formation and has been included in some descriptions of the Mille Lacs Group, contains detrital zircons with ages as young as 1844 Ma. Terry Boerboom and others have raised the possibility that the Little Falls Formation correlates with the post-Sudbury Animikie Group, indicating that the (Late) Animikie Basin was much larger than indicated by its present outcrop (Boerboom, 2011; Boerboom et al., 2014).

25. Rocks of the South Cuyuna range are not included in the North Range Group. Even the iron formations in the two ranges are different; the Trommald Iron Formation contains much more manganese than the South Cuyuna range iron formations.

26. John Craddock and others (2013) have argued from ages of dikes and detrital zircon that the Great Stratigraphic Gap was only 117 million years long rather than earlier suggestions of 250 to 350 million years.

27. The Paint River Group, consisting largely of greywacke and shale, was originally thought to overlie the Baraga Group. Discovery of the Sudbury ejecta layer at the base of this sequence suggests that it is actually part of the Menominee and Baraga groups.

28. If this sounds confusing, it is. The term Animikie Basin is indeed used for this small, late part of the Animikie Superbasin, and that is why the term "superbasin" has been used here (to distinguish the two). To help remind you that the Animikie Basin (and its Animikie Group) is a late part of the sequence, it is referred to here as the (Late) Animikie Basin. The (Late) Animikie Basin was cut into two parts by the Duluth Complex when it was emplaced at about 1100 Ma as part of the Midcontinent Rift. In

the northern half, the Gunflint Range iron formation rests on Kakabeka Quartzite or directly on Archean basement (fig. 6.11G in color section) and is overlain by shales of the Rove Formation (fig. 6.11H in color section). In the southern half, the Biwabik iron formation (Mesabi Range) rests on a thin layer of Pokegama Quartzite and is overlain by Virginia Formation shales (fig. 6.9). The far southern end of the Animikie Basin contains the Emily iron formation and Thomson Formation, which probably correlate with the Biwabik iron formation and Virginia Formation. Although the (Late) Animikie Basin is largely undeformed, its southern margin in the area of the Emily iron formation is deformed and, as mentioned in note 24, there is even the possibility that the (Late) Animikie Basin extended southward to include the deformed Little Falls Formation. Ages of the Rove and Virginia Formations are in Fralick et al. (2002), Heaman and Easton (2005), Addison et al. (2005), and Addison and Brumpton (2012).

29. The Animikie and Baraga Groups differ in one important respect. The base of the Baraga Group is above the Sudbury impact layer in most areas and does not include iron formations. In contrast, the base of the Animikie Group is below the iron formations; in addition to the Biwabik and Gunflint iron formations, it even includes the underlying Kakabeka and Pokegama Quartzites. These iron formations correlate with the Ironwood and Riverton iron formations on the east side of the rift in the Baraga Group. The simplest way to repair this problem would be to put the base of the Animikie Group (west of the rift) at the top of the iron formations, and include the iron formations and underlying rocks in the Menominee Group. There is one possible complication, however. The Gunflint, Biwabik, Ironwood and Riverton are almost certainly the same age because they are all overlain by the Sudbury debris layer. The position of the Negaunee is less clear because of two conflicting dates (Schneider et al, 1874 Ma; Pietrzack-Renaud, 1891 Ma). However, both ages are older than the Sudbury layer and confirm that the iron formation was part of the Menominee Group rather than an early phase of Baraga-Animikie Group deposition.

30. Ophiolites are rocks that were originally part of the ocean crust and immediately underlying mantle. They consist of ocean floor basalts and underlying dikes that fed the basalt flows. These rest on mafic and ultramafic rocks of the lower ocean crust and upper mantle. Packages of ocean crust and mantle are moved onto continents and island arcs by obduction, during which ocean plates are thrust onto other plates. See Schulz (1984) for a description of the Wisconsin ophiolites and Dilek and Newcomb (2003) for a summary of ophiolite concepts and geology.

31. Calc-alkaline rocks commonly form in subduction zone environments. See chapter 5, note 29 for a discussion of the petrochemistry of igneous rocks.

32. Although the Marshfield and Pembine-Wausau terranes started life as independent entities, they collided before about 1833 Ma to form the Wisconsin magmatic terranes. We know this because the Eau Pleine fault that separates them is cut by felsic intrusions with ages between about 1853 and 1833 Ma.

33. See Whitmeyer and Karlstrom (2007) and NICE (2007) for details of the Yavapai and Mazatzal terranes.

34. These intrusions are characterized by abundant potassium and sodium (alkalis) relative to their aluminum content and range from granite (with excess silica that formed quartz) to syenite (with silica content too low to form quartz). Similar alkaline intrusions are widespread in central North America and are referred to as A-type granites (Smith 1983; Anderson et al. 1988; Holm et al. 2005; Dewane and Van Schmus 2007).

They can form by melting of lower crust or by extreme differentiation of basaltic magma (Frost and Frost 2011).

35. Metamorphic intensity refers to the degree to which a rock has been changed by metamorphism. The summary by Cannon and Schulz is based on their own work and that of Terrence Boerboom, Val Chandler, Daniel Holm, Mark Jirsa, John Klasner, Bruce Marshak, David Schneider, Paul Sims, Eric Tohver, and others.

36. The dome-and-keel belt contains two types of domes separated by a line known as the Great Lakes Tectonic Zone. South of the line are granite/gneiss domes such as the McGrath, Watersmeet and Carney Lake, which contain Archean rocks as old as 3800 Ma. North of the line are domes consisting of a mixture of granite and volcanic rock called greenstone with ages of about 2700 Ma. The line between these two types of domes is known as the Great Lakes Tectonic Zone, and it marks the boundary between two Archean-age terranes that form the basement rocks below the Paleoproterozoic cover sequence. These are the Superior Province on the north (especially the Abitibi-Wawa Belt) and the Minnesota River Valley terrane on the south, which are discussed further in the next chapter.

37. Profits from these operations allowed Joel Hirschhorn to acquire the 6000 paintings, sculptures, and other works of art, which were donated to the United States to found the Hirschhorn Museum and Sculpture Garden on the National Mall in Washington, DC.

38. At Elliot Lake, at least 200 million tons of mine waste and mill tailings were left on the surface without much long-term planning. Weathering of this material created acid mine drainage and released radioactive daughter products. During later reclamation, the material was placed in flooded zones where the overlying standing water prevented oxidation of pyrite and served as a shield from radiation. Barium was also used to co-precipitate radium that was generated in the wastes (Feasby 1997).

39. A common variant of BIF is GIF, or granular iron formation, which consists of iron-rich sediment fragments probably formed by wave action in shallow water. By far the best place to see BIF is Jasper Knob at the top of the hill just east of the Cleveland Cliffs research lab in Ishpeming, Michigan.

40. These special conditions probably apply to the global ocean rather than an isolated arm of the sea in the Great Lakes region because iron formations of the same (1850–1890 Ma) age are found in other parts of the world, especially Labrador.

41. Bill Cannon (1976) and Natalie Pietrzak-Renaud (2013) have described post-depositional changes in the Marquette Range iron ores related to later fluid flow and metamorphism. Birger Rasumssen and others (2016) measured ages of 1800 to 1770 Ma for monazite in Marquette Range ores, suggesting that some changes took place during Penokean or Yavapai metamorphism. But, Ken Farley and Ryan McKeon (2015) reported ages of 1060 to 770 Ma for hematite in enriched ore from the Gogebic range, supporting the possibility of fluid flow caused by uplift of the Midcontinent Rift, as originally suggested by Glenn Morey (1999).

42. Miners at that time simply discarded arsenic-rich wastes on the surface, where some material remains today (Sprague et al. 2016). See Jambor (1971) and Andrews et al. (1986) for descriptions of the Cobalt deposits.

43. The only VMS deposit of this group that has been mined so far is Flambeau at Ladysmith, Wisconsin, where the uppermost part of the deposit was removed and shipped directly to a smelter without any processing. This upper part had extremely

high concentrations of copper, which had been formed by weathering of the deposit when it was exposed at the ancient surface, probably during the weathering that produced the paleosols beneath the Baraboo Interval Quartzites. The weathering leached copper from the top of the deposit and precipitated it a few meters lower, gradually enriching that part of the deposit (as the overlying part was removed by erosion) by a process known as supergene enrichment. The much larger Crandon deposit, discovered in the 1970s, has not been mined because of environmental opposition. In 2003 the deposit was sold to the local Sokaogan Ojibwe tribe for 16.5 million dollars derived from casino revenue. The Pembine volcanic belt dies out toward the east, although the small sliver of it that extends into Michigan contains the Back Forty VMS deposit (fig. 6.12).

44. See Holland (2002) and Blaustein (2016) for reviews of the Great Oxidation Event.

45. See Kasting and Howard (2006) for a summary of MIF sulfur isotope compositions through time. Some periods during Archean time, without evidence of MIF might be oxygen oases as discussed in chapter 7 (box 7.6). For more information on MIF of sulfur isotopes, see Domagal-Goldman et al. (2011).

CHAPTER 7

Making a Craton

Archean Greenstone Belts and Granites

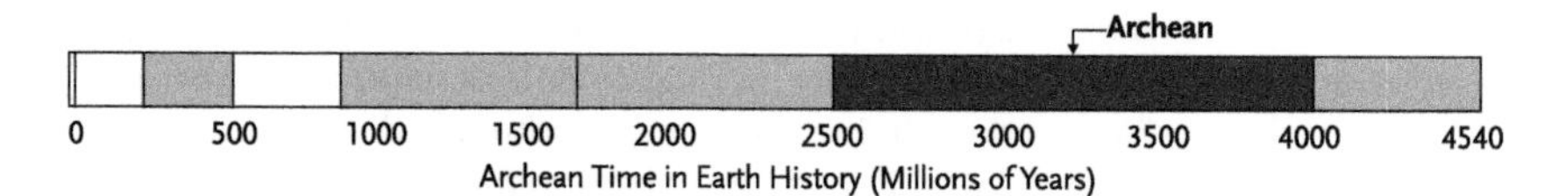

7.1. The Superior Craton Is the Core of the Precambrian Canadian Shield

Our time traveler has now arrived in Archean time and is trying to find a place to set up shop on the Superior craton. But, there is no geologic information. Throughout the last chapter, the Superior craton was described as a large homogeneous blob. Before landing, our traveler needs to know what's going on down there. What parts of the craton will form first and which last, which are likely to be stable and which will be covered by lava flows, and even where to find minerals that might supply the last big time jump into the Hadean. Unraveling the history of the Superior craton will also help in future exploration of other planets because it is here that we and the time traveler can learn about how and why cratons begin to aggregate into larger land masses.

We can start by referring to the exposed part of the Superior craton as the Superior Province, the name used in most studies.[1] Rocks in the Superior Province are almost entirely Archean in age, ranging from about 2500 to 4000 Ma. Many geologists who work with younger rocks look toward ancient Archean terranes with the awe a child might feel for a centenarian. Old rocks, like old people, have had a lot of experiences and their appearances often show it. In the case of Archean rocks, the common expectation is that deep burial and metamorphism have obliterated evidence of their earlier history, including their protolith, a term that refers to the original, premetamorphic rock (which might have been volcanic, intrusive, or sedimentary). Although some Archean rocks have indeed been strongly metamorphosed and lost evidence of their ancestry, surprisingly large parts of the Superior Province are made up of lightly metamorphosed rocks that retain their original rock textures and structures. This is fortunate because it allows us to learn more about Archean history and the early stages of formation of Earth's cratons and continents.

It might be surprising that we would be taking up the origin of continents in a section that deals with Archean time. After all, the Archean Eon ended a full 2 billion years after Earth formed. What about the preceding Hadean Eon? Was Earth such a slow starter that it took a billion years to get into the continent-forming business? Sadly, the answer is yes. As we will see in the next chapter, Hadean time was just too tumultuous to produce numerous, long-lived continents.

In our quest to understand the origin of continents, we are fortunate to have the Great Lakes region as our focus of study because it contains the southern part of the Superior Province, the largest Archean craton on Earth. The Superior Province has an area of about 1.4 million km^2, more than twice the area of France, and accounts for almost a quarter of all Archean and older rocks that are exposed on our planet. The range of ages found in rocks of the Superior Province is impressive, almost as long as the time span covered in all six previous chapters. And, as you may have noticed, the span of Superior Province ages extends a few hundred million years back into the Hadean Eon, a topic we will review in the next chapter.

The Superior Province (along with rest of the Superior craton, which is buried beneath younger sediment) makes up the core of the present-day North American continent. As we learned in the last chapter, it was the principal continental fragment that constituted early Laurentia. As you may recall from figure 6.17, the Superior craton is bounded on all sides by highly deformed, deeply eroded mountain belts of Paleoproterozoic age, including the Southern/Penokean Province that we dissected in chapter 6. These are orogenic zones that formed when smaller cratons, or continental fragments, collided with the Superior craton to form Laurentia. Although the collisions associated with this assembly chewed a little against the edges of the Superior craton, much of it has remained undisturbed. The stability of the Superior craton is due in part to the fact that it is underlain by a 200- to 250-km-thick keel that has protected it from later deformation. The term Canadian Shield is often used to refer to this highly stable region (including the later Proterozoic additions), and later in this chapter we will look briefly at just how the keel formed.

Before we go any further, take a quick look at the Superior Province as shown in figure 7.1. It is not the single homogeneous blob that we worked with in chapter 6; instead, it is made up of long units that are referred to formally

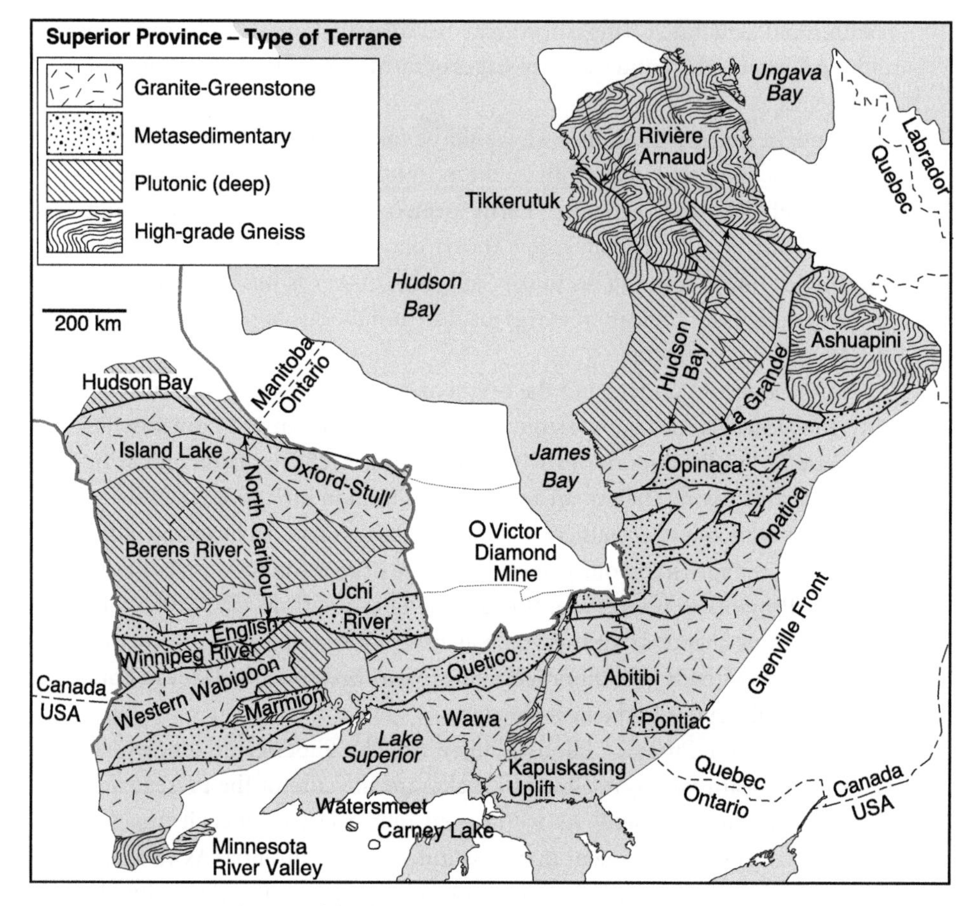

Figure 7.1. Distribution of the four main types of terranes in the Superior Province. (Modified from Card and Ciesielski 1986; Card 1990; Percival et al. 2006, 2012; Stott et al. 2010.)

as terranes and informally as subprovinces or belts. These terranes have evocative names derived, in part, from the aboriginal people of northern North America, and they are definitely distinct geologic units.

In figure 7.1, the terranes are divided into four main types based on the dominant rocks that they contain, as discussed below. Recognition of these terranes began in 1964 when C. H. Stockwell outlined their general east-west orientation in Ontario. In 1970 A. H. Lang, Alan Goodwin, and others used aeromagnetic maps to show the extent of terranes and their boundaries, and in the 1980s, Ken Card summarized the early work and extended it to the east in Quebec. Expanded mapping and age measurements summarized by John Percival and Gregory Stott resulted in the terranes that are recognized today.

So what are these terranes and what story do they tell about the Superior Province and the origin of continents? To answer this, we must first look at the main rock units in this immense area and how they were recognized during mapping and then assembled into geologic maps.

7.2. Mapping the Superior Province Involved Geology, Geochronology, and Geochemistry

Imagine yourself with the assignment of composing the first report on the geologic history of the Superior Province. The area is huge, it has few roads, and much of it is covered with glacial deposits, lakes, beaver dams, and moose pasture. It is also relatively flat, a setting that yields few exposures of rock and precious little three-dimensional information about how rock units are related to one another at depth. But that's not the first problem. The first problem is even more basic: what types of rock to map. What are the main geologic units that make up these vast terranes? This was the situation faced by early geologists of the Geological Surveys of Ontario and Quebec, as well as the Geological Survey of Canada.[2]

Their efforts involved three types of research, including careful field mapping to determine the main rock types and units and their relations to one another, detailed measurements of the ages of the rocks and rock units, and finally chemical and isotopic analyses to show how the rocks formed. These three steps overlapped in actual practice, but we will summarize them as a sequence in the following sections. All this work was aided by geophysical surveys similar to those that we used in chapter 5 to learn the shape of the Midcontinent Rift (fig. 5.1).

7.2.1. The Superior Province Consists Largely of Supracrustals and Intrusive Rocks

Mapping in the Superior Province showed that much of it consists of metamorphosed volcanic and sedimentary rocks, which are referred to in Archean parlance as supracrustals to indicate that they formed at and near Earth's surface. These are surrounded by intrusive rocks that were emplaced at depth and are referred to as plutonic rocks. Many parts of the supracrustals are not very metamorphosed and provide valuable information about near-surface conditions during Archean time. Other supracrustals have undergone intense metamorphism, which has largely obliterated the original rocks but provides information about temperatures and pressures reached during later burial and metamorphism.

These volcanic, sedimentary, igneous, and metamorphic rocks are the building blocks of the Superior Province. Because they have been metamorphosed to varying degrees, they are referred to with the *meta* prefix.[3]

Metavolcanic rocks consist of volcanic flows and fragmental rocks that have undergone metamorphism. Basalt is by far the most abundant volcanic rock, and many basalts have the very characteristic pillows (fig. 7.2A, B in color section) that form when mafic lava is extruded underwater, where it cools quickly to form a carapace that is expanded by new lava.[4] The abundance of pillows in these lavas confirms that they were emplaced in a submarine environment. Smaller amounts of intermediate and felsic volcanic rocks are also present, along with an unusual ultramafic rock called komatiite (fig. 7.2C in color section), which we will get back to a little later. Fragmental rocks range from coarse-grained volcanic breccias to fine-grained tuffs, many of which show sorting and layering, indicating that they were deposited in water (fig. 7.2D, E in color section). Many metavolcanic rocks contain green metamorphic minerals like chlorite, which give the rocks their distinctive name greenstones.

Metasedimentary rocks consist largely of clastic (fragmental) grains that were eroded from the volcanic rocks. The grains range from cobbles that were deposited in streams (fig. 7.2F in color section) to finer-grained material deposited by density currents that formed fans of sediment in deeper water (fig. 7.2G in color section). Limestones, evaporites and other chemical sediments typical of shallow continental-shelf passive-margin settings are very rare. The most widespread chemical sediments are iron formations and volcanogenic massive sulfide deposits, both of which are discussed later in this chapter. The Archean iron formations are not as extensive as the large Paleoproterozoic banded iron formations described in chapter 6, in part because continental shelf environments were smaller and less numerous.[5]

Plutonic rocks are intrusive igneous rocks that were emplaced at relatively deep levels in the Archean crust. In Archean terranes, they are mostly felsic varieties that are referred to collectively as granite, even though they have a wide range of compositions (box 7.1). Mafic and ultramafic intrusive rocks are also present but are less common. Many of these intrusions are coarse-grained and contain numerous fragments (xenoliths) of the surrounding wallrocks (fig. 7.2H in color section).

The fourth rock type, *high-grade gneiss*, consists of metamorphosed varieties of the first three rock types. The term *high-grade* indicates that the rocks have undergone relatively intense metamorphism involving high temperatures and deep burial, and the term *gneiss* refers to their crudely layered texture (fig. 7.3C, D in color section).[6] Some high-grade gneiss provides clues about the original rock that was metamorphosed (protolith). The most common protolith is plutonic granite, which has been metamorphosed to form granite gneiss.

BOX 7.1. WHAT IS GRANITE?

In Archean terranes, the term *granite* refers to a large range of coarse-grained, felsic, intrusive igneous rocks. These rocks are sometimes referred to with the acronym TTG, which is short for tonalite-trondjhemite-granodiorite, all of which are felsic intrusive rocks. Archean "granites" get this awkward moniker because they differ significantly in chemical composition from "granites sensu stricto" found in younger terranes. In particular, TTG granites are richer in sodium and poorer in potassium than granites sensu stricto. Although this might seem like a very fine point, it turns out to be important to Archean geology. According to Nick Arndt, granite sensu stricto in younger terranes can form by partial melting of basalt at temperatures of about 700 to 900°C, whereas TTG rocks require temperatures of 900 to 1100°C. This means that the tectonic/geologic environment in which TTG magmas formed was hotter and is one of several lines of evidence that the Archean Earth was hotter and more volcanically and tectonically active.

7.2.2. Supracrustals and Intrusive Rocks Form Separate Terranes with Different Ages

Various combinations of these four rock types, metavolcanic, metasedimentary, plutonic, and high-grade gneiss, make up the four main types of terranes found in the Superior Province (fig. 7.1). The name given to each terrane reflects its dominant rock type, although all of the terranes are mixtures of various rock types, as explained below.

Granite-greenstone terranes consist of greenstone belts surrounded by "granite," as defined above. Greenstone belts, which are the supracrustal rocks mentioned above, consist of metavolcanic rocks with interlayered metasediments. They are the remains of ancient volcanic mountain belts and plateaus that were tens of kilometers long and thousands of meters thick. Granite surrounds the greenstone belts in most terranes and is usually more abundant. In the enormous Abitibi granite-greenstone terrane, which makes up most of the southern part of the Superior Province (fig. 7.1), granites make up about 60 percent of the area compared to only 40 percent for supracrustals.[7] The Abitibi terrane is one of the youngest and least deformed terranes in the Superior Province; in other volcanic (granite-greenstone) terranes such as the Oxford-Stull and Uchi (fig. 7.1), supracrustals are even less abundant and the proportion of granites is greater, probably reflecting a deeper level of erosion (which would expose more of the deep intrusive rocks).

The role of granite in these belts varies. In some belts, like the Marmion terrane and Island Lake domain (part of the North Caribou terrane), many granites are older than the supracrustal greenstone belts, and they form the basement on which the supracrustal rocks were deposited. In other belts, such as the Uchi and Abitibi, large volumes of granite intruded the supracrustals and are therefore younger. In a classic study of metamorphism in the Abitibi terrane, Wayne Jolly showed that many supracrustal rocks were most strongly metamorphosed at their margins with the granite intrusions, confirming that the granites intruded the supracrustals.[8]

Metasedimentary terranes consist of thick sequences of clastic sediments. These supracrustals differ from the greenstone belts in having relatively few associated volcanic rocks. The English River and Quetico-Opinaca metasedimentary terranes (fig. 7.1), which are usually referred to as belts, extend all the way across the Superior craton (and are much larger than the small zones of metasedimentary rocks that are interlayered with volcanic rocks in individual greenstone belts). Metasedimentary belts are also intruded by granitic intrusive rocks. And they have undergone relatively intense and regionally extensive metamorphism that contrasts with lower-intensity metamorphism in many granite-greenstone terranes.

Plutonic terranes consist largely of plutonic igneous rocks with very little in the way of supracrustals or high-grade gneiss. According to Ross Stevenson and his coworkers, the central part of the North Caribou terrane (previously known as the Berens River subprovince) consists almost entirely of granitic intrusive rocks, with less than 5 percent supracrustals. These are generally considered to be deeply eroded granite-greenstone terranes from which most of the supracrustals have been removed.

High-grade gneiss terranes lack abundant or obvious supracrustals, either metavolcanic or metasedimentary. They consist instead of a mixture of granite gneiss (metamorphosed granite), highly metamorphosed volcanic and sedimentary rocks, and some plutonic rocks. Good examples include rocks of the Ashuanipi part of the Rivière Arnaud terrane, as well as the Minnesota

River Valley terrane and its eastern extension in the Watersmeet and Carney Lake domes in Michigan (fig. 7.1). These terranes are even more deeply eroded and metamorphosed than the plutonic terranes. Studies based on the temperatures and pressures of metamorphism in these rocks show that the Ashuanipi and Hudson Bay terranes were originally at depths of 20 to 25 km in the crust. The small Kapuskasing uplift, which separates the Wawa and Abitibi terranes, exposes similar deep rocks.[9]

From the very beginning, high-grade gneiss terranes were suspected of being the oldest parts of the Superior Province. Perhaps they represented highly metamorphosed early phases of volcanism, sedimentation, and continent formation that were deformed and metamorphosed during late Archean agglomeration of terranes to form the Superior Province. To test this, however, it was necessary to determine ages and origins for the rocks, and that was the second big push in research.

7.2.3. U-Pb Analyses of Zircons Provide Ages of Superior Province Rocks and Terranes

As mapping proceeded, it became clear that isotopic ages were needed to sort out the terranes. There was no guidance from fossils, or course, and contact relations between different rock units gave only relative ages and were sometimes ambiguous. Early age measurements using the K-Ar system were bedeviled by low closure temperatures, as discussed in chapter 6. The best method turned out to be U-Pb isotopic analyses of zircon ($ZrSiO_4$), which has a high closure temperature. Although zircons did present some challenges, including rims that grew during later events, loss of original U or Pb, and zircons inherited from older rocks, they proved to be very useful.[10]

The Minnesota River Valley high-grade gneiss terrane provides a good example of the development of isotopic age measurement methods in Archean rocks. In 1963 E. J. Catanzaro reported a surprisingly old age of 3300 Ma for a group of zircons found in gneiss from the Minnesota River valley. By 1970, Sam Goldich and Carl Hedge had pushed the age back to 3550 Ma, and in 1974 they revised it to 3800 Ma, in part using the Rb-Sr isotope system. These results led to two important conclusions. First, rocks in the Minnesota River Valley terrane were among the oldest on the planet. Second, the range of age measurements was very large, and methods had to be improved, in terms of both the precision of the analysis and the size of the sample (number of zircon grains) that had to be analyzed.

A big breakthrough in Superior Province age measurements happened in 1975 when Tom Krogh set up the Jack Satterly Geochronology Laboratory at the Royal Ontario Museum and University of Toronto. Along with Jim Mortenson at the Geological Survey of Canada, the Satterly lab pioneered the development of new analytical methods that yielded high-precision U-Pb ages for individual zircon grains. Fernando Corfu, who joined Krogh during the early stages of this research, has written an interesting summary of the history of technique development and its impact on zircon geochronology. The result was a veritable gusher of ages for the many units that had been mapped in the Superior Province. With analytical uncertainties as low as just a few million years, these new ages allowed events to be measured with a precision almost as good as that obtained from very young rocks.

Just how much progress has been made is shown by a recent study of rocks in the Minnesota River Valley terrane by Pat Bickford and associates. You can see in figure 7.3A (in color section) that there isn't much rock to work with in the valley. The only rocks that are exposed are at the bottom of the ancestral River Warren valley; almost everything else is covered with glacial deposits. Even so, the exposed rocks differ enough that they are considered to represent two different Archean crustal fragments, the Montevideo and Morton blocks, which are intruded by the Sacred Heart granite. Precise ages of these two blocks would obviously help us decide whether they are siblings or unrelated fragments with very different histories. It turned out that they are siblings. This was investigated using an even more recent technique in which individual spots in zircons are analyzed (rather than an entire single grain) using microbeam techniques. This method yielded ages of 3485±6 Ma and 3499±6 Ma for the Montevideo granite gneiss and an age of 3524±9 Ma for the Morton granite gneiss. These ages are very similar and suggest that the Minnesota River Valley terrane consists of two closely related fragments of a single craton with an age of about 3500 Ma.[11]

Several other events were indicated by the zircon analyses, but one, in particular, is noteworthy for us. Figure 7.3B (in color section) is an image of a single grain of zircon from the Montevideo gneiss that has been cut and polished. Although this grain is only a few tens of micrometers in size, you can see that it consists of several layers or zones that formed as it grew. Spot analytical techniques yield an age of 3499 Ma for the central part of the grain, whereas the outer part of the grain, the rim, has an age of 2603 Ma. The simplest interpretation of these measurements is that the zircon grain formed at about 3500 Ma but grew an outer rim during a later event at about 2600 Ma. The younger age is the same as the 2604 Ma age of zircons in the Sacred Heart granite. The texture of the rocks shows that they underwent metamorphism accompanied by melting and intrusion of granite (fig. 7.3C, D in color section), and the zircons show that this happened during intrusion of the granite. Keep this late Archean (~2600 Ma) age in mind as we will see a lot of it later in this chapter.

U-Pb analyses have been used to determine ages for granites, greenstones, and metamorphic rocks throughout the Superior craton. Figure 7.4, which is based on decades of age measurements, shows the maximum crystallization ages for intrusive rocks in each of the terranes. The oldest terrane found so far is the Tikkerutuk domain, which is part of the Rivière Arnaud terrane in northern Quebec near the far northern margin of the Superior Province. This terrane is at least in part late Eoarchean in age (4000 to 3600 Ma), and marks the Tikkerutuk crust as truly among the oldest on the planet. Also falling in the Eoarchean range is the Carney Lake gneiss in northern Michigan with an age of 3750 Ma reported by Robert Ayuso and others. The Carney Lake Gneiss and its nearby companion the Watersmeet dome were mentioned in chapter 6 as dome-and-keel structures that formed during the Yavapai orogeny (fig. 6.15).

Paleoarchean intrusive rocks ranging in age from 3600 to 3200 Ma have been found in other terranes, including the Minnesota River Valley, Winnipeg River, and parts of the Hudson Bay terranes.[12] Studies have also indicated that some terranes and subterranes should be combined into larger units. One of these is the North Caribou superterrane, with ages in the 3200 to 2800 Ma range, which resulted from Mesoarchean assembly of the Oxford Stull, Island Lake, and Uchi domains around the previously recognized Berens River central plutonic subprovince (fig. 7.4). Most of the rest of the Superior Province consists of rocks that formed largely during the last part of the Archean, the Neoarchean, from 2800 to 2500 Ma.

7.2.4. Nd Model Ages Provide Information on the Ancestry of Igneous and Sedimentary Rocks

As age determinations for single zircon grains and spots in the grains accumulated, more and more old ages were found in relatively young rocks, many of them from zircons that had been picked up from older crust during magma formation and transport. Extremely old zircons that were found in metasedimentary rock units had probably been eroded from old source terrains. These ancient zircon ages were tantalizing hints that there were much older Archean and possibly even Hadean rocks in the Superior Province. But where were these rocks? Had they been melted and recycled to form some of the granites or were they hidden in the high-grade gneiss terranes that were recycled preexisting crust? Some of the best answers to these questions

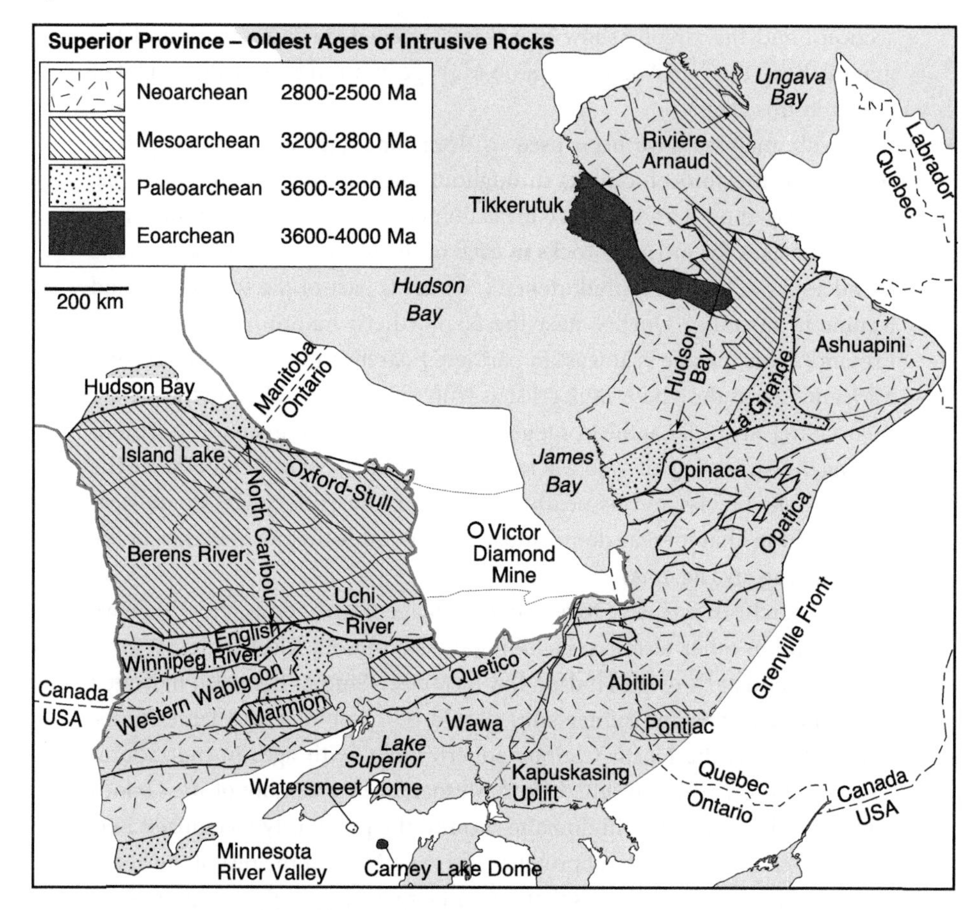

Figure 7.4. Ages of the oldest intrusive igneous rocks exposed today in Superior Province terranes. (From sources shown in figure 7.1 and Ayuso et al., 2018.)

came from isotopic age measurements and model ages obtained from the samarium-neodymium (Sm-Nd) and lutetium-hafnium (Lu-Hf) isotopic systems.

Nd model ages provide information on the age and complexity of the source region from which magmas are derived by partial melting (as well as eroded terranes from which clastic sediments are derived). The ages are referred to as "model ages" because they do not usually indicate the actual age of the igneous, metamorphic, or sedimentary rock. Instead they represent an estimate of the average age of the crustal source region from which the rocks or magmas were derived by partial melting.[13] Earth's crust and mantle have very different Nd isotope compositions. This means that Sm-Nd-model ages

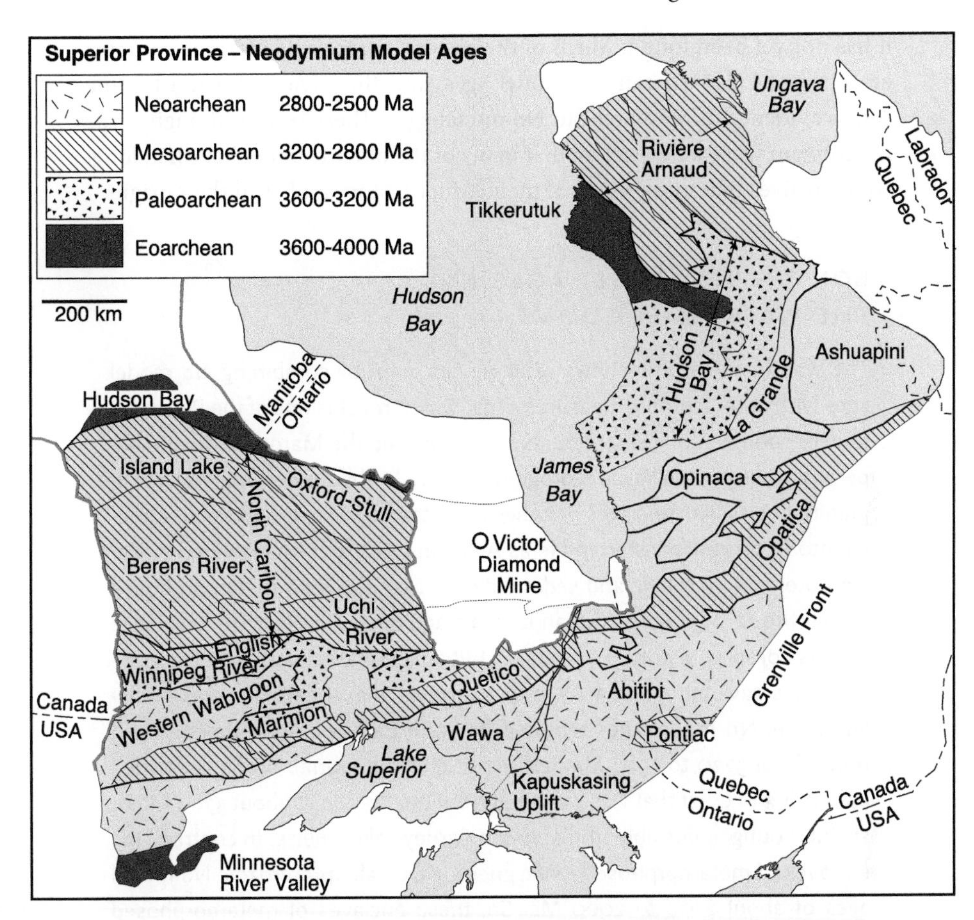

Figure 7.5. Neodymium model ages of intrusive igneous rocks exposed today in the Superior province. Note that Nd model ages are not available for some terranes. (From the sources listed in figure 7.1.)

of igneous rocks are very sensitive to the presence of older crust that may be hidden at depth or otherwise undetectable by mapping. The Watersmeet dome shows how these model ages can help unravel complex problems (box 7.2).

For granites the Nd model age is usually considered to be an estimate of the point in time when the rock that was melted to form the granite magma (possibly the lower crust) was separated from the mantle. As you can see in figure 7.5, Nd model ages of granitic rocks in the western Hudson Bay terrane indicate Eoarchean ages of 3600 to 4000 Ma, hundreds of millions of years older than the oldest zircon ages obtained so far. This suggests that much older crust was probably involved in the formation of these terranes, even though it has not yet been found. Much of the eastern Hudson Bay terrane contains evidence of Paleoarchean Nd model ages, and the Rivière Arnaud terrane shows evidence of Mesoarchean Nd model ages. These results, though cryptic, confirm what we expected: that many of the older terranes have had long, complex lives that mask much of their earlier history and even their origin.

BOX 7.2. ND MODEL AGES AND THE WATERSMEET DOME

The Watersmeet dome shows what we can learn by combining Nd model ages with other isotopic measurements. Watersmeet is one of the Paleoproterozoic dome-and-keel structures that deformed the Marquette Range Supergroup during the Yavapai orogeny (fig. 6.15). The dome consists now of granite gneiss with enclaves or zones of metamorphosed mafic gneiss folded into it. It probably started life as Archean crust that was covered by Paleoproterozoic volcanic and sedimentary rocks, and then metamorphosed during the Yavapai orogeny. Can isotopic age measurements find evidence for this complex history? For a start, U-Pb analysis of zircon from the granite gneiss gave an age of 3562 Ma, confirming that the granite was Archean in age. Nd model ages reported by Karin Barovich and others ranged from about 3720 to 3520 Ma, indicating that the granite probably formed by melting of crust that was extracted from the mantle at about 3700 Ma or slightly younger, possibly similar to the Carney Lake Gneiss. In contrast, the enclaves of metamorphosed mafic gneiss gave Paleoproterozoic Nd model ages of about 2100 to 2000 Ma. So, these enclaves of metamorphosed mafic gneiss probably started life as mafic volcanic rock (basalt) that came from melting of the mantle during early Paleoproterozoic time, and that was deposited on the Archean basement rocks during formation of the Animikie Superbasin. That left the question of how a Paleoproterozoic volcanic rock got itself enveloped in an Archean granite to make a complex granite gneiss with enclaves of mafic gneiss. The answer had to be that the mafic volcanic rocks were mixed together and metamorphosed with the Archean granite basement rocks during formation of the dome-and-keel structures. Proof of this was provided by 1750 Ma Yavapai ages that were obtained from these same rocks by Rb-Sr isotopic methods.

7.3. Formation of the Superior Province Was a Multistage Process

Now that we have information on types and ages of rocks and terranes, it's time to work out how they formed. How did the terranes themselves form and how did they amalgamate to form the Superior craton? We can start by assembling the basic building blocks of the craton, namely, the granite-greenstone and metasedimentary belts, and go on to look at how these blocks collided, amalgamated, and were uplifted and eroded to form the Superior craton.

7.3.1. The First Step Involved Formation of Granite-Greenstone Terranes

Formation of granite-greenstone belts is a three-part problem. First, we need to make greenstone belts, then we need to smash a bunch of them together to make a terrane, and finally we must figure out how to intrude the greenstone belts with a sea of granite. By far the best area to go to in search of answers is the Abitibi terrane (fig. 7.6), which has all the right qualifications. It is one of the largest granite-greenstone belts in the world, with an area of about 85,000 km^2, and it is less deformed and metamorphosed than most other terranes. The Abitibi terrane terminates on the west against the Kapuskasing uplift, but Abitibi-like rocks are also found to the west in the Wawa terrane (fig. 7.1) and the two are often considered to be a single, giant granite-greenstone terrane that was cut by the Kapuskasing uplift. The combined Abitibi-Wawa terrane also has some of the largest and most numerous gold and base metal ore deposits in the world, lending practical importance to efforts to understand its history, as discussed at the end of this chapter.

Much of what we know about the Abitibi greenstone belt comes from detailed studies by geologists of the Geological Survey of Canada and the Ontario Geological Survey, first summarized in their outstanding 1991 volume on the geology of Ontario and updated in numerous reports since then.[14] In northeastern Ontario, where the Abitibi has been studied in greatest detail, it turns out to be a composite of seven distinct volcanic complexes and two younger sedimentary sequences. The geologic map of this area by Phil Thurston and others shows seven different volcanic complexes with lengths of 10 to 100 km and thicknesses of up to 7 km (fig. 7.6).

These volcanic complexes formed over an impressively short period of time. John Ayer and Phil Thurston have shown in separate summaries that the entire suite of volcanic complexes in the Abitibi formed over a period of

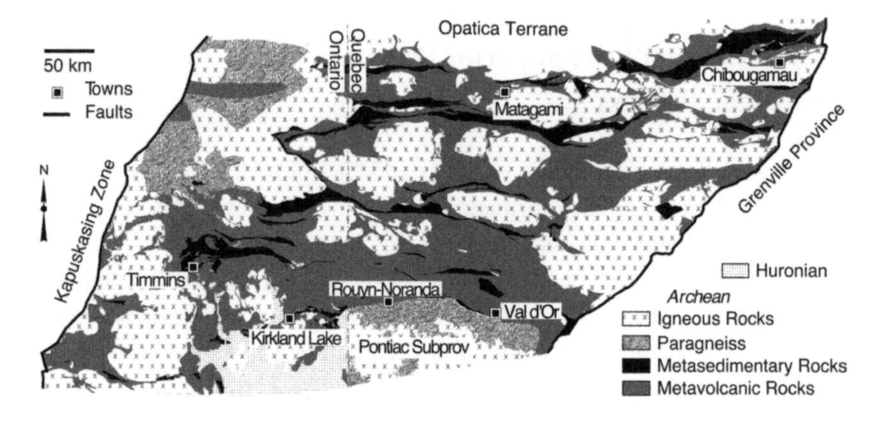

Figure 7.6. Generalized geologic map of the Abitibi granite-greenstone terrane showing the distribution of greenstone (metavolcanic) belts and granite, as well as the younger Pontiac metasedimentary rocks that were derived from erosion of the volcanic rocks. (Modified from Thurston et al. 2008.)

only about 50 million years, between 2747 and 2697 Ma, and that individual volcanic complexes formed over periods ranging from only 1 to as many as 15 million years. You might wonder, could Earth really have piled up so much volcanic rock in such a short period of time? Surprisingly, the answer appears to be yes. For instance, the modern Hawaiian Island chain, which extends about 500 km from Kauai to Hawaii, formed in only about 5 million years, and similar volumes of volcanic rock have been measured in young volcanic arcs of the western Pacific. In fact rates of Archean volcanism may have been even higher because Earth was considerably hotter. Evidence in support of this is seen in estimates made by Ayers and Thurston, who showed that, within individual volcanic cycles, about 10 percent of the actual time was take up by volcanism. The remaining 90 percent of the time involved deposition of sedimentary rocks, largely iron formations and volcanogenic massive sulfide deposits.

How did these volcanic complexes form and smash together to form greenstone belts? In particular, were the volcanic complexes formed by plate tectonic processes similar to the ones we know today or were they formed by uniquely Archean processes that have no direct modern analogue? Seismic surveys have provided some evidence for plate tectonics in the Superior Province. One of the most important salvos came in 2003 from D.J. White and others, who observed a slab of remnant material interpreted to be ocean

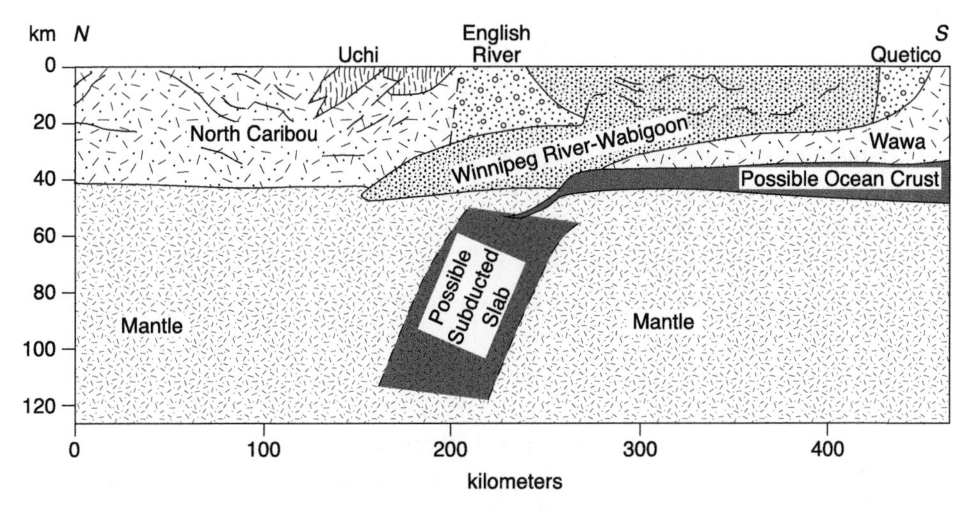

Figure 7.7. North-south cross section from the Wawa to the North Caribou terrane based on a seismic survey showing a slab interpreted to be ocean crust beneath the Wabigoon terrane. It could have been emplaced during subduction that formed the Abitibi-Wawa terrane. Rock units are labeled. (Modified from White et al. 2003.)

crust from the Abitibi-Wawa terrane immediately below the adjacent Wabigoon terrane to the north (fig. 7.7). Not everyone accepts this evidence, however (box 7.3); in the Superior Province, the strongest arguments against plate tectonics have come from Jean Bédard, who favors a process called "mantle wind" in which currents in the upper mantle push terranes around the surface. In this process, partial melting takes place at the base of volcanic piles rather than at subduction zones, a point we will revisit in the discussion of granites below.

BOX 7.3. WHEN DID PLATE TECTONICS START IN EARTH HISTORY?

Ever since plate tectonics was recognized in the 1970s, there has been a debate about when it began. Was plate tectonics active throughout most of Earth history or did it start at some later point? Current opinions for the start of plate tectonics range from about 4300 to 800 Ma, an impressively large range. These differing opinions are based on both field observations and theoretical calculations. Field observations show that modern terranes like the island of Cyprus consist of ocean crust and underlying mantle (known as ophiolites) that have been pushed up onto land (obducted), and we don't see rocks like these in Archean terranes. Archean terranes also lack metamorphic rocks that form at high pressures but low temperatures, such as might have slid into the mantle along subduction zones. However, these rocks could have been removed from Archean terranes by erosion. More disturbing is the abundance of intermediate volcanic rocks (andesites) in modern convergent-margin volcanic belts and their marked scarcity in Archean greenstone belts. Early theoretical calculations also suggested that ocean crust during Archean time was too thick and buoyant to subduct. Nevertheless, seismic studies are finding evidence for slabs of crust (fig. 7.7), which could have been subducted beneath Archean terranes, and revised theoretical studies have found ways in which a new "steep and deep" type of subduction could have moved lithospheric plates into the mantle. The book is far from closed on this controversy. In two recent reviews, Nick Roberts and his coworkers concluded that plate tectonics began as early as about 3200 Ma, but Phil Thurston favored the mantle wind hypothesis.[15]

Studies of individual Archean greenstone (volcanic) belts show that they have physical characteristics similar to several different types of modern volcanic complexes.[16] Some are volcanic plateaus, like the Hawaiian Island chain, that formed above mantle plumes; others look like volcanic arcs, such as the Aleutian Islands, that formed at convergent margins above subduction zones, and a third group even looks like volcanic rifts similar to the Midcontinent Rift. Furthermore, igneous rocks in these volcanic complexes can be divided into the same two compositional series, tholeitic and calc-alkaline, that we have seen in younger igneous rocks (box 5.5).[17] Archean volcanic plateaus contain abundant tholeiitic rocks similar to those in modern volcanic systems, and the Archean volcanic arcs are dominantly calc-alkaline, similar to those in modern volcanic rocks.

Archean volcanic rocks do differ from those in modern settings in one important way; they contain komatiites. Komatiites are unusually magnesium-rich volcanic rocks that form by means of melting in the mantle at temperatures of about 1600°C, conditions much hotter than those that form basalt magmas in the mantle.[18] They are found almost exclusively in Archean and early Proterozoic rocks, and they are best explained as a product of an early Earth that was considerably hotter than it is now and that would have partially melted mantle rocks at correspondingly higher temperatures. Komatiites are easy to recognize in the field because they contain magnesium-rich olivine (Mg_2SiO_4), which crystalizes into unusual radiating forms known as spinifex, after an Australian grass that has this form (fig. 7.2B in color section). Komatiites are most common in tholeiitic volcanic plateaus rather than the calc-alkaline volcanic arcs.

For those who favor Archean plate tectonics, these relations indicate that greenstone (volcanic) belts are a collage of volcanic complexes, with tholeiitic plateaus forming above plumes and calc-alkaline arcs forming at oceanic convergent-margin subduction zones. Figure 7.8A shows how the plateaus and arcs could have formed and then been pushed together by continuing plate tectonic movement. Plate motion would cause individual volcanic complexes to merge into larger, deformed complexes, gradually building a greenstone belt. Exactly how the mantle wind would do this remains poorly known, although the final result would obviously be the same.

Once we have a greenstone belt, we have to face the next question. How can we account for all the granite that pervades the greenstone belts? Ages compiled for the volcanic rocks and granites in the Abitibi Belt by Ben Frieman show that some granites were emplaced as much as 30 to 40 million years after the close of volcanism, indicating that granite formation was not part of the melting process that formed the volcanic rocks. What was melted to form all of this granite magma? It turns out that Earth's recipe for making granite is relatively simple. The only things you need are basalt, water, and heat. As basalt is heated in the presence of water, it begins to melt. But the first melt that forms does not have the same composition as basalt. Instead this early "partial melt" has a composition that is similar to the TTG granites described in box 7.1. (This should come as no surprise; partial melts always differ in composition from their parent rock, as when ultramafic mantle rocks partially melt to form mafic basalt magma.)

Finding a good recipe for granite is only half the job, however. We also need to identify a realistic geologic setting in which basalt can melt in the presence of water. One obvious possibility would be for heat from the mantle to melt the bottom of the volcanic complexes that make up the greenstone belts, perhaps as they sink into the mantle (fig. 7.8B). This process has sometimes been called sagduction because it acts like subduction but involves the sagging of small volumes of the crust rather than downward movement of a continuous slab.[19] However, there is not much water at the base of the crust, and the tholeiitic volcanic complexes that were formed by plumes have large amounts of komatiite rather than basalt (and komatiite will not partially melt to form granite). For that reason, many people favor the alternative possibility, namely, that the granite magmas formed in convergent margin settings. There the downgoing slab would have contained abundant water gained from

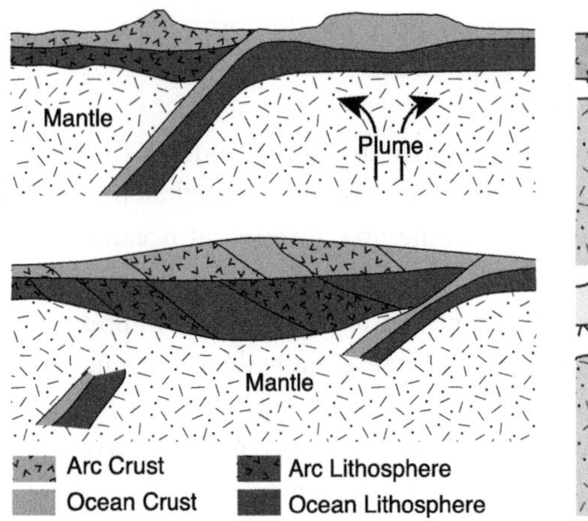

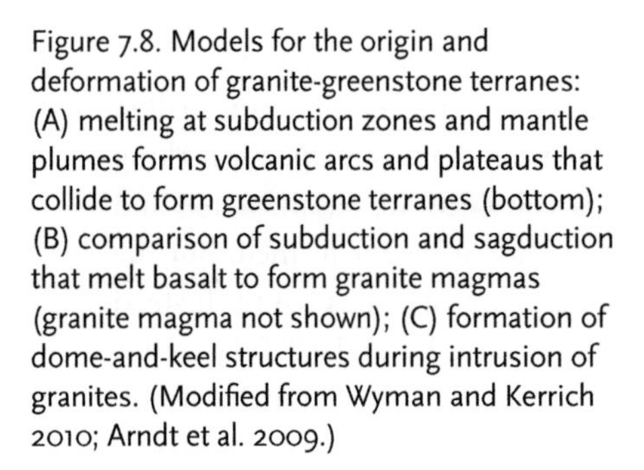

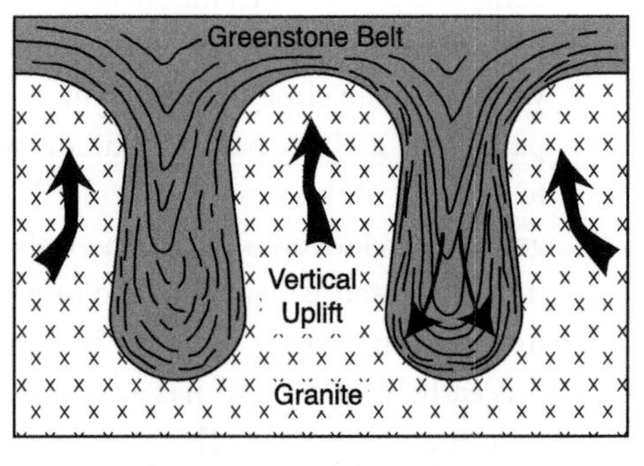

Figure 7.8. Models for the origin and deformation of granite-greenstone terranes: (A) melting at subduction zones and mantle plumes forms volcanic arcs and plateaus that collide to form greenstone terranes (bottom); (B) comparison of subduction and sagduction that melt basalt to form granite magmas (granite magma not shown); (C) formation of dome-and-keel structures during intrusion of granites. (Modified from Wyman and Kerrich 2010; Arndt et al. 2009.)

reactions between basalt magma and ocean water during mid-ocean ridge volcanism. As the slab descended, this water would have steamed off and moved upward into basaltic volcanic rocks in the overlying crust, causing partial melting to form granite magma.

However it formed, the volume of granitic magma was truly enormous, and it had a great impact on the ultimate form of the terranes. Because the granite magmas were less dense than the basaltic volcanic rocks, they rose into the upper part of the crust and shouldered the volcanic rocks aside. This produced the "dome-and-keel" structure that characterizes Archean granite-greenstone terranes. In this structure, the granites form domes and the supracrustal volcanic and sedimentary rocks (greenstones) form keels that consist of steeply dipping layers that extend downward (fig. 7.8C). We have encountered this type of structure before; in chapter 6, Paleoproterozoic sediments in the Marquette trough were compressed into vertical folds surrounded by domes of granite gneiss. Although the exact processes that formed the two terranes differ somewhat, they still reflect a rise of hot underlying rocks and magmas toward the surface.

7.3.2. Terranes Were Amalgamated to Form the Superior Province

We now face the "final question" of how and in what order the many terranes in the Superior Province converged into a single craton. Many parts of the Superior Province have been studied in only a general way. So the story might change. At present it looks something like figure 7.9, which is based largely on summaries by John Percival and his associates.

Before we start, here is an important reminder. Amalgamation of the terranes to form the Superior Province took place near the end of Archean time (Neoarchean time) at about 2700 Ma. But many of the terranes that were incorporated into the Superior Province, especially the Rivière Arnaud, Hudson Bay, and North Caribou, contain rocks that are almost a billion years older than that. These rocks have had a complex history, which clearly involved other, older amalgamation events that formed even older continen-

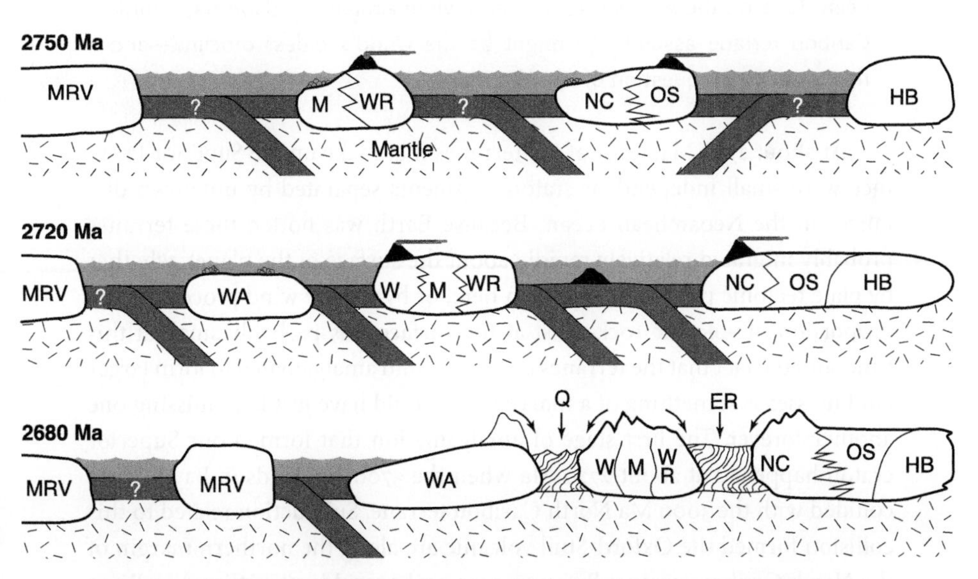

Figure 7.9. Cross sections of the western part of the Superior Province during late Archean amalgamation of terranes, the first step in forming cratons. Volcanism is shown above water for clarity, although most was submarine (as indicated by pillow forms). North is to the left. The ocean is gray. ER = English River, HB = Hudson Bay, M = Marmion, MRV = Minnesota River Valley, NC = North Caribou, OS = Oxford-Stull, Q = Quetico, W = Wabigoon, WA = Abitibi-Wawa, WR = Winnipeg River. Curved arrows show sediment being eroded into the metasedimentary belts. (Modified from Percival et al. 2006, 2012; Hammer et al. 2010.)

tal fragments. So, as we put these pieces together to form the Superior Province, don't forget that the pieces themselves hold even older histories (box 7.4).

BOX 7.4. ARCHEAN OROCLINES?

One of the interesting aspects of the Rivière Arnaud–Hudson Bay–North Caribou terrane assemblage is its change in orientation from east-west in Ontario to north-south in northern Quebec. Curved mountain belts, called oroclines, are found in many younger rocks. One of the best examples is the Andes Belt, which changes from a southwest trend in Peru to a south trend in Chile. Rob van der Voo and others have tried to determine whether these oroclines are originally straight mountain belts that were later bent or simply originally curved belts. Paleomagnetic studies in several Phanerozoic belts, notably the Cantabria-Asturias orocline in northern Spain described by Gabriel Gutierez-Alonso and others, have shown that later bending is required to account for their form. Although studies of this type have not yet been done on the Archean rocks, the Rivière Arnaud–Hudson Bay–North Caribou terrane assemblage might be the world's oldest orocline—and then there is the question of what bent it.

At about 2750 Ma, most of the terranes that make up the Superior Province were small independent craton fragments separated by unknown distances in the Neoarchean ocean. Because Earth was hotter, these terranes probably migrated relatively rapidly about the surface of the planet, whether by plate tectonic processes, as shown here, or by mantle wind processes. The proportion of volcanic rock crust at the surface was probably small at this time, and the fact that the terranes converged and amalgamated to form larger land masses is something of a marvel; they could have just kept missing one another forever. The first stage of amalgamation that formed our Superior craton happened at about 2720 Ma when the 3700 Ma Hudson Bay terrane collided with the 3000 Ma North Caribou terrane. Subduction related to this collision formed the Oxford-Stull volcanic arc along the northern margin of the North Caribou terrane. Between 2720 and 2700 Ma, the Winnipeg River terrane began to collide with the south side of the North Caribou terrane. North-directed subduction beneath the North Caribou terrane formed the Uchi volcanic belt, and zircon summaries by Ben Frieman and others show that sediments shed into the intervening basin formed the English River sedimentary belt. Intense metamorphism of the English River sediments happened when the North Caribou-Uchi and Winnipeg River terranes collided.

At about the same time, the western Wabigoon terrane began to collide with the south side of the Winnipeg River terrane.

Finally, between about 2700 and 2690 Ma, the Abitibi-Wawa terrane docked against the southern side of the newly formed craton, and sediments shed into the intervening basin formed the Quetico metasedimentary belt. The last event in the assembly involved collision of the Minnesota River Valley terrane, arriving from the south, an event that probably involved emplacement of the 2600 Ma Sacred Heart granite. The Archean Marshfield terrane in Wisconsin, discussed in chapter 6, was still out in the Archean ocean during all these events and did not show up in the area until Paleoproterozoic time.

The boundary between the Minnesota River Valley and Wawa-Abitibi terranes was part of the Paleoproterozoic story that we reviewed in chapter 6. If you look at figure 6.15, you can see that this boundary shows up in the domes of Archean basement rocks that poked through the Marquette Range Supergroup sediments, where it is called the Great Lakes Tectonic Zone (GLTZ). Domes of granite-greenstone terrane are found north of the GLTZ and domes of granite gneiss are found south of the GLTZ. Because granite gneiss of the Minnesota River Valley terrane was more easily deformed, it made the more prominent domes like the McGrath, Carney Lake and Watersmeet.

7.4. The Superior Province Was a Key Player in the Archean Supercontinent Story

Once the Superior craton formed, did it merge with other cratons to form supercratons, continents, and supercontinents? We already know the answer for the end of Archean time. The Superior craton joined Hearne, Karelia, Wyoming, and Kola to form the Superior supercraton (fig. 6.1), which then rifted apart, opening basins that host the Huronian and Marquette Range supergroups described in chapter 6. The question we have now is whether even older supercontinents or supercratons formed during earlier parts of Archean time.

As we move farther back in Archean time, the supercraton story begins to blur. There are fewer and fewer possible cratons and even fewer paleomagnetic measurements providing an estimate of their Archean history, and this has placed an even greater premium on geologic correlations. And that, in turn, focused attention on the best exposed and most thoroughly studied Archean cratons, the Superior of Canada, Transvaal of South Africa, and Pilbara of western Australia. Suggestions were made for a possible correlation between the Transvaal and Pilbara cratons in the 1970s, and Eric Cheney proposed that they were joined in a supercontinent known as Vaalbara. Subsequent paleomagnetic measurements have provided some support for Vaalbara,[20] although strictly speaking its small size only qualifies it as a supercraton like Superia. More recently, Ashley Gumsley, Wouter Bleeker, and others have shown that both the Transvaal and Pilbara cratons were probably part of Superia, with the Kaapvaal connected to both the Superior and Wyoming cratons and the Pilbara craton hanging off the end of the train connected only to the Kaapvaal craton.

Other older Archean supercratons that have been suggested include Sclavia, consisting of the Slave and Wyoming cratons of North America; the Dharwar craton of India; and the Zimbabwe craton, Zimgarn, consisting of the Zimbabwe and Yilgarn cratons. Something is wrong here. Unless the Wyoming and Zimbabwe cratons had Archean doppelgangers, they should not be part of more than one supercraton. There are at least two possible explanations for these anomalies. First, we are talking about an extended period of time throughout the Archean; cratons may have moved around the globe rapidly during that period, showing up in two places at different blinks of Archean time. Second, the histories of individual cratons over this period of time are still being worked out, and they might consist of more than one part that had different locations at different times.

A final part of the Archean craton mystery is whether any of these supercratons went on to form a continent or supercontinent, and that brings up the putative supercontinent known as Kenorland. Kenorland is so shrouded in mystery that it is not even clear if or when it might have formed. David Evans pointed out that various studies have proposed assembly dates of 2700 or 2500 Ma for Kenorland, with breakup dates sometime in the Paleoproterozoic. It might be that as we learn more about Superia, it will take the place of Kenorland. Or perhaps Kenorland was an even earlier supercraton.[21]

Regardless of all this uncertainty, it does appear that cratons were merging during Archean time, and this raises the possibility that one or more could have been large enough and in a sufficiently polar latitude to start a global glaciation event (although that might not have been needed in view of the faint sun problem). The Tayla Conglomerate in the Dharwar craton in southern India, which was recently described by Dick Ojakangas and his colleagues, is a tantalizing glimpse of what may have been one such event at about 2700 Ma.

7.5. Cratons Were Preserved Because They Are Underlain By Anomalously Thick Lithosphere

One remaining challenge is to explain why the Superior craton has lasted so long and been so little disturbed by later tectonic events. This stability was an early aspect of the Superior craton, as can be seen from the fact that it was not deformed significantly when craton fragments crashed into it during Paleoproterozoic Penokean, Yavapai, and other orogenies. One reason for its resiliency involves the thickness of the Superior craton crust, which ranges from about 33 to 43 km, but there is more.

Archean cratons such as Superior are directly underlain by a zone of mantle, called the subcontinental lithospheric mantle. This subcontinental lithospheric mantle is at least 200 km thick and is more buoyant and rigid than the mantle that underlies younger continents. This means that the Superior craton was effectively about 240 km thick and highly resistant to any deformation by a mere ocean or even continental plates migrating around the world. Most people agree that the subcontinental lithospheric mantle beneath Archean cratons is the residue left over from melting that formed the overlying crust.[22] But even the advocates of plate tectonic processes do not agree on where the melting took place. On one side of the plate tectonic debate are Nick Arndt and others, who suggest that the dominant location was plumes. On the other side are Hugh Rollinson and others, who advocate melting beneath hot mid-ocean ridges. The two possibilities are not that different; one is a point of rising mantle and the other is a line.

7.6. The Superior Province Contains Important Deposits of Metals and Diamonds

Mineral deposits are unusually numerous in the Superior Province and are the basis for local economies in most northern communities. According to a recent study at the University of Toronto, the average mine employs several hundred people and results in about five times that many jobs in surrounding communities. Many northern economic centers, including Timmins, Ontario, and Rouyn-Noranda, Quebec, began life as mining towns and still rely heavily on mineral exploration and production.

The most important mineral deposits in the Abitibi Belt are volcanogenic massive sulfide (VMS) copper-zinc deposits, orogenic gold deposits,

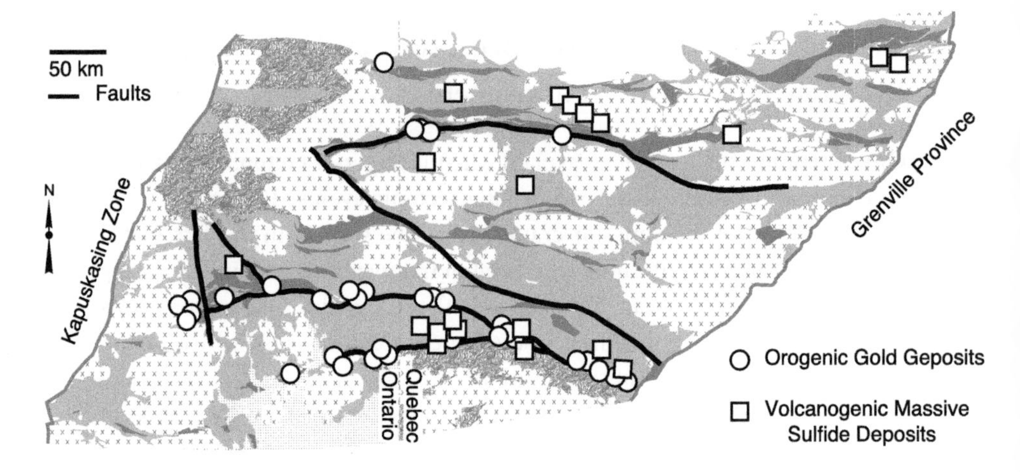

Figure 7.10. Location of Archean-age VMS and orogenic gold deposits in the Abitibi greenstone-granite terrane. (Modified from Dubé and Gosselin 2007; Gaboury and Pearson 2008.)

and kimberlite diamond deposits. Despite their tongue-twisting names, these deposits formed by relatively simple processes that are clearly related to the overall evolution of the Superior craton. The VMS and orogenic gold deposits are very abundant in the Abitibi Belt (fig. 7.10), and research on them has provided basic information on ore-forming processes that has aided mineral exploration in other Precambrian terranes around the world.

7.6.1. VMS Deposits Formed at Submarine Hot Springs

VMS deposits consist of copper, zinc, lead, and iron sulfide minerals that were deposited around hot springs on the ancient seafloor. Most submarine hot springs of this type are associated with volcanic rocks, which accounts for the term *volcanogenic* in VMS. The term *massive* is used because the ore consists entirely of sulfide minerals; it is not mixed with non-ore minerals that are common in many other types of ore deposits. This means that VMS deposits are among the richest ores on the planet (fig. 7.11A in color section). These deposits are very rich, but also relatively small. This makes them difficult to find, but it does allow more compact, environmentally friendly mining.

Archean VMS deposits are the fossil remains of systems very similar to the black smoker hot springs found at modern mid-ocean ridges.[23] In older rocks, mounds of sulfide deposited by the hot springs have been solidified and preserved under later lava flows to become today's VMS ore deposits. VMS deposits have been found in rocks of all ages, but they are particularly abundant in Archean greenstone belts, possibly because the ocean crust was hotter then. The Abitibi terrane contains a large number of deposits (fig. 7.10), including two of the largest: Kidd Creek near Timmins, Ontario, with more than 12 million tons of copper + lead + zinc (box 7.5); and Horne, with 2.5 million tons of metal, near Rouyn-Noranda, Quebec. Because supracrustals in Archean terranes are strongly folded by the dome-and-keel structure, most VMS deposits are tilted so that the originally horizontal strata are strongly inclined. Kidd Creek, for instance, is almost vertical, and has been mined downward to a depth of 9600 feet, making it one of the deepest mines in the world. VMS deposits have been a major target for mineral exploration in the Canadian Shield because they are conductive and can be detected by airborne electromagnetic geophysical surveys, which can "see through" the glacial deposits.

BOX 7.5. A MAJOR DISCOVERY!

The discovery of Kidd Creek in 1964 was one of the most exciting events in the long history of mineral exploration in Canada, both because the deposit was so huge and because the discovery involved all sorts of chicanery. For a start, officers of the Texas Gulf Sulfur Company loaded up on shares of the company right before announcing their discovery, thus profiting from a 400 percent increase in the share price. This led to a protracted investigation and court case involving the US Securities and Exchange Commission and the adoption of much tighter rules about insider trading. In the second case, well-known Canadian explorers George and Viola MacMillan set up exploration drills along the border of the Texas Gulf property and made misleading press releases. The price of shares in their company, Windfall Oil and Mines, rose by more than ten times before crashing when it became clear that worthless rock was being drilled. The resulting investigation by the Ontario Securities Commission improved standards for exploration financing. Although George died before a reckoning, Viola served a short prison sentence but went on to become a respected philanthropist and to receive the Order of Canada. Much of this is wonderfully described by Morton Shulman in his book *Billion-Dollar Windfall*, a must read for anyone interested in the history of minerals, exploration, and the northland.

7.6.2. Orogenic Gold Deposits Formed During Metamorphism of the Greenstone Belts

Orgenic gold deposits consist largely of gold and quartz in sheeted veins that cut metamorphosed volcanic (and to a lesser extent, metasedimentary and plutonic) rocks (fig. 7.11B in color section).[24] The deposits get their name from the fact that they formed during the orogenesis (deformation and metamorphism) that accompanied amalgamation of greenstone-granite terranes. (These events are much younger than the VMS deposits, which formed during volcanism that built the greenstone belts.) The gold-bearing veins occupy subsidiary faults, or splays, near large terrane-scale faults, and they are unusually widespread in the southern part of the Abitibi Belt (fig. 7.10).[25] Gold was deposited in the faults by hydrothermal solutions (hot water with dissolved CO_2) released by deeply buried volcanic rocks as they were being metamorphosed (as metamorphism progresses, it drives water and CO_2 out of low-temperature clay and carbonate minerals). Most veins are surrounded by carbonate minerals, including calcite, dolomite, and siderite ($FeCO_3$), which formed when CO_2 in the hydrothermal solutions reacted with the wallrocks.

Numerous observations confirm that the gold-bearing hydrothermal solutions generated pressures greater than the pressure of overlying rocks (fig. 7.11C in color section) and thus were able to push open the faults they passed through (much like fracking fluids in oil and gas production). Gold was deposited in the faults when the hydrothermal solution reacted with iron-bearing volcanic wallrocks or when it boiled to release CO_2 (fig. 7.11D in color section).

Large orogenic gold deposits have vertical extents of up to 2000 m and contain millions of ounces of gold. One of the largest deposits in the Superior Province, the Hollinger-McIntyre near Timmins, produced about 32 million ounces of gold worth around 40 billion dollars in 2017 prices. Other large deposits in the Abitibi greenstone belt include the Dome, Kirkland Lake, Sigma-Lamaque, and Kerr-Addison, all with production of more than 10 million ounces of gold. Large orogenic gold deposits are not as abundant elsewhere in the Superior Province, although the Campbell–Red Lake mine in the North Caribou terrane has produced at least 25 million ounces.

7.6.3. Kimberlites and Their Diamonds Were Intruded from the Mantle

Diamond mines are new to the Superior craton and nearby parts of northern Canada; the first deposit began production only in 1998, as outlined in the interesting history written by Bruce Kjarsgaard and Al Levinson. People had suspected that diamond deposits were present in Canada for many years, and their appetites had been whetted by diamonds found in glacial gravels (which must have been scraped from concealed deposits by the glaciers). Most of the diamond mines active in Canada today are in the Slave Province (fig. 6.17), but several kimberlite fields have been found in the northern part of the

Superior Province, especially in and near the James Bay Lowlands. The first of these kimberlites to be mined is the Victor mine (fig. 7.11E in color section).

Diamond deposits like the Victor are found in curious igneous rocks known as kimberlites, which consist largely of olivine and other minerals typical of Earth's mantle. Kimberlites commonly form cylindrical and dikelike intrusions that contain fragments of mantle rock ranging from boulders to powder in a matrix rich in olivine and other mantle minerals. Some kimberlite intrusions even reached the surface where they scattered a ring of debris. Some, but not all, kimberlites contain small amounts of diamonds. Diamonds are simply carbon in an unusual crystal form that is only stable at the high pressures found at depths of more than about 150 km in the mantle (fig. 7.10F in color section). At Earth's surface, diamonds should change to graphite, the stable form of carbon at low pressure, but this does not happen as long as the diamonds are not heated for a prolonged period. The fact that many diamonds are found in kimberlites near the present-day surface has been interpreted to mean that the kimberlites were intruded into the upper crust very rapidly, possibly within hours or days.

Kimberley Webb and others who worked on the Victor deposit have shown that its host kimberlite intrusion is not Archean in age; it is only 170 million years old (Jurassic), as discussed in chapter 5. Although most of the kimberlite intrusions in the Lake Superior region are relatively young, they are most abundant in areas with thick Archean crust and subcontinental lithospheric mantle.[26] The problem for diamond explorers is that there are many kimberlites in Archean terranes and only a few of them contain enough diamonds to be mined. Apparently, as indicated recently by Stephane Faure and others, the diamond-bearing kimberlites form only in areas that overlie edges of the keels of lithosphere that extend deep into the mantle. The younger age of Superior Province kimberlites probably reflects incipient but largely unsuccessful rifting during Mesozoic time.

7.7. Evidence for Archean Life Is Found in Stromatolites and Microfossils

The Superior Province has been an important part of the continuing search for evidence of Archean life. William Schopf and others have summarized the case for fossils that might represent Archean life and divided them into two categories: stromatolites, which we first encountered in chapter 4; and microfossils. The Superior Province has both.

The best-known stromatolites in the Superior Province are in the Marmion terrane near Steep Rock Lake, Ontario. These stromatolites take several forms, including giant domes almost 15 m high (fig. 7.12 in color section). They are found in the ~2800 Ma Steep Rock Group, a sequence of shallow shelf sediments that includes iron formation and the Mosher Limestone, one of the oldest limestones in the world and the host for the stromatolites. Philip Fralick and Robert Riding concluded that the Steep Rock sediments represent an oxygen oasis (box 7.6) that was isolated from the anoxic Archean ocean and atmosphere. Abundant oxygen in the oasis formed by cyanobacteria would have produced dissolved carbonate (the oxidized form of carbon), which would have facilitated deposition of limestones. Older stromatolites are found in other Archean passive-margin shelf sequences, especially in South Africa and Australia. Among the oldest of these is the 3430 Ma stromatolite reef in the Pilbara craton in western Australia, which was described by Abigail Allwood and others.

BOX 7.6. OXYGEN OASES

Although Earth's atmosphere was reducing prior to the Great Oxidation Event, discussed in chapter 6, there may have been special places or times when oxygen was plentiful. We know that anomalous environments can form on Earth today. Even though the present-day atmosphere is oxygen rich, there are plenty of anoxic (low-oxygen) lakes and even parts of the ocean where the available oxygen has been consumed. So maybe the reverse happened in the Archean ocean and locally plentiful photosynthetic cyanobacteria produced enough oxygen in isolated areas to overwhelm the reducing gases dissolved in the water. These areas would become oxygen oases, and they are the best theory we have to account for the presence of limestones and some iron formations that require oxygen to form. Stephanie Olson, Lee Kump, and James Kasting investigated this possibility theoretically and concluded that "oxygen oases are an expected consequence of oxygenic photosynthesis beneath an essentially O_2-devoid atmosphere" It has also been suggested that a high enough level of oxygen production might have allowed an oasis to expand and include the entire planet for at least a short period of time, thus forming transient oxygen-rich periods during the anoxic Archean.

The Archean microfossil story is confused by a lack of consensus about what exactly constitutes a microfossil. The most recent controversy began in 2004 when Harald Furnes, Neil Banerjee, and others reported the presence of microscopic filaments of titanite (a calcium titanium mineral) in pillow basalts of the 3470 Ma Barberton Greenstone Belt in South Africa. Later, Nathan Bridge along with Banerjee and others reported similar features in volcanic rocks from the Blake River Group, one of the volcanic units in the Abitibi Belt (fig. 7.6). It was proposed that these filaments are the remains of small tunnels left by microorganisms that burrowed through the outer part of the pillow basalts after they were emplaced. Early enthusiasm for these features has been dampened by ultradetailed analytical studies, particularly those reported by Eugene Grosch, Nicola McLaughlin, and their associates, which find that they lack chemical or isotopic traces of life.

Despite this setback, interest in Archean microfossils is very strong. Several localities, especially in the ~3000 Ma Pilbara craton in western Australia, have yielded material that is more widely accepted. Kathleen Grey and Kenichiro Sugitari have reported the presence of spheres and filaments from the Farrel Chert in the Pilbara. And Greg Retallack and others followed with a report that claims to show evidence for microfossils in a paleosol in the Farrel Chert, which would be the first evidence for life of this age in a nonmarine setting. The Superior Province got back into the story when possibly Hadean microfossils were reported from Quebec, as discussed in the next chapter.

7.8. Did It All Begin in the Archean . . . or the Hadean?

Clearly, things were pretty inhospitable at the beginning of the Archean. The atmosphere was anoxic, there were few continents, and volcanism was very active and unusually hot. But was that the dawn of cratons? Or did cratons and maybe even life actually start in the Hadean? There is growing interest in that possibility. So we should take a look at what has been learned about this long-ago part of Earth's history and about how the Superior Province has contributed to the investigation.

Notes

1. Strictly speaking, Superior craton (or supercraton) refers to the entire block of crust, and Superior Province refers to that part of it that is exposed (not covered with younger rocks).

2. At the same time, US geologists, including Frank Grout, Charles van Hise, and C. K. Leith, were also puzzling over their smaller piece of the Superior Province.

3. Even the lightly metamorphosed rocks have undergone at least some metamorphism and, strictly speaking, should be referred to as metavolcanic or metasedimentary. Many geologists ignore this, and it is not uncommon to hear someone talk about Archean rocks as if they were unmetamorphosed and formed yesterday.

4. Compositions of igneous rocks were reviewed in chapter 5. Most igneous rocks

are silicates that are made up mostly of silicon, aluminum, and oxygen. Mafic silicate rocks like basalt contain relatively large proportions of iron and magnesium, whereas felsic rocks like granite contain sodium and potassium.

5. Banded iron formation in volcanic environments is commonly referred to as Algoma type to distinguish it from the thicker and more extensive Superior-type (named for Lake Superior) banded iron formation that is found in Proterozoic passive-margin shelf environments (Gourcerol et al. 2016). Where Algoma-type iron formations are found on shelf environments, the shelves are much smaller, as indicated by a shorter distance between shallow-water clastic sediments and deep-water sediments.

6. Gneiss is a metamorphic rock, usually consisting of quartz, feldspar, and mica, with a poorly defined layered or foliated structure. *Foliation* refers to the planar alignment of minerals.

7. Despite the dominance of granites in granite-greenstone terranes, they are sometimes referred to simply as greenstone belts. Strictly speaking, this term should be reserved for belts consisting entirely of metavolcanic and related sedimentary rocks, with no granites, although this not always done. For a comprehensive review of greenstone belts and granite-greenstone terranes, see Thurston (2015).

8. See also Thurston and Breaks (1978). Even in terranes dominated by young, intrusive granites, evidence from zircon age measurements and Nd-model ages suggests that older granites were present before the last stage of granite intrusions.

9. These deep zones were uplifted by a complex combination of processes involving erosion and faulting over an extended period. See Percival and West (1994) for a summary of the structural history of the Kapuskasing zone.

10. In looking for the oldest zircon ages in an igneous rock, we must be careful to avoid inherited zircons (xenocrysts). When magmas are forming, some zircons can be transferred from the source that is melting to the new magma. Because of its high closure temperature, these zircons retain the age of the older source rock.

11. The Minnesota River Valley terrane also includes the Benson block that contains ~2600 Ma granite and gneiss (Schmitz et al., 2006). The ± symbol used for these ages indicates the analytical uncertainty for the analysis. Because it is so small for most zircon U-Pb analyses, it is often omitted. We have followed that practice here.

12. The Carney Lake dome is probably an extension of the Minnesota River Valley (MRV) terrane, suggesting that Eocambrian rocks might be present in the MRV and Watersmeet dome as well.

13. According to James Gleason, Nd model ages used in Precambrian studies, are derived from samarium-neodymium isotope analysis. Samarium has a naturally occurring radioactive isotope, ^{147}Sm, which decays to ^{143}Nd with a half-life of 106 billion years. All naturally occurring isotopes of neodymium are stable, but because ^{143}Nd is radiogenic, its abundance changes with time through radioisotopic decay. Thus, ratios of radiogenic ^{143}Nd to nonradiogenic ^{144}Nd ($^{143}Nd/^{144}Nd$ ratio) will change through time. Because the ratio changes very little with time (because of the long half-life) other isotopic systems are used more commonly for geologic age measurements. However, Nd-isotope measurements are widely used in isotope tracer studies and for determining model ages (known as "crust formation" ages). Sm-Nd model ages are calculated in a three-step process. For igneous rocks, the first step is to obtain the $^{143}Nd/^{144}Nd$ ratio of a representative sample using isotope ratio mass spectrometry. The next step is to use the half-life of ^{147}Sm to calculate a line showing how the $^{143}Nd/^{144}Nd$ ratio would change

backward through time (considered to be how the $^{143}Nd/^{144}Nd$ ratio has changed since the Nd was removed from the mantle and entered the crust). This line is then compared to two theoretical lines that show the change through time of theoretical mantle source reservoirs for the crust. For igneous rocks, the most commonly used theoretical lines are CHUR (undifferentiated mantle) and DM (differentiated mantle). The age in the past at which the calculated sample line intersects either CHUR or DM is known as the Nd model age, and it gives an average age for when the crust that melted to form the magma originally separated from the mantle (represented by CHUR or DM). For basalts that are derived by direct partial melting of the mantle, the model age will be close to the actual crystallization age (because the evolution lines intersect at the time the basalt formed). For many igneous, metamorphic, and sedimentary rocks, the Nd model age is commonly much older than the crystallization, formation, or depositional ages, respectively. In some cases, the theoretical curves do not intersect, or yield an impossible age (e.g., older than the age of the Earth). In these cases, disturbed Sm/Nd ratios in the sample are usually implicated. For this reason, model ages should be used with caution. If you want to read more about this method, and its interpretation, see Bennett and DePaolo (1987) or Arndt and Goldstein (1987).

14. These more recent contributions were based on earlier work by Morley Wilson, Willet Miller, H. C. Gunning, J. W. Ambrose, F. F. Grout, Frank Pettijohn, and many others.

15. Additional information on the history of plate tectonics and its relation to the Archean can be found in Langford and Morin (1976); Poulsen et al. (1992); Davies (1992); Condie and Kröner (2008); Stern (2008, 2013); Thurston (2015); Hawkesworth et al. (2016).

16. Individual volcanic complexes in the Abitibi-Wawa Belt differ considerably. The Shebandowan belt has a large proportion of intermediate rocks and probably formed at least in part by subduction-related processes (Lodge 2016). The Blake River group is a plateau-caldera system similar to Hawaii with a principal caldera almost as large as the Olympic Mons caldera on Mars; it formed at least partly over a plume (Pearson and Daigneault, 2009). See also Polat (2009); Wyman and Kerrich (2010); Thurston (2015).

17. A third, less common, magma series known variously as alkalic, alkaline, or peralkaline, contains enough sodium and potassium to combine with most of the silica in the magma, producing felsic rocks that have little or no quartz. Rocks of this type are found in the Kirkland Lake area of the Abitibi Belt.

18. Experiments indicate that komatiite magmas form by partial (about 50%) melting of mantle material at temperatures of about 1600°C. This compares to basalts, which are produced at temperatures of only about 1100 to 1250°C (Arndt et al., 2008; Mole et al., 2014).

19. See Johnson et al. (2016) for a discussion of sagduction related to mafic and ultramafic rocks.

20. See Wingate (1998) and Strik et al. (2003) for contrasting paleomagnetic tests of the Vaalbara hypothesis.

21. Grant Young (2015) has suggested that the breakup of Kenorland happened even later, shortly before 2100 Ma, and that uplift associated with this breakup accounts for the Great Stratigraphic Gap discussed in chapter 6.

22. Characteristics of the subcontinental lithospheric mantle beneath Archean cra-

tons have been studied in two ways. First, seismic surveys show that it is less dense than mantle that underlies younger continents. Second, pieces of the subcontinental lithospheric mantle have been brought up as fragments by magmas rising from the mantle to form volcanoes, and studies of these fragments (called xenoliths) show that they consist largely of the mineral olivine. Olivine is hard to melt and would be left behind if the mantle underwent some melting. This suggests that the olivine-rich subcontinental lithospheric mantle is the residue from partial melting that formed the volcanic and maybe some of the intrusive rocks that make up the overlying craton. Support for this interpretation comes from the fact that the olivine-rich xenoliths are the same age as the Archean volcanic rocks; in other words, the partial melting that formed them coincided with Archean volcanism. See Griffin et al. (2009) for a recent review.

23. Modern VMS-forming submarine hot springs were discovered in 1977 along the spreading ridge in the Atlantic Ocean. The hot springs flow onto the seafloor at temperatures of about 350°C at depths great enough to prevent boiling. The hot water cools quickly and precipitates small grains of sulfide minerals. The sulfide grains look like black smoke, which led to the appellation "black smokers." The small grains of sulfide minerals agglomerate into chimneys and other forms that collapse into a mound of sulfide minerals. The water that flows out of the vents is largely seawater that was heated by volcanic and intrusive rocks at depth and leached metals along its path. The hot spring vents are surrounded by curious organisms that metabolize hydrogen sulfide rather than oxygen, and fossils of these creatures have been found in ancient deposits. Images of the modern hydrothermal vents and biota are available on the web sites of oceanographic institutions, especially Woods Hole in Massachusetts (http://www.whoi.edu/main/topic/hydrothermal-vents), and Deborah Kelley has written a readable short summary of their geology and biology (2001). For more information, see Arndt et al. (2015).

24. The terms *ribboned* and *sheeted* refer to the striped appearance of the veins, which consist of numerous parallel white quartz veins separated by thin layers of dark wallrock. These are thought to result from repeated fracturing events caused by hydrothermal fluids that are injected into the cracks in the wallrock, causing it to split. Injection happens because the fluids have been isolated by fault movement and have developed high pressures that exceed the pressure of surrounding wallrocks (overpressured fluids).

25. Because they are found largely in greenstone belts, they are sometimes referred to as greenstone gold deposits. See Arndt et al. (2015) for additional information on the origin of orogenic gold deposits. Goldfarb et al. (2005) provides a more detailed summary of the nature and origin of orogenic gold deposits, and Robert et al. (2005) discusses their setting in the Superior Province.

26. Recent exploration has located kimberlite fields in the Pontiac and Opatica-Opinaca terranes, as well as farther south near Wawa and north of Lake Superior. The Renard kimberlite in northern Quebec, which is also being mined, has an age of about 640 my. Pearson and Wittig (2008) discuss the formation of Archean continental lithosphere and its relation to diamonds and kimberlites. See Kesler and Simon (2016) for the geology of diamond-bearing kimberlites.

CHAPTER 8

Making the Crust

Solidification of the Hadean Magma Ocean

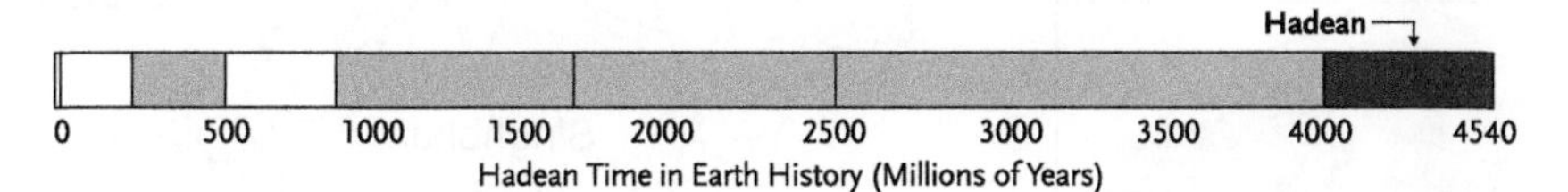

8.1. New Rock Evidence and New Controversies Have Focused Attention on the Hadean

The Hadean Eon endured for 500 million years, starting when Earth formed at 4540 Ma and continuing to 4000 Ma, the start of the Archean Eon. As the name implies, the Hadean was an unwelcoming destination for our time traveler. It was so tough, in fact, that early opinion among many geologists was that there were no Hadean rocks to be found. If they ever formed, they would have been obliterated by later events. As people looked more carefully, however, evidence of Earth's Hadean history began to turn up. The increasing rate of Hadean finds is shown by the number of geologic research papers with Hadean in the title, which increased from 16 in the 1990s, to 119 in the 2000s and more than 200 by 2018.

Some of the most important recent Hadean finds have been in Canada, including our old friend the Superior Province, and they are leading to new ideas about Hadean history. Other aspects of the Hadean story, including the origin of the Moon and life itself, are also getting a fresh look.

8.2. Hadean Rocks Have Been Found on Earth, but Lots More Are on the Moon

The first discovery of a Hadean rock on Earth was made by Samuel Bowring and Ian Williams who were working in the Slave craton in northern Canada (fig. 8.1). The Slave craton is one of the continental fragments that merged

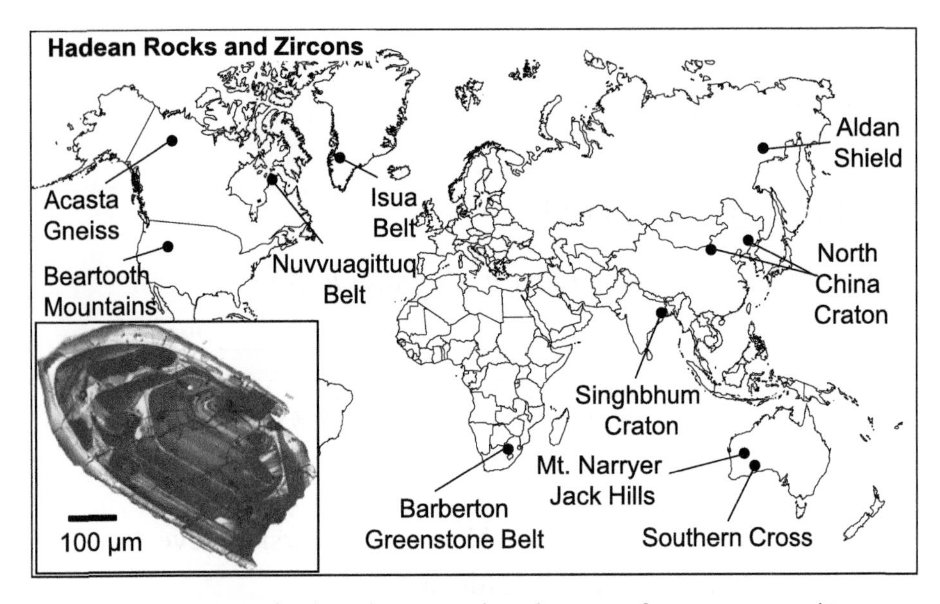

Figure 8.1. Location of rocks and zircons with Hadean ages (from sources noted in the text and Iizuka et al. 2006; Zhang et al. 2014; Glukhovskii et al. 2017; Miller et al. 2017.) Inset shows a grain of zircon from Jack Hills, Australia, Earth's oldest mineral (Courtesy of John Valley.)

with the Superior craton to form Laurentia (fig. 6.17), and it contains several rock units that appear to be especially old. One of these, the Acasta Gneiss Complex, contains zircons with ages ranging from about 3600 to 4031 Ma. The rocks containing these old zircons are metamorphosed granitic intrusions, and the zircon ages represent several different intrusive events, as described by Ann Bauer and others. The oldest zircons just squeaked into the Hadean. Although it was almost 500 million years younger than the birth of the planet, the Acasta Gneiss achieved the title "Earth's oldest rock." Even so, a lot had happened before the Acasta Gneiss formed; Robert Emu and others have reported that the gneiss contains evidence of evolved crust that had probably passed through more than one crust-forming event.

The Superior Province got into the oldest rock game with a recent discovery in the Nuvvuagittuq Supracrustal Belt (NSB), located in the Tikkerutuk terrane, part of the Superior Province in northeastern Quebec (fig. 8.1). The NSB consists largely of metamorphosed basalts and granites and probably constitutes an ancient granite-greenstone terrane. Jonathan O'Neil and others measured an Nd model age of 4280 Ma for the basalts. This would mark the NSB as Earth's oldest rock if it represented the time that the basalt was extruded and cooled, although there is considerable controversy about this age.[1] Despite this dispute, the northern part of the Superior Province looks like good hunting ground for very old rocks and we will likely hear more about it.

If you want to find really old rocks, however, you must go to the Moon, where the search has been more successful. This might be surprising in view of the relatively limited "fieldwork" done there by the Apollo astronauts. Success was more likely, however, because the Moon did not have a plate tectonic history that disturbed its early crust. Instead its crust formed by crystallization of molten rock that originally formed a magma ocean on the lunar surface; this crust was disturbed by later (though still very old) volcanism and impact cratering. The oldest rock found so far on the Moon appears to have formed during crystallization of its magma ocean, and it makes up much of the lunar highlands. Marc Norman and his colleagues reported an age of 4460 Ma for this rock. This age is about 100 million years younger than the age of Earth, and it raises the question of what the Moon was doing during these early days.

Although the lack of plate tectonics has made the last 4 billion years of Moon history pretty boring, its early years were anything but. A widely accepted theory for the origin of the Moon involves a large impactor, or bolide, which hit Earth shortly after it formed. Originally, it was thought that the Moon consisted at least in part of fragments of the bolide, but Alex Halliday showed that the Moon has a composition similar to that of Earth's mantle and therefore must include pieces of Earth.[2]

The search for old rocks continues among lunar samples returned by the Apollo astronauts as well as on Earth. However, the real interest now is in old minerals. Because minerals are smaller, they have a greater likelihood of being preserved. Not just any mineral can be used, however. It must contain radioactive isotopes that can yield an isotopic age, and it must be highly resistant to later change. The winner, of course, is zircon!

8.3. The List of Locations on Earth with Hadean Zircons Continues to Grow

One of the first age-related surprises provided by zircons came from the sensitive high-resolution ion microprobe (SHRIMP),[3] a huge microanalytical device at the Australian National University. In 1983 Derek Froude, working with a group that included Bill Compston, who built the SHRIMP, reported ages of 4100 to 4200 Ma for zircons from Mt. Narryer in the Yilgarn craton of western Australia (fig. 8.1). They went on to find even more very old zircons at a nearby area called the Jack Hills, and in 2001 Simon Wilde, working with John Valley, reported an amazingly old Hadean age of 4374 Ma for one of the zircons. These are the oldest natural materials known on Earth. Both the Mt. Narryer and Jack Hills areas are part of the Yilgarn craton, which, as you will recall from the last chapter, has been linked to the Superior craton in some Archean supercontinent reconstructions.[4]

These old zircons stimulated a search for Hadean zircons elsewhere in the world, a project that was aided by development of laser-based analytical methods that were faster and simpler than SHRIMP-based methods. As analyses accumulated, Hadean-age zircons turned out to be widespread. Stephen Wyche and others found zircons as old as 4350 Ma in the nearby Southern Cross granite-greenstone terrane in Australia (fig. 8.1). Then, Tsuyoshi Iizuka and others reported 4200 Ma zircons from the Acasta Gneiss in the Slave craton. Analisa Maier and others found similar Hadean zircons in the Beartooth Mountains in the western United States and suggested that they might be part of the same terrane that contained the Acasta Gneiss. Shortly afterward two studies, one headed by Yuansheng Geng and another by Pei-Long Cui, reported Hadean zircons in the North China craton. By conservative estimate, that meant Hadean crust is present in three (now) widely separated parts of the world. As you can see in figure 8.1, the list has continued to grow and now includes parts of Siberia and India.

The growing list suggests that zircons were an important part of the Hadean crust, and this makes it all the more important to come up with a realistic geologic process that can form them. We need to remind ourselves that almost all these old zircons are detrital grains in younger sediments. So their host sedimentary rocks cannot tell us anything about the origin of the zircons. Instead we must look at the zircon grains themselves, and it turns out that they have a lot to tell us. The bottom line, as indicated in the early studies of Roland Maas and Malcolm McCulloch and corroborated by lots of later work, is that most of the zircons grew in a granite magma. This is big news because granites form in continental crust,[5] and their presence seems to be telling us that continental crust existed during Hadean time. But granite crust with zircons presents a problem because the Hadean is supposed to have begun with a mafic magma ocean of probable basalt composition. How can we make a granite crust in a world that is overwhelmingly basalt and should have crystalized to form a basalt crust?

8.4. Hadean Granite Magmas Could Have Formed by Meteorite Impacts, Subduction, or Sagduction

We already know that you can make granite magma by melting basalt in the presence of water. So once the magma ocean began to develop a basalt crust, it might have been possible to melt that crust to form granite magma, which would have been a first step in making a crust that is more complex than just basalt.

In the previous chapter, we reviewed two possible ways to melt granite, subduction and sagduction. Both were possible in the Hadean, but a third possible mechanism needs to be considered: melting caused by meteorite impacts. Impact melting is on the list because the inner Solar System underwent intense meteorite bombardment during the first billion years of its history, with a spike in activity during an especially intense phase at about 3900 Ma called the Late Heavy Bombardment.[6] Simone Marchi, Oleg Abramov, and others have shown that this bombardment reprocessed Earth's surface, forming large impact melt sheets (like the one at Sudbury), which buried and mixed with preexisting crust. Gavin Kenny and others suggested that these melt sheets could have formed granites with zircons during late stages of progressive crystallization, much like the crystallization process that formed late granite at the top of the Sudbury Igneous Complex.[7]

Whereas impact melting could have taken place whenever a meteorite struck solid basalt crust, subduction and sagduction require that the basalt crust sink into hotter parts of the upper Earth. Mark Harrison reviewed the factors that favor sinking and pointed out that a critical requirement is development of density differences in the Hadean crust. Subduction would take advantage of these density differences by causing a plate of crust to sink into the mantle, but how would sagduction work? Balz Kamber, Tony Kemp, and their coworkers have proposed ways to do this. They start with formation of a large volcanic plateau on the crust, which then sags downward, initiating melting (fig. 8.2). For both sagduction and subduction, melting would probably have happened at relatively shallow depths because the planet was hotter.[8]

Regardless of which mechanism we choose, they all need water to be involved in the melting process, as we saw in chapter 7. An environment as hot as the Hadean might not seem like a good place to find enough water to do the job, so that's where we should look next.

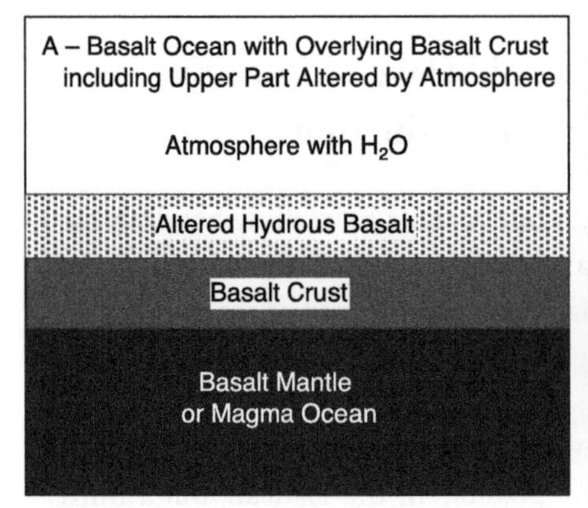

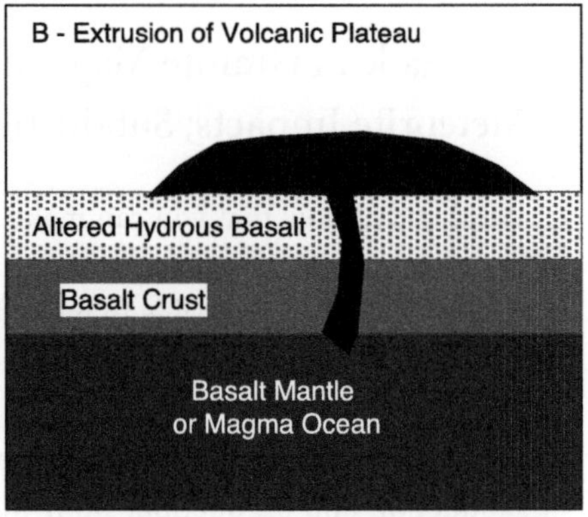

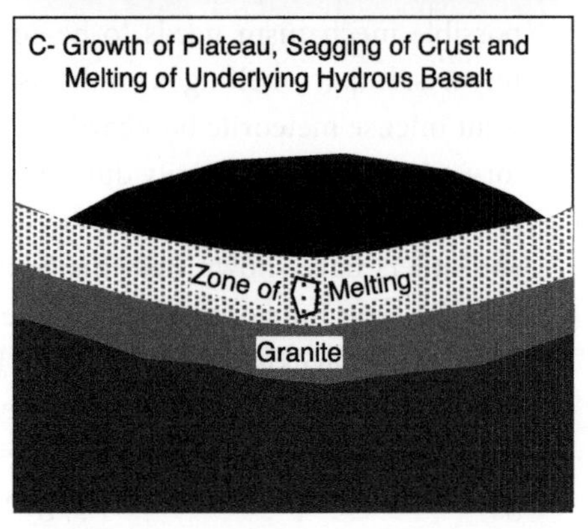

Figure 8.2. Forming granite magmas by sagduction melting of basalt: (**A**) original configuration of basalt crust with upper altered hydrous zone formed by reaction with the atmosphere or ocean; (**B**) a volcanic plateau is extruded onto the crust and then depresses the hydrous upper part of the crust into the mantle; (**C**) causing it to start melting. (Modified from Kamber et al. 2005; Kemp et al. 2010.)

8.5. Water Might Be the Key to Understanding Hadean History

Water was available to react with rocks and magmas during the Hadean. Most of it probably came along with the material that collected to form Earth, as reviewed recently by Ron Cowen, and it was driven toward the surface (degassed) during accretion. Early surface conditions were probably so hot that the water was present in the atmosphere as vapor. As things cooled, atmospheric water would have condensed during a period known imaginatively as the Great Rain, and collected into early oceans.

BOX 8.1. WHAT WAS THE COMPOSITION OF EARTH'S FIRST ATMOSPHERE?

Heat caused by the accretion of dust and small planets to form Earth would have driven gases outward to form the first atmosphere. The composition of this atmosphere can be estimated by heating meteorite materials, which represent the dust and small planets. Early experiments involved heating carbonaceous chondrite meteorites, a type that contains abundant carbon and some hydrogen. To no one's surprise, CO_2 and H_2O were the main gases released, leading to suggestions that the early atmosphere was CO_2-rich steam (water vapor). But carbonaceous chondrite meteorites are relatively rare and are thought to have constituted only about 5 percent of the original mass of the Earth. Other types of meteorites lack abundant carbon, and they release gases dominated by H_2 when heated. Thus, it is more likely that the earliest atmosphere was made up of H_2, N_2, and CH_4 with smaller amounts of CO_2 and H_2O.

The Hadean water story has become surprisingly important. In the early 2000s, John Valley, Stephen Mojzsis, William Peck, and others pointed out in several studies that the oxygen isotope composition of the 4300 Ma Jack Hills zircons is typical of oxygen that has been in contact with cool water. They interpreted this to mean that an ocean was present on Earth only 200 million years after the planet formed. This means that the basalt magma ocean must have crystallized to form a crust and then the water in the atmosphere must have condensed during the Great Rain to form the oceans, all within a very short time.[9]

The hypothesis that liquid water was present, even if intermittently, on the Hadean Earth, has become known as the Waterworld hypothesis, and it has stimulated a lot of new research.[10] One line of research has recognized an obvious complication to the Waterworld story, and that is the ~3900 Ma Late Heavy Bombardment, which could have vaporized the ocean and would certainly have caused some prodigious melting, possibly forming granite melt sheets. However, Elizabeth Bell and Mark Harrison studied the Jack Hills zircons with this in mind and found that only the outer (youngest) growth zones of some zircons are the same age as the Late Heavy Bombardment; most of the zircon grains are indeed older than the bombardment and probably originated in a more typical granite magma.

Recognition that Earth may have had an early ocean, whether permanent or intermittent, has also stimulated thinking about its possible role as a cradle for early life, and that brings us back to the Superior Province.

8.6. The Hadean Ocean Might Have Contained Life

The scarcity of sample material makes the search for Hadean life even more challenging than in the Archean. Nevertheless, optimism has been stimulated by recent observations of two types suggesting that our time traveler might have had some company.

The first observations involve possible microfossils. These were reported by Matthew Dodd and others from iron-rich sedimentary rocks in the NSB in Quebec, discussed above as the site of possible old rocks. These sediments are thought to be precipitates from a submarine hot spring, and the putative microfossils found in the precipitates consist of tiny iron oxide tubes and filaments that have the same general shape as microorganisms found at modern submarine hot springs. Although not everyone accepts these features as microfossils, they could indicate that submarine life existed at least as far back as 3770 Ma and possibly in the Hadean, depending on resolution of the debate about the age of the NSB rocks, as discussed above.

The second observation involves carbon isotopes. Carbon is the basis for life on Earth, and its isotopic composition is affected when it becomes part of a living organism. This provides a useful way to fingerprint small patches of carbon that are found in ancient, often metamorphosed rocks. These patches might represent an unfortunate organism of some sort that was deformed and cooked beyond recognition. Or they might just be small grains of graphite, a mineral that can form inorganically from carbon in rocks. In an early application of this method in 1996, Stephen Mojzsis and others claimed that carbon patches in iron formations from the 3800 Ma Isua complex in Greenland had isotopic "biosignatures" reflecting formation by living organisms. Elizabeth Bell and others used a similar method to investigate tiny 3-micrometer patches of carbon that are completely enclosed within Jack Hills zircon grains and reported isotopic biosignatures that reflect life. The carbon, which was described as poorly crystalline graphite, was found in only 2 of the 656 zircon grains studied. Exactly how the carbon got into the zircon is not clear.

Even to a skeptic, these observations raise the possibility that our time traveler might have found life on the early Earth, possibly during Hadean time. Where and how it formed remain unclear, however, although the ocean is the most likely place for things to have started, particularly deep-sea hot or cold springs.[11] Whether it persisted once it formed is also unclear. Perhaps it formed and was obliterated one or more times by large impacts that vaporized the ocean. At this point, we cannot peer any farther back in time and will have to wait for new field and laboratory discoveries to clarify how Earth and our Great Lakes region really began. In the meantime, keep your eyes open for the rock sample that will solve these mysteries.

Notes

1. The 4280 Ma basalt "age" is an Nd model age that probably represents the age of the mantle from which the basalt was extracted by partial melting, as discussed in chapter 7. Debate about the significance of this age also involves the granitic material in the NSB. Nicole Cates, working with Karen Ziegler, Axel Schmitt, and Stephen Mojzsis (2013), concluded that zircons in the granitic material defined a maximum age of only 3750 to 3780 Ma and that this age should be applied to the basalts.

2. For oxygen isotope results on the origin of the Moon, see Wiechert et al. (2001). For Earth to throw off enough material to make the Moon requires that it was spinning so fast that Earth's day was only a few hours long, and this has required some modifications to the lunar origin theory as discussed by Tim Elliott and Sarah Stewart (2013) and Maria Ćuk and others (2016).

3. The SHRIMP can measure the chemical and isotopic composition of spots no larger than about 30 micrometers.

4. Some reconstructions correlated the Superior Province with the Yilgarn craton, possibly including it in a Vaalbara supercontinent (Chamberlain et al. 2015).

5. A granitic source magma for the zircons is supported by their isotopic and trace element compositions, abundant inclusions of minerals commonly found in granites, and low temperatures of crystallization, which are typical of granite magmas (Harrison 2009; Mueller and Wooden 2012; Bauer et al. 2017; Burnham and Berry 2017).

6. The Late Heavy Bombardment was originally recognized in 1974 by Fouad Tera, Dimitri Papanastassiou, and Gerald Wasserburg in a clustering of ages of impact melts brought back from the Moon by the Apollo astronauts. Although no new age measurements have become available since the Apollo programs, later theoretical and modeling studies based partly on new geophysical data have suggested that the spike in activity related to the Late Heavy Bombardment may have been less distinct, with a longer period of heavy bombardment (Marchi et al. 2014; Kamata et al. 2015; Shibaike et al. 2016).

7. Burnham and Berry (2017) showed that Ce and Eu in the Jack Hills zircons indicate that their parent magma formed by melting of deep lower crustal rocks, which could have been melted by an impact.

8. In addition to the heat released during formation of Earth's core, there would have been much more heat from radioactive decay, especially ^{235}U, which has a shorter half-life than ^{238}U and was almost twice as abundant at this time in Earth history. Liquid water is not essential to the basalt-melting process. As you can see in figure 8.2, the

upper part of the basalt crust could have reacted with either the ocean or the atmosphere. Both processes would have formed new (alteration) minerals containing water, which would have reacted with the rock during burial and heating to form granite.

9. According to John Valley, the zircon grain shown in figure 8.1 has a U-Pb age of 4400 Ma and is the oldest known piece of the Earth. Nano-geochronology analyses of the grain rule out Pb-mobility biasing of the Haden age and confirm chemical homogenization of the silicate Earth and formation of a crust before 4400 Ma, supporting evidence for a cool early Earth that was habitable for life by about 4300 Ma.

10. Not all research has supported the Waterworld hypothesis (Nemchin et al., 2006).

11. One attractive feature of springs as a place for life to start is the presence of bubbles from escaping gas, which might form a spherical template for enclosed entities such as cells. If you want to read back through the literature on springs and the origin of life, a good place to start is Mike Russell's commentary on progress (Russell, 2017).

CHAPTER 9

Sustaining the Continent

Our Geologic Future

Now that we have traveled as far back in time as Earth will allow, it's time to use this knowledge to look forward. After all, one of the reasons we study geology is to use information on the past to predict Earth's future. This sort of prediction is important to us and our residence on Earth. If we are to have a sustainable relation to our planet, we need to understand the long-term processes that affect its welfare and our own. Geology and its allied sciences are clearly at the heart of many of these long-term issues, especially plate tectonics and climate change. So, we should close out our tour of Great Lakes rocks by a brief review of some of the geologic challenges that we face in sustaining our planet . . . and ourselves.

We geologists have gotten pretty good at predictions, at least in our collective opinion. We are especially good at really long-term predictions, and less certain as the time frame gets shorter. For instance, probably millions of years from now, subduction will increase along the margins of the Atlantic Ocean, causing Europe and North America to move toward one another again, possibly forming a new Euramerica. The increased subduction will cause renewed volcanism along the eastern side of North America, but it will probably not extend as far west as the Great Lakes region. It might, however, cover our region with volcanic ash, as it did during Ordovician time.

Over the shorter term, one of our best hints as to what might happen comes from earthquakes. It turns out that the continental interior around the Great Lakes region experiences quite a few earthquakes, and they show the pattern of faulting that will govern future movements of the crust. Many of these earthquakes are twitches of the crust related to continuing uplift of the northern part of the continent as it recovers from the weight of the glaciers, but some others are a result of larger-scale tectonic processes that are moving the North American plate across the globe. These modern earthquake movements, and the faults they form at the surface, go by the name neotectonics, and they are a window on the future.

9.1. Earthquakes and Neotectonics Outline the St. Lawrence Rift

Earthquakes have shaken many parts of the Great Lakes region (fig. 9.1), and they outline the important fault zones that are active today. In the east, most of the earthquakes occur along a regional fault system known as the St. Lawrence Rift (fig. 9.1). Some of these earthquakes have been large, including magnitude 6.1 and 5.2 events in the Timiskaming, Ontario area and magnitude 5.2 and 5.0 events in the Val-des-Bois area in Quebec. Just outside the Great Lakes region to the east, there have been several magnitude 6 earthquakes and one of magnitude 7 near Saugenay, Quebec, also along the St. Lawrence Rift. Things have been quieter to the west, although there are a few hotspots along what appears to be the western extension of the rift. One of these is an earthquake cluster at Cleveland on the south shore of Lake Erie, including one magnitude 5 event. Just to the southwest of Cleveland, the Anna area in Ohio has had several events, including one of magnitude 5.4.

The main part of the St. Lawrence Rift system is a fault valley, or graben, that hosts the St. Lawrence River.[1] The rift has two important subsidiaries, the Saguenay Graben, northeast of the city of Quebec, and the larger Ottawa-Bonnechere Graben, which splits from the St. Lawrence Rift near Ottawa and extends westward, where it splits again to form the Timiskaming Graben on the north and a poorly defined rift that continues through North Bay, Ontario (fig. 9.1). It was this graben system that early French explorers used as a short cut to the upper Great Lakes. The western extension of the main rift system probably extends through Cleveland and Anna in Ohio. Michael Hansen has suggested that the earthquake cluster at Cleveland is related to a subsidiary fault system similar to the Saugenay Graben.[2]

According to Alain Tremblay, Rolly Rimando and others, the rift system first appeared during late Proterozoic rifting associated with the dismemberment of Rodinia and Pannotia, which opened the Iapetus Ocean, and it was reactivated during early Mesozoic rifting, which opened the present Atlantic Ocean. As we saw in chapter 4, the grabens preserve remnants of the Paleozoic sediments that covered parts of the Canadian Shield.

As we move farther west, earthquakes are fewer and smaller, and they

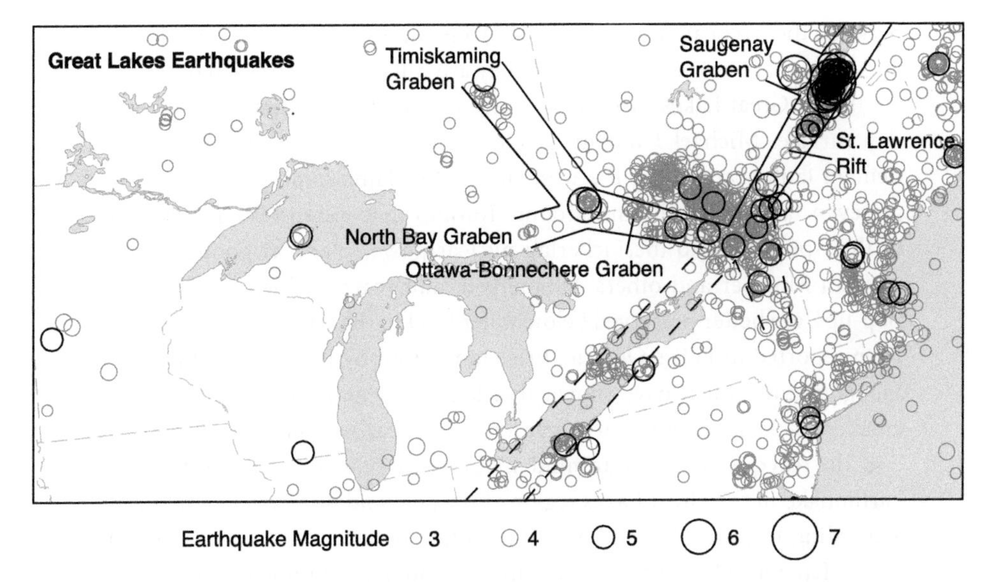

Figure 9.1. Distribution of earthquakes between 1627 and 2015 and related fault systems. Earthquakes with magnitudes of 6 and 7 are shown with darker symbols. (Modified from Earthquakes Canada: http://www.earthquakescanada.nrcan.gc.ca/historic-historique/images/caneqmap_e.pdf; Doughty et al. 2013; Ebel and Tuttle 2002.)

include explosions set off for mining and cryoseisms, which form when expansion of ice causes a rupture of the ground. According to Kazuya Fujita and Norman Sleep, the largest earthquake experienced in Michigan was a magnitude 4.6 tremor near Coldwater in 1947. In Illinois there have been several magnitude 5 and 4 events just southwest of Chicago, and in Wisconsin the largest event was the magnitude 4 event at Milwaukee. In Minnesota one of the largest earthquakes, a magnitude 4.6 event in 1975 at Morris, is part of a cluster of earthquakes in the central part of the state. These earthquakes are not related to the St. Lawrence Rift, and their causes are less obvious. In the southern part of Michigan, for instance, it has long been common to ascribe earthquakes to reactivation of faults along the margins of the buried Midcontinent Rift. However, closer examination of many of these earthquakes, such as the 1994 Central Michigan earthquake studied by Trent Faust and others, shows that they are not directly on rift-bounding faults. Farther to the west, in Minnesota, the 1975 and 1993 earthquakes around Morris, one seismically active area in the state, occurred along a line of earthquakes known as the Morris fault. Val Chandler has pointed out that this fault is the boundary between the Archean Minnesota River Valley and Wawa-Abitibi terranes discussed in chapter 8, suggesting that they are still adjusting to each other.

Some Great Lakes earthquakes have formed fault scarps at the present-day surface. Michael Doughty and his coworkers have described a well-defined fault scarp along the west side of the Timiskaming Graben. There are even fault scarps in and around Toronto that were of interest because of their proximity to the Pickering nuclear power plant, as pointed out by Arsaian Mohajer and others. Submerged fault scarps, which are known as pop-ups, are better preserved below water level in the Great Lakes. They have been reported in Lakes Ontario, Erie, and Huron by Blasco and others, and Doughty and others have shown that landslides and other features in the smaller glacial lakes in Ontario give a record of earlier landslides.

The real question in your mind might be just how soon to expect another earthquake in the Great Lakes region and where and how strong it might be. According to John Ebel and Martitia Tuttle, the level of seismic activity in the St. Lawrence Rift system is slightly lower than in the Appalachians to the east, although the level in the seismically active Ottawa-Bonnechere Graben is almost as high as around the New Madrid earthquake zone in Missouri, the scene of one of North America's largest earthquakes.[3] There appear to be two driving forces for earthquakes in the Great Lakes region: plate tectonic forces that are pushing North America westward and glacial rebound following disappearance of the Pleistocene ice sheets. Neither is very strong, which means that the risk of earthquakes is comparatively low in terms of both frequency and magnitude.

For Great Lakes earthquakes to attract more attention, they need to take on a larger project, like splitting the continent. That project cannot involve the Midcontinent Rift in view of the lack of earthquakes along its margins. Furthermore, the Midcontinent Rift had a great chance to reopen in Paleozoic time during the breakup of Pannotia and Pangea, and it dropped the ball. Instead the St. Lawrence Rift opened closer to the eastern margin of Laurentia. In its present form, the St. Lawrence Rift seems to give out to the southwest somewhere near Anna, Ohio. But you can extend the line of the rift farther southwest and come near the New Madrid area of Missouri. The New Madrid area is at the north end of a series of rifts and faults that extend down the Mississippi River valley toward the Gulf of Mexico. Although it is definitely a long shot, the St. Lawrence Rift could grow to the southwest and try to split North America in two.

9.2. Meteorites Will Be Infrequent and Probably Small . . . We Hope

Another long-shot geologic possibility is a meteorite impact. You might say that meterorites are the province of astronomy, which is correct, but once they hit the ground, we geologists take over. In fact meteorite impacts have been a part of the geologic history of the Great Lakes region since at least middle Proterozoic time, as we saw in chapters 4 and 6, and they are still happening. There are several celestial objects out there waiting to strike Earth, including gassy comets, rocky asteroids, and meteoroids, which are asteroids smaller than about a meter. On their way into the atmosphere, all of them burn to form a meteor. If the light is brighter than the planet Venus, the meteor is a fireball. At that point geologists and astronomers part ways, with astronomers using the term bolide for fireballs that explode in the atmosphere and geologists using it for impactors that form a crater—in other words, the big ones.

According to the American Meteor Society, worldwide fireball events occur at the rate of several thousand per day. A few of these reach the surface and produce meteorites that are sought by a small army of enthusiastic collectors. Their search is aided by weather radar, which detects the passing meteorites and shows approximate landing zones. You can get an idea of the frequency of material raining down from space in figure 9.2, which shows the global distribution and magnitude of reported fireballs over a 30-year period. On the basis of this map, we might expect about 1 fireball per decade in the Great Lakes region.[4] But what about larger events? Peter Brown and his coworkers estimated that objects in the 10-m size range strike Earth each year and that a strike like the Tungaska event,[5] which exceeded 100 m, should occur every 1000 years. But, you persist, what about really big events such as those that killed dinosaurs? At that point, we get into the long-standing question of whether all or some of the largest events are periodic in nature. In *Dark Matter and the Dinosauers*, Lisa Randall and Matthew Reece reviewed the record and proposed that an approximately 35-million-year periodicity in big events is related to as yet undetected dark matter at the center of our galaxy. Fortunately, we are not near the end of the most recent 35-million-year period.

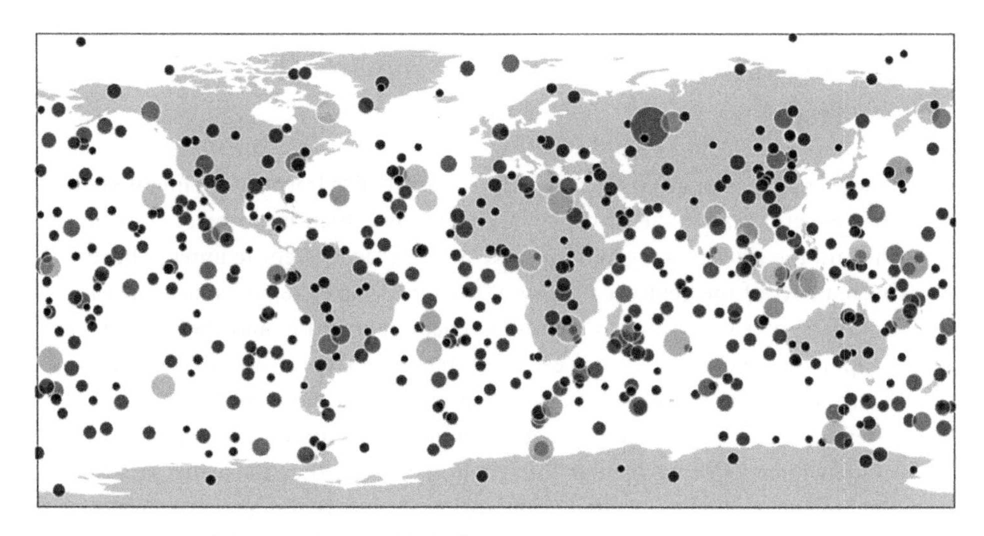

Figure 9.2. Global distribution of fireballs reported by US government sensors between April 15, 1988, and February 1, 2018. The size of each circle indicates fireball energy. (Modified from Center for Near Earth Object Studies: https://cneos.jpl.nasa.gov/fireballs/.)

9.3. Climate Change Is Likely, with Warmer Weather and More Storms

For short-term geologic events that are almost certain to happen, you cannot find a better candidate than climate change. One of the most recent reviews of climate change in the Great Lakes region was made by the Great Lakes Integrated Sciences and Assessments Program. It reported that average temperatures in the region have increased by 1.1°C since 1900 and that this trend is expected to continue with a further 4.2 to 6.1°C increase in average temperatures by 2100, based largely on global predictions by the United Nations Intergovernmental Panel on Climate Change (IPCC). The increase will be accompanied by stronger storms, including an increase in the number of days with unusually heavy precipitation. Changes will be greatest in the northern parts of the region and cooler parts of the year, with a continuation of the already obvious trend toward less ice and snow. (Average ice cover on the Great Lakes decreased by about 70 percent from 1973 to 2010, and average thickness of snow in the US part of the region decreased by about 2 inches from 1975 to 2004.) The decline in Great Lakes ice cover will continue, leading to greater evaporation of the lakes, greater stratification, and warmer lakes with more algal blooms, such as those seen recently on Lake Erie. Whether lake levels will rise or fall is unclear, which is not surprising in view of the many variables that must be included in the estimates.

These changes are definitely affecting Great Lakes region lifestyles. Nikolay Damyanov and his associates have reported a decrease in the number and length of season for outdoor hockey rinks and skating ponds in Canada. The length of the ski season in most areas is decreasing, although greater winter evaporation caused by decreased ice levels on the Great Lakes is increasing the regional extent and in some cases the total amount of lake effect snow along the lakeshores.

Great Lakes fauna and flora will also respond to climate change. Fish populations will shift from cold to warm water species, with lake trout, Chinook salmon, and whitefish giving way to bass and perch. For land-based wildlife, migration northward, following their natural habitat, will be complicated by the barrier formed by the lakes themselves. Large and more mobile animals will probably make the change; smaller and less mobile ones might not. Similar changes will be seen in the forests, with maple, beech, birch, and aspen being replaced by oak and hickory. Disappearance of maple trees would damage (or at least displace) the tourism industry's focus on Fall colors, not to mention the maple syrup industry and even the Canadian flag.

But, what about the return of the glaciers? They left the Great Lakes region only about 10,000 years ago, and the Sangamonian (interglacial) Stage ended only 75,000 years ago. In the last 4 million years, each retreat has been followed by an advance, and it is reasonable to expect that our present interglacial period will end with another. The wild card, of course, is the enormous contribution of anthropogenic greenhouse gas emissions that we are adding to the atmosphere during the current interglacial. This has turned the timing of the next ice advance, if it actually happens, into a contest between the composition of the atmosphere and the Milanković cycles that controlled glacial processes before we came on the scene. Most climate scientists agree that Milanković will win and we will have another glacial advance, although there is considerable uncertainty about exactly when.

It should come as no surprise that the key factor contributing to this uncertainty is CO_2 and the greenhouse effect. Paul Tzedakis and his associates have shown that another return of the glaciers could happen within about 1500 years if atmospheric CO_2 levels are similar to preindustrial levels of about 280 ppmv.[6] However, our additions of fossil fuel CO_2 have driven atmosphere concentrations above 400 ppmv, and they are almost certainly on their way to much higher values. This will be especially true if we burn all the fossil fuels on the planet, which some of us seem to be intent on doing. We know that the carbon cycle will work to pull this CO_2 out of the atmosphere by dissolving it in the ocean, precipitating carbonate minerals, weathering rocks, and making more trees, and the question is just how long this will take. A common response is that CO_2 returns to the hydrosphere and lithosphere over tens to hundreds of years. If so, we could expect the atmosphere to purge itself of the pulse of fossil fuel CO_2 within a century or so after our fossil fuel orgy. However, David Archer and others have argued that even though a significant amount of CO_2 does clear out of the atmosphere rapidly, as much as 10 percent of it remains in the atmosphere for much longer time periods. On the basis of their estimates, it will take tens to hundreds of thousands of years for CO_2 to return to a level at which continental glaciers could advance again.

This result might seem reassuring if it were not for what will happen while we wait for the glaciers to return. According to some climate models, melting of much of Earth's polar ice will cause sea level to rise tens of meters, threatening coastal areas around the world. Sea level changes won't have a direct effect on the Great Lakes region, of course, but the increased precipitation and decreased snow cover will intensify erosion, leading to important changes in our drainages patterns.

9.4. Increased Precipitation Will Lead to Stream Capture and Lake Filling

In most nonmountainous areas of the world, such as our Great Lakes region, erosion has continued for a relatively long time and the drainage system has begun to equilibrate with local base level. That is not true for us. Here we have a water-dominated system that is trying to come to terms with the topography and debris left by a recently departed ice-dominated system. To make the job more complex, isostatic rebound is changing the elevation and tilt of the land, which changes the base level for streams, and even the direction in which they might flow. As the new water-dominated drainage system adjusts to the ice-dominated topography, the most obvious changes will involve filling of lakes, headward erosion of streams and waterfalls, and a process known as stream capture.

We saw in chapter 2 that lakes are temporary features of the landscape and that many small glacial lakes have filled with sediment to form the widespread wetlands that litter the Great Lakes region. Even the largest lakes have an uncertain future in the Great Lakes region. Lake Pepin in the Mississippi

River between Wisconsin and Minnesota is a good example (fig. 9.3). According to Dylan Blumentritt and Herbert Wright, Lake Pepin formed when the Mississippi was dammed by a delta formed where the Chippewa River flowed into the Mississippi. How this could happen takes us back to chapter 3 and the history of glaciers in Lake Superior. At that time, the Chippewa River was carrying lots of sediment from the glaciers and ice front lakes in the Superior lobe of the ice sheet (fig. 3.8B).[7] Where the Chippewa River entered the Mississippi, this sediment formed a delta that gradually advanced across the Mississippi, forming a dam. The water that collected behind this dam formed Lake Pepin. Now that the Chippewa River is no longer draining a glacier it has a lower sediment load and cannot replenish this delta. So you might expect the Mississippi River to erode the Chippewa Delta, releasing water from Lake Pepin. But, ironically, concern now centers on the possibility that the lake will fill with sediment.

The offending sediment is being carried into the lake by the Minnesota River (fig. 9.3), which joins the Mississippi upstream from Lake Pepin. We last saw the Minnesota River in chapter 3 when it was glacial River Warren draining glacial Lake Agassiz (fig. 3.14). At that time, there was a huge amount of water flowing through glacial River Warren, and it cut a deep and wide valley across Minnesota. The present Minnesota River, which inherited the valley from glacial River Warren, is much too small for its valley, and its tributary streams flow down steep bluffs along the side of the valley to reach the Minnesota River. Much of the bluff material consists of easily eroded glacial till that was being eroded and carried down the Minnesota River, into the Mississippi River and finally into Lake Pepin.[8]

Production of all of this sediment involves headward erosion and stream capture, the other big processes that are modifying our Great Lakes drainage system. In the Minnesota River, for instance, tributary streams flowing down the bluffs have a steeper gradient than they did when they flowed into the higher water levels of glacial River Warren. The increased gradient gives the streams greater erosive power, and this causes them to extend their drainage basins upstream by means of headward erosion. But, as a stream's drainage basin extends upstream, it enlarges and encroaches on adjacent drainage basins. The stream that has the greatest erosive power (because of either a greater gradient or weaker rock to erode) will capture parts of adjacent drainage basins. Patrick Belmont and others have described how this happened with two tributaries of the Minnesota River. The Le Sueur River originally flowed into the Minnesota River but was completely captured by the Blue Earth River (fig. 9.4).[9] Stream capture was common along moraines left by the

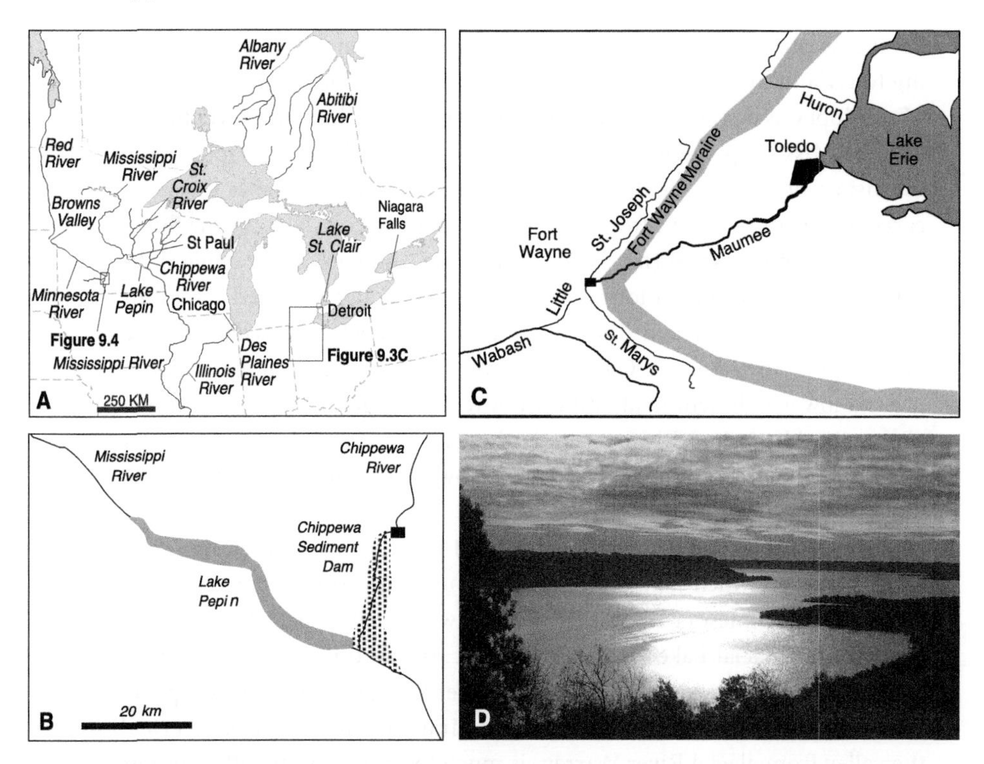

Figure 9.3. (**A**) Location of areas experiencing stream capture, lake filling, and migrating waterfalls (including the location of figure 9.4); (**B**) Lake Pepin, showing sediment dam deposited by the Chippewa River; (**C**) Stream capture by the Maumee and Huron Rivers; (**D**) View of Lake Pepin, looking south from Frontenac State Park. The steep, 100 meter bank along the east side of the lake is part of the original valley of Glacial River Warren.

glaciers (fig. 9.3C). In Michigan, the early Huron River on the east side of the moraine had a steeper gradient into the Detroit River, and captured streams on the west side of the moraine. Similarly, in Ohio, the Maumee River, with a steeper gradient into Lake Erie, captured the St. Joseph and St. Marys Rivers that had originally drained into the Wabash River.

We can use these insights to look into the future, trying to identify lakes that might fill or streams that might be captured. The most imperiled of the Great Lakes chain is Lake St. Clair. A glance at figure 2.8C makes it obvious that the delta at the north end of the lake will continue to grow, ultimately filling the lake. Slowing this process would require stopping the flow of sediment down the St. Clair River, which is not a realistic option. Other river migration and capture events are also staring us in the face. One of the most obvious

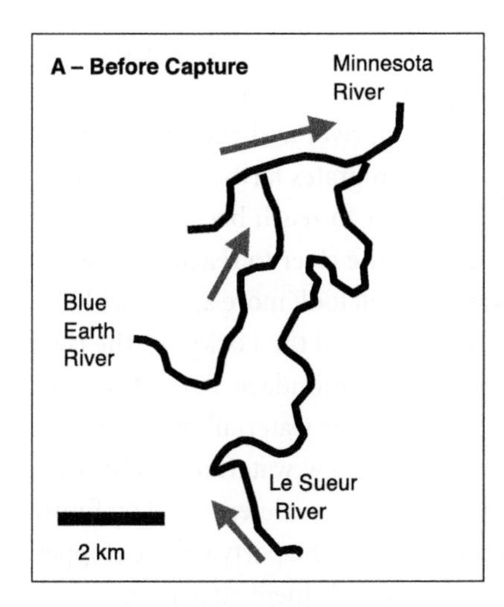

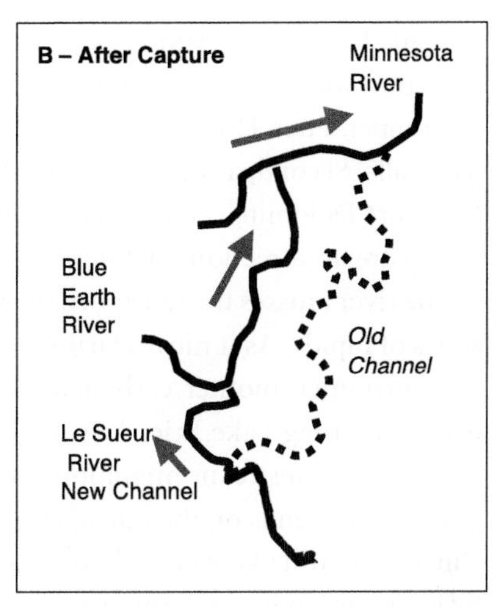

Figure 9.4. Capture of the Le Sueur River by the Blue Earth River in central Minnesota. (Modified from Belmont et al. 2011.)

is the contest between the Red River of the North and the Minnesota River, whose headwaters meet at Browns Valley, Minnesota, on the South Dakota border (fig. 9.3). These two rivers meet at the drainage divide between rivers that flow to the Mississippi River and those that flow north to Hudson Bay. Herbert Wright has shown that the present divide at Browns Valley is another delta, this one deposited by the Little Minnesota River, which blocked the upper part of Glacial River Warren after its flow declined. The part of Glacial River Warren north of the delta was forced to flow northward and became the upper part of the Red River. Today the headwaters of the Red River (Bois de Sioux River) and Minnesota River are cutting toward each other through this delta. When they meet, the river with the steepest slope will begin to capture the upper reaches of the other river. At present it looks like the Red River, which has a shorter path to the sea, might be the winner, but there is an additional complication. Glacial rebound is greater in the north, and this means that uplift in Manitoba might eventually cause the Red River to reverse flow toward the south.

Rivers might also capture lakes. In chapter 3, we learned that Niagara Falls has migrated more than 11 km up the Niagara River in about 12,000 years (fig. 2.13A), and this raises the obvious question of whether it will continue migrating upriver to drain Lake Erie. If the falls meet Lake Erie, will there be a single, catastrophic draining of the lake or just a slow dribble? Although this would make a great disaster movie, don't hold your breath. In

1974 Shailer Philbrick summarized the likely future trend of the Niagara River and concluded that it would be a multistage process ending in something of a whimper. First, Horseshoe Falls will migrate upriver and cut off the American Falls. Second, as the new Horseshoe Falls migrates farther upstream the Lockport Dolomite cap rock layer will be harder to reach because it dips in the upstream direction, making it too deep for the river to reach by erosion. As the river crosses the Lockport, the waterfall will look more and more like a series of rapids. As it moves farther upstream beyond the Lockport, the river will encounter another carbonate cap rock, the Onondaga Formation, just before reaching Lake Erie. This might form another waterfall, which would start its own upstream migration. Just when this new waterfall might reach Lake Erie depends on the rate of erosion, which will be determined by future climate. We will likely never know because no doubt the party will be stopped by intervention in the form of a dam or other containment structure.

The ultimate capture, though, would be by the Des Plaines River at the Chicago Sanitary and Ship Canal, which would divert flow from Lake Michigan into the Mississippi River and Gulf of Mexico. Here the contest will be between drainage flowing into the St. Lawrence system and that flowing into the Mississippi system. Stream gradients from Chicago to sea level are about the same for the two systems, making it less clear which one would win, although human intervention will make it hard for the Mississippi system to triumph. Finally, there is the challenge posed by the Albany and Abitibi rivers, which drain the area north of Lake Superior into Hudson Bay. These rivers have a much shorter path to sea level than either the St. Lawrence or the Mississippi and would seem to pose a big threat to the upper Great Lakes if they cut through to Lake Superior. However, they have a major impediment in the area of high, hard rocks along the north shore of Lake Superior, which will likely slow their progress considerably. And their northward flow might eventually be slowed by greater isostatic uplift around Hudson Bay. Finally, as we learned in chapter 2, upper parts of these rivers have already been diverted to flow into Lake Superior rather than Hudson Bay.

9.5. Our Short-Term Geologic Future Contains Two Final Challenges: Too Much Heat and Too Few Mineral Resources

To top off our geologically short-term concerns, we face two other, possibly severe challenges. The first of these is simply a hotter world produced by climate change. *Simply* might not be the appropriate term, however, because the world might become very hot. There is a precedent for this, of course, in the Paleocene-Eocene Thermal Maximum discussed in chapter 3. During this period, about 55 million years ago, global temperatures were as much as 12°C higher than today, and there was at least one rapid-warming event during which average global temperatures increased by about 5°C within only a few thousand years.[10] Even closer to us in terms of geologic time is the mid-Pliocene warm period about three million years ago when giant camels roamed Arctic forests, as described by Natalia Rybczynski and associates. According to estimates by the IPCC and other groups, we can anticipate similar high temperatures in the short-term geologic future. Henry Pollack's *A World Without Ice* provides a look at our march toward this hotter world.

High temperatures of this type will obviously raise sea level, but the real insult might come from the temperatures themselves. It would be worst for mammals because they must dissipate heat generated by metabolic processes into the environment. As the temperature of the surrounding air rises, this becomes more and more difficult, leading to heat stress and death. Some high-humidity, extremely hot tropical regions in today's world approach these conditions, as do some deep mines, but the situation would become much more widespread with global warming. Steven Sherwood and Matthew Huber have estimated that this effect would begin to be felt more widely at a temperature increase of 7°C and that large parts of the world would become uninhabitable if temperatures warmed by 12°C. These are extreme numbers, of course, but even incremental approaches to these conditions would place enormous burdens on society to either move people to cooler climates or expend large amounts of energy and materials on cooling.

And that brings up our second challenge: mineral resources. As global populations and standards of living increase, we are calling on Earth to supply more and more of the mineral resources required for modern living. It is true that "if it can't be grown, it has to be mined," and a glance around confirms that the cars, buildings, and roads that distinguish us from stone age people came from Earth's mineral resources. We are developing the ability to derive energy from the sun and wind, but we cannot get materials that way. They must be mined. Recycling cannot solve the problem because standards of living are increasing in many areas, requiring more and more material, and because material is degraded as it is recycled. So we will require a continuing supply of mineral resources as long as we maintain an advanced society.

Because of both energy and space considerations, mineral resources cannot be derived from common rock. Instead they have to come from the sorts of ore deposits discussed throughout this book. And, as we have seen, different materials must come from different types of ore deposits in different geologic environments. There is, of course, a finite inventory of these mineral deposits on Earth.[11] Estimating the size of this inventory for all mineral commodities is an enormous job on which geologists have made only a little progress. And just how long it will last depends on both the size of the inventory and future demand.

The best progress in estimating our global resource inventory has been made for copper and its ore deposits. In a recent study that I did with Nick Arndt and others, we showed that land-based deposits might meet about 4000 years' worth of current copper demand and that resources in the oceans might add another 5000 years of supply. We also evaluated the possibility of using rocks that are slightly enriched in copper (containing far less copper than ore deposits do) and estimated that current needs might be supplied for as much as 450,000 years from this source. All these estimates have to be tempered by the sobering fact that we have not found these deposits yet and that many are probably too deep or too disturbed or too close to our cities or too this or too that to be exploited. And demand is likely to increase. So our estimated lifetimes are likely much shorter. That's it. After exhausting these deposits, there will be no more new copper to be had and society will be able to sustain our current life style only with dramatically improved recycling or materials from another planet or asteroid.

It is interesting that estimated time frames for the exhaustion of global mineral deposits are similar to those for the return of the glaciers. The fatalistic reader might conclude that humans were meant to be only an "interglacial" species. But our ancestors were active along the margins of the glaciers in North America and far back into the Sangamonian interglacial in Europe, Asia, and Africa. So we will likely persist. Whether our Great Lakes region will be wiped clean by another glacial advance in thousands of years or flooded by an advancing sea in millions of years remains to be seen. In the meantime, we should try to take care of it so that the next generation can continue to enjoy its beauty and puzzle over its geologic evolution.

Notes

1. A graben is a linear valley with a down-dropped central zone bounded by faults on both sides. It forms during extension, when the crust is being pulled apart.

2. Moid Ahmad and Jeffrey Smith (2008) have shown that at least one recent earthquake at Cleveland was initiated by the injection of cooling water from the Perry Nuclear Power Plant.

3. The New Madrid area in far southeastern Missouri is an area of frequent earthquakes, including three on December 16, 1811, of estimated magnitude 7.4, 7.5, and 7.9.

4. The 734 fireballs reported by US government sensors over a 30-year period between April 15, 1988, and February 1, 2018, included about 30 over North America. If the Great Lakes region comprises about 10 percent of North America, that would mean about 3 events over a period of around 30 years or 1 per decade.

5. The Tunguska Event in 1908 in central Siberia is the largest known impact event in recorded history. That bolide exploded in the atmosphere, releasing energy estimated to have been in the 10- to 15-megaton range and flattened trees over a 2000-km^2 area. More information on near-Earth objects and the hazards they present can be found in Yeomans (2013).

6. The abbreviation "ppmv" refers to parts per million by volume, a common measure of the concentration of atmospheric constituents.

7. The enormous potholes in the adjacent St. Croix River (fig. 2.12D) testify to the large sediment load these rivers carried as they drained proglacial lakes. Several other riverine lakes formed by similar damming processes along the Mississippi River and up the St. Croix.

8. Karen Gran (2017) has pointed that this sediment is an important concern in river management.

9. Even larger stream capture events took place during glacial retreat. Eric Carson and others (2013) have described how the Mississippi River originally flowed eastward through the lower part of the Wisconsin River, probably draining into the St. Lawrence. It was diverted westward when ice blocked its eastward flow, leading to capture by a south-flowing river that became what we know today as the Mississippi.

10. See Pagani et al. (2006) for a short description of the PETM and DeConto et al. (2012) for a discussion of possible sources for CO_2 that caused it.

11. Earth is forming new mineral deposits all the time, but we are consuming them much more rapidly than they form. So the inventory is essentially fixed.

REFERENCES

Abramov, O., Kring, D. A., and Mojzsis, S. J. 2013. The impact environment of the Hadean Earth: Chemie der Erde. Geochemistry, v. 73, p. 227–248.

Addison, W. D., and Brumpton, G. R. 2012. Field trips 1 & 13: Sudbury impactoclastic debrisites at Thunder Bay. Proceedings of the 58th ILSG Annual Meeting, part 2.

Addison, W. D., Brumpton, G. R., Vallini, D. A., McNaughton, N. J., Davis, D. W., Kissin, S. A., Fralick, P. W., and Hammond, A. L. 2005. Discovery of distal ejecta from the 1850 Ma Sudbury impact event. Geology, v. 33, p. 193–196.

Ahmad, M. U., and Smith, J. A. 1988. Earthquakes, injection wells, and the Peerry Nuclear Power Plant, Cleveland, Ohio. Geology, v 16, p. 739–742.

Aldrich, L. T., Davis, G. L., and James, H. L. 1965. Ages of minerals from metamorphic and igneous rocks near Iron Mountain, Michigan. Journal of Petrology, v. 6, p. 445–472.

Alexander, E. C. 2013. Mystery Cave, Minnesota: A window into the paleohydrology of the Upper Mississippi Valley. Abstracts with Programs, Geological Society of America, v. 45, p. 640.

Alexander, E. C., and Lively, R. S. 1995. Karst: Aquifers, caves, and sinkholes In Text supplement to the Geologic Atlas, Fillmore County, Minnesota. Minnesota Geological Survey County Atlas Series C-8, part C, p. 10.

Alexander, E. C., Jr., and Wheeler, B. J. 2015. A proposed hypogenic origin of iron ore deposits in southeast Minnesota karst. 14th Sinkhole Conference, NCKRI Symposium 5, p. 167–176.

Algeo, T. J., Berner, R. A., Maynard, J. B., and Scheckler, S. E. 1995. Late Devonian oceanic anoxic events and biotic crises: "Rooted" in the evolution of vascular land plants? GSA Today, v. 5, p. 445ff.

Allen, D. J., Hinze, W. J., Dickas, A. B., and Mudrey, M. G., Jr. 1997. Integrated geophysical modeling of the North American Midcontinent Rift system. Geological Society of America Special Paper 312, p. 47–67.

Alley, R. B. 2005. Ice sheet and sea level changes. Science, v. 310, p. 457.

Allwood, A. C., Walter, M. R., Kamber, B. S., Marshall, C. P., and Burch, I. W. 2006. Stromatolite reef from the Early Archaean Era of Australia. Nature, v. 441, p. 714–718.

Ames, D. E., Davidson, A., and Wodicka, N. 2008. Geology of the giant Sudbury polymetallic mining camp, Ontario, Canada. Economic Geology, v. 103, p. 1057–1077.

Anderson, J. L., Cullers, R. L., and Van Schums, W. R. 1980, Anorogenic metaluminous and peraluminous granite plutonism in the Mid-Proterozoic of Wisconsin, USA: Contributions to Mineralogy and Petrology, v. 74, p. 311–328.

Anderson, R. R. 1997. Keweenawan Supergroup clastic rocks in the Midcontinent Rift of Iowa. In Ojakangas, R. J., Dickas, A. B., Green, J. C., eds., Middle Proterozoic to Cambrian rifting, central North America. Geological Society of America Special Paper 312, p. 211–230.

Anderson, T. W., and Lewis, C. F. M. 2012. A new water-level history of Lake Ontario Basin: Evidence for a climate-driven Holocene lowstand. J Paleolimnol., v. 47, p. 513–530.

Andrews, A. J., Owasiacki, L., Kerrich, R., and Strong, D. F. 1986. The silver deposits at Cobalt and Gowganda, Ontario, I: Geology, petrography, and whole-rock geochemistry. Canadian Journal of Earth Sciences, v. 23, p. 1480–1506.

Annin, P. 2009. Great Lakes water wars. Island Press. 303 p.

Arbogast, A. F. 2012. Sand dunes. In Schaetzl, R., Darden, J., Brandt, D. Michigan geography and geology. Pearson Learning Solutions. p. 274–287.

Arbogast, A. F., and Garmon, B. Undated. The emerging science of coastal sand dune age and dynamics: implications for regulation and risk management in Michigan. http://www.environmentalcouncil.org/mecReports/EmergingScienceofDuneAge.pdf

Arbogast, A. F., and Loope, W. L. 1999. Maximum-limiting ages of Lake Michigan coastal dunes: their correlation with Holocene lake level history. Journal of Great Lakes Research, v. 29, p. 372–382.

Archer, D., Eby, M., Brovkin, V., Ridgwell, A., Cao, L., Mikolajewicz, U., Daldeira, K., Matsumoto, K., Munhoven, G., Montenegro, A., and Tokos, K. 2009. Atmospheric lifetime of fossil fuel carbon dioxide. Annual Reviews Earth and Planetary Science, v. 37, p. 117–134.

Armstrong, D. K., and Dodge, J. E. P. 2007. Paleozoic geology of southern Ontario. Ontario Geological Survey Miscellaneous Release: Data 219. 27 p.

Arndt, N. T. 2013. The formation of massif anorthosite: Petrology in reverse. Geoscience Frontiers, v. 4, p. 195–198.

Arndt, N. T., Coltice, N., Helmstaedt, H., and Gregoire, M. 2009. Origin of Archean subcontinental lithospheric mantle: Some petrological constraints. Lithos, v. 109, p. 61–71.

Arndt, N. T., Fontbote, L., Hedenquist, J. W., Kesler, S. E., Thompson, J. F. H., and Wood, D. G. 2017. Future global mineral resources. Geochemical Perspectives, v. 6, 166 p.

Arndt, N. T., and Goldstein, S. L. 1987. Use and abuse of crust formation ages. Geology, v. 15, p. 893–895.

Arndt, N. T., Kesler, S. E., and Ganino, C. 2015. Metals and society. Springer. 205 p.

Arndt, N. T., Lesher, C. M., and Barnes, S. J. 2008. Komatiite. Cambridge University Press. 488 p.

Arndt, N. T., Lewin, É., and Albarède, F. 2012. Strange partners; formation and survival of continental crust and lithospheric mantle. Geological Society of London Special Publication 199. p. 91–103.

Austin, J. A., and Colman, S. M. 2007. Lake Superior summer water temperatures are increasing more rapidly than regional air temperatures: A positive ice-albedo feedback. Geophysical Research Letters, v. 34, p. L06604.

Awramik, S. M., and Sprinkle, J. 1999. Proterozoic stromatolites: The first marine evolutionary biota. Historical Biology, v. 13, p. 241–253.

Ayer, J., Amelin, Y., Corfu, F., Kamo, S., Ketchum, J., Kwok, K., and Trowell, N. 2002. Evolution of the southern Abitibi Greenstone Belt based on U-Pb geochronology: autochthonous volcanic construction followed by plutonism, regional deformation, and sedimentation. Precambrian Research, v. 115, p. 63–95.

Ayuso, R. A. Schulz, K. J., Cannon, W. F., Woodruff, L. G., Vazquez, J. A., Foley, N. K., and Jackson, J. 2018. New U-Pb zircon ages for rocks from the granite-gneiss terranes in northern Michigan: Evidence for events at 3750, 2750 and 1850 Ma: Institute on Lake Superior Geology Abstract Volume, p. 78.

Baedke, S. J., and Thompson, T. A. 2000. A 4,700-year record of lake level and isostasy for Lake Michigan. Journal of Great Lakes Research, v. 26, p. 416–426.

Bailey, J., Lafrance, B., McDonald, A. M., Fedorowich, J. S., Kamo, S., and Archibald, D. A. 2004. Mazatzal-Labradorian age (1.7–1.6 Ga) ductile deformation of the South Range Sudbury impact structure at the Thayer Lindsley mine, Ontario. Canadian Journal of Earth Sciences, v. 41, p. 1491–1505.

Bakker, K. 2006. Eau Canada: The future of Canada's water. University of British Columbia Press. 440 p.

Balco, G., and Rovey, C.W., II. 2010. Absolute chronology for major Pleistocene advances of the Laurentide ice sheet. Geology, v. 38, p. 795–798.

Banerjee, N. R., Simonetti, A., Furnes, H., Muehlenbachs, K., Staudigel, H., Heaman, L., and Van Kranendonk, M. J. 2007. Direct dating of Archean microbial ichnofossils. Geology, v. 35, p. 487–490.

Bankey, V., Cuevas, A., Daniels, D., Finn, C. A., Hernandez, I., Hill, P., Kucks, R., Miles, W., Pilkington, M., Roberts, C., Roest, W., Rystrom, V., Shearer, S., Snyder, S., Sweeney, R., Velez, J., Phillips, J. D., and Ravat, D. 2002. Digital data grids for the magnetic anomaly map of North America. U.S. Geological Survey Open-File Report 02–414. U.S. Geological Survey, Denver.

Barghoorn, E. S., and Tyler, S. A. 1965. Microorganism from the Gunflint Chert. Science, v. 147, p. 563–577.

Barnes, D. A., Harrison, W. B., III, and Shaw, T. H. 1996. Lower-Middle Ordovician lithofacies and interregional correlation, Michigan Basin, U.S.A. Geological Society of America Special Paper 306.

Barnett, P. J. 1992. Quaternary geology of Ontario. In Geology of Ontario. Ontario Geological Survey Special Volume 4. p. 1011–1090.

Barnosky, A. D., Matzke, N., Tomiya, S., Wogan, G., Swartz, B., Quental, T., Marshall, C., McGuire, J. L., Lindsey, E. L., Maguire, K. C., Mersey, B., and Ferrer, E. A. 2011. Has the Earth's sixth mass extinction already arrived? Nature, v. 471, p. 51–57.

Barovich, K. M., Patchett, P. J., Peterman, Z. E., and Sims, P.K. 1989. Nd isotopes and the origin of 1.9–1.7 Ga Penokean continental crust of the Lake Superior region. Geological Society of America, v. 101, p. 333–338.

Barovich, K. M., Patchett, P. J., Peterman, Z. E., and Sims, P. K. 1991. Neodymium isotopic evidence for Early Proterozoic units in the Watersmeet gneiss dome, northern Michigan. U.S. Geological Survey Bulletin 1904G. p. 1–8.

Bartoli, G., Honisch, B., and Zeebe, R. E. 2011. Atmospheric CO_2 decline during the Pliocene intensification of northern hemisphere glaciation. Paleoceanography, v. 26, PA 4213.

Bauer, A. M., Fisher, C. M., Vervoort, J. D., and Bowring, S. A. 2017. Coupled zircon Lu-Hf and U-Pb isotopic analyses of the oldest terrestrial crust, the >4.03 Ga Acasta Gneiss Complex. Earth and Planetary Science Letters, v. 458, p. 37–49.

Bayley, R. W., Dutton, C. E., and Lamey, C. A. 1966, Geology of the Menominee iron-bearing district Dickinson County, Michigan and Florence and Marinette Counties, Wisconsin: U.S. Geological Survey Professional Paper 513, 96 p.

Beaty, C. B., 1978. The causes of glaciation: major glaciations have been caused not by dramatic changes in the earth's climate but by the conjunction of several discrete factors at times when the continents were located at high latitudes. American Scientist, v. 66, p. 452–459.

Bechle, A. J., Kristovich, A. R., and Wu, C. H. 2015. Meteotsunami occurrences and causes in Lake Michigan. Journal of Geophysical Research: Oceans, v. 120, p. 8422–8438.

Bédard, J. H. 2006. A catalytic delamination-driven model for coupled genesis of Archaean crust and sub-continental lithospheric mantle. Geochimica et Cosmochimica Acta, v. 70, p. 1188–1214.

Behrendt, J. C., Green, A. G., Cannon, W. F., Hutchinson, D. R., Lee, M. W., Milkereit, B., Agena, W. F., and Spencer, C. 1988. Crustal structure of the Midcontinent Rift system: Results from GLIMPCE deep seismic reflection profiles. Geology, v. 16, p. 81–85.

Bell, E. A., Boehnke, P., Harrison, T. M., Mao, W. 2015. Potentially biogenic carbon preserved in a 4.1 billion-year-old zircon. Proceedings of the National Academy of Science, v. 112, http://www.pnas.org/content/112/47/14519.full.pdf

Bell, E. A., and Harrison, T. M. 2013. Post-Hadean transitions in Jack Hills zircon provenance: A signal of the Late Heavy Bombardment? Earth and Planetary Science Letters, v. 364, p. 1–11.

Bello, D. 2009. Peat and repeat: can major carbon sinks be restored by rewetting the world's drained bogs? Scientific American. https://www.scientificamerican.com/article/peat-and-repeat-rewetting-carbon-sinks/

Belmont, P., Gran, K., Jennings, C. E., Wittkop, C., and Day, S. S. 2011. Holocene landscape evolution and erosional processes in the Le Sueur River, central Minnesota. In Miller, J. D., Hudak, G. J., Wittkop, C., and McLaughlin, P. I., eds. Archean to Anthropocene: Field guides to the geology of the mid-continent of North America. Geological Society of America Field Guide 24. p. 439–455.

Bennett, G. 2006. The Huronian Supergroup between Sault Ste. Marie and Elliot Lake. Institute on Lake Superior Geology Field Trip Guidebook 52, pt. 4.

Bennett, G., Dressler, B. O., and Robertson, J. A. 1991. The Huronian Supergroup and associated intrusive rocks. Ontario Geological Survey Special Volume 4, pt. 2. p. 549–591.

Bennett, V. C., and DePaolo, D. J. 1987. Proterozoic crustal history of the western United States as determined by neodymium isotopic mapping. Bulletin of the Geological Society of America, v. 99, p. 674–685.

Benton, M. J. 2005. When life nearly died: The greatest mass extinction of all time. Thames and Hudson. 336 p.

Berger, A., and Loutre, M-F. 2005. Glaciation causes: Milankovitch theory and paleoclimate. Encyclopedia of Quaternary Science. p. 1017–1022.

Bickford, M. E., Wooden, J. L., and Bauer, R. L. 2006. SHRIMP study of zircons from Early Archean rocks in the Minnesota River valley: Implications for the tectonic history of the Superior Province. GSA Bulletin, v. 118, p. 94–108.

Bjørnerud, M. 2016. Proterozoic geology of the Baraboo Interval. Geological Society of American Field Guide 43. p. 21–33.

Blasco, S. M., Lewis, C. F. M., Jacobi, R. D., Covill, R. D., Keyes, D., Armstrong, D., and Harmes, R. A. 2003. Bedrock pop-ups in western Lake Ontario, eastern Lake Erie, and eastern Lake Huron; evidence of neotectonic activity on the lakebed of the southern Great Lakes. Geological Society of America Abstracts with Programs 35. p. 76–77.

Blaustein, R. 2016. The Great Oxidation Event. BioScience, v. 66, p. 189–195.

Bleeker, W. 2018. Archean BIF clasts vs. Paleoproterozoic jasper clasts? The proof is in the pudding (stone): Institute on Lake Superior Geology Proceedings, v. 64, Part I, Program and Abstracts, p. 9–10.

Blewett, W. L., Lusch, D. P., and Schaetzl, R. J., 2012. The physical landscape: A glacial legacy. In Schaetzl, R., Darden, J., and Brandt, D. eds. Michigan geography and geology, Pearson Learhing Solutions, p. 249–273.

Blumentritt, D. J., and Wright, H. E. 2009. Formation and early history of lakes Pepin and St. Croix of the upper Mississippi River. Journal of Paleolimnology, v. 41, p. 545–562.

Boerboom, T. 2011. Transect from Archean basement to the Animikie Basin, east-central Minnesota. Institute on Lake Superior Geology Field Guidebook, Trip 8. p. 131–175.

Boerboom, T. J., Wirth, K. R., and Evers, J. F., 2014, Five newly acquired high-precision U-Pb ages in Minnesota, and their geologic implications: Institute on Lake Superior Geology Abstract Volume, v. 60, p. 13–14.

Bond, D. P. G., and Wignall, P. B. 2014. Large igneous provinces and mass extinctions: An update. Geological Society of America Special Paper 505. p. 29–55.

Boomer, I., Aladin, N., Plotnikov, I., and Whatley, R. 2000. The paleolimnology of the Aral Sea: a review. Quaternary Science Reviews, v. 19, p. 1259–1278.

Bornhorst, T. J., and Barron, R. J. 2011. Copper deposits of the western Upper Peninsula of Michigan. Geological Society of America Field Guide 24. p. 83–99.

Bornhorst, T. J., and Lankton, L. D. 2012. Copper mining: A billion years of geologic and human history. In Schaetzl, R., Darden, J., and Brandt, D., eds. Michigan geography and geology. Pearson Learning Solutions. p. 150–173.

Bornhorst, T. J., and Mathur, R. 2017. Copper isotope constraints on the genesis of the Keweenaw Peninsula native copper district, Michigan, USA. Minerals, v. 7. http://www.mdpi.com/2075-163X/7/10/185

Bornhorst, T. J., Paces, J. B., Grant, N. K., Obradovich, J. D., and Huber, N. K. 1988. Age of native copper mineralization, Keweenaw Peninsula, Michigan. Economic Geology and the Bulletin of the Society of Economic Geologists, v. 83, p. 619–625.

Bornhorst, T. J., and Williams, W. C. 2013. The Mesoproterozoic Copperwood sedimentary rock-hosted stratiform copper deposit, Upper Peninsula, Michigan. Economic Geology, v. 108, p. 1325–1346.

Bottomley, D. J., Gregoire, D. C., and Raven, K. G. 1994. Saline ground waters and brines in the Canadian Shield: geochemical and isotopic evidence for a residual evaporite brine component. Geochimica et Cosmochimica Acta, v. 58, p. 1483–1498.

Bowring, S. A., and Williams, I. S. 1999. Priscoan (4.00–4.03 Ga) orthogneisses from northwestern Canada. Contributions to Mineralogy and Petrology, v. 134, p. 3.

Boyce, F. M., Donelan, M. A., Hamblin, P. F., Hurthyu, C. R., and Simons, T. J. 1988. Thermal structure and circulation the Great Lakes. Atmosphere-Ocean, v. 27, p. 607–642.

Brannen, P. 2017. The ends of the world. Harper Collins. 336 p.

Brannon, J. C., Podosek, F. A., and McLimans, R. K. 1992. Alleghenian age of the upper Mississippi Valley zinc-lead deposit determined by Rb-Sr dating of sphalerite. Nature, v. 356, p. 509–511.

Braun, J. 2010. The many surface expressions of mantle dynamics. Nature Geoscience, v. 3, p. 825–833.

Breckenridge, A. 2013. An analysis of the late glacial lake levels within the western Lake Superior Basin based on digital elevation models. Quaternary Research, v. 80, p. 383–395.

Breckenridge, A., and Johnson, T. C. 2009. Paleohydrology of the upper Laurentian Great Lakes from the late glacial to early Holocene: Quaternary Research, v. 71, p. 397–408.

Bridge, N. J., Banerjee, N. R., Mueller, W., Muehlenbachs, K., and Chacko, T. A. 2010. Volcanic habitat for early life preserved in the Abitibi Greenstone Belt, Canada. Precambrian Research, v. 179, p. 88–97.

Briggs, D. E. G. 2015. The Cambrian explosion. Current Biology, v. 25, p. 864–868.

Broecker, W. S., and Farrand, W. R. 1963. Radiocarbon age of the Two Creeks forest bed, Wisconsin: Geological Society of America Bulletin, v. 280, p. 795–802.

Brown, A. C. 1971. Zoning in the White Pine copper deposit, Ontonagan County, Michigan. Economic Geology, v. 66, p. 543–573.

Brown, M. J., and Hampton, B. A. 2016. U-Pb detrital geochronology and Hf isotopic analyses from Mesoproterozoic Keweenawan Supergroup strata of the Midcontinent Rift, northern Michigan. Geological Society of America Abstracts with Programs 48.

Brown, P., Spalding, R. E., ReVelle, D. O., Tagliaferri, E., and Worden, S. P. 2001. The flux of small near-Earth objects colliding with the Earth. Nature, v. 420, p. 294–296.

Brummer, J. J., MacFadyen, D. A., and Pegg, C. C. 1992. Kimberlite discoveries, sampling, diamond content, ages and emplacement. Exploration & Mining Geology, v. 1, no. 4.

Bruxer, J., and Southam, C. 2010. Analysis of Great Lakes volume changes resulting from glacial isostatic adjustment. http://iugls.org/DocStore/ProjectArchive/DVR_StClairDataVerificationAndReconciliation/DVR03_Bruxer_GIAVolumeChanges/Reports/DVR03-R1_Bruxer.pdf

Buchan, K. L., and Ernst, R. E. 2004. Diabase dyke swarms and related units in Canada and adjacent regions. Geological Survey of Canada Map 2022A.

Budai, J. M., and Wilson, J. L. 1991. Diagenetic history of the Trenton and Black River formations in the Michigan Basin. Geological Society of America Special Paper 256. p. 73–88.

Buggisch, W., Joachimski, M. M., Lehnert, O., Bergstrom, S. M., and Repetski, J. E. 2011. Did intense volcanism trigger the first Late Ordovician icehouse? REPLY. Geology, v. 39, p. 238.

Bulbeck, R. C., Diment, W. H., Deck, B. L., Baldwin, A. L., and Lipton, S. D. 1971. Runoff of deicing salt: Effect on Irondequoit Bay, Rochester, New York. Science, v. 172, p. 1126–1132.

Buol, S. W., Southard, R. J., Graham, R. C., and McDaniel, P. A. 2011. Soil genesis and classification. 6th ed. John Wiley and Sons. 351 p.

Burgess, P. M. 2008. Phanerozoic evolution of the sedimentary cover of the North American craton. Sedimentary Basins of the World, v. 5, p. 31–63.

Burgess, S. D. 2016. Initial pulse of Siberian traps sills as the trigger of the end-Permian mass extinction. Nature Communications 164, https://www.nature.com/articles/s41467-017-00083-9

Burke, G. V. 1988. Agriculture's struggle for survival in the Great Clay Belt of Ontario and Quebec. American Review of Canadian Studies, v. 18, p. 455–464.

Burnham, A. D., and Berry, A. J. 2017. Formation of Hadean granites by melting of igneous crust. Nature Geoscience, v. 10, p. 457–461.

Cannon, W. F. 1992. The Midcontinent Rift in the Lake Superior region with emphasis on its geodynamic evolution. Tectonophysics, v. 213, p. 41–48.

Cannon, W. F. 1994. Closing of the Midcontinent Rift: A far-field effect of Grenvillian compression. Geology, v. 22, p. 155–158.

Cannon, W. F. 1999. Hard iron ore of the Marquette Range, Michigan. Economic Geology, v. 71, p. 1012–1028.

Cannon, W. F., and Gair, I. E. 1970. A revision of stratigraphic nomenclature of middle Precambrian rocks in northern Michigan: Geological Society of America Bulletin, v. 81, p. 2843–2846.

Cannon, W. F., Green, A. G., Hutchinson, D. R., Lee, M. W., Milkereit, B., Behrendt, J. C., Halls, H. C., Green, J. C., Dickas, A. B., Morey, G. B., Sutcliffe, R., and Spencer, C. 1989. The North American Mid-continent Rift beneath Lake Superior from Glimpse seismic reflection profiling. Tectonics, v. 8, p. 305–332.

Cannon, W. F., and Hinze, W. J. 1992. Speculations on the origin of the North American Midcontinent Rift. Tectonophysics, v. 213, p. 49–55.

Cannon, W. F., LaBerge, G. L., Klasner, J. S., and Schulz, K. J. 2007. The Gogebic iron range: A sample of the northern margin of the Penokean fold and thrust belt. U.S. Geological Survey Professional Paper 1730. 44 p.

Cannon, W. F., and Mudrey, M. G., Jr. 1981. The potential for diamond-bearing kimberlite in northern Michigan and Wisconsin. U.S. Geological Survey Circular 842. 32 p.

Cannon, W. F., Schulz, K. J., Ayuso, R. A., and Mroz, T. H., 2018, Archean and Paleoproterozoic geology of the Felch district, central Dickinson County, Michigan: Institute on Lake Superior Geology, 64th Annual Meeting, Field Trip Guidebook, p. 1–38.

Cannon, W. F., Schulz, K. J., Wright Horton, J., Jr.and, Kring, D. A. 2010. The Sudbury impact layer in the Paleoproterozoic iron ranges of northern Michigan, USA. Geological Society of America Bulletin, v. 122, p. 50–75.

Cannon, W. F., Woodruff, L. G., Schulz, K. J. 2012. The Hiawatha graywacke of the Iron River–Crystal Falls district, Michigan: A megaturbidite triggered by seismicity related to the 1850 Ma Sudbury impact. Institute on Lake Superior Geology Proceedings, v. 59, p. 14–15.

Card, K. D. 1978. Metamorphism of the middle Precambrian (Aphebian) rocks of the eastern Southern Province: Metamorphism in the Canadian Shield, Geological Survey of Canada paper p. 78–10, 269–282.

Card, K. D. 1990. A review of the Superior Province of the Canadian Shield, a product of Archean accretion. Precambrian Research, v. 48, p. 99–156.

Carlson, A. E. 2013. The Younger Dryas climate event: Encyclopedia of Quaternary Science, v. 3, p. 126–134.

Card, K. D., and Ciesielski, A. 1986. Subdivisions of the Superior Province of the Canadian Shield. Geoscience Canada, v. 13, p. 5–13.

Carlson, H., and Caballero, R. 2017. Atmospheric circulation and hydroclimate impacts of alternative warming scenarios for the Eocene. Climate of the Past, v. 13, p. 1037–1048.

Carson, E. C., Rawling, J. E., Attig, J. W., and Bates, B. R. 2012. Stream capture and the genesis of the upper Mississippi drainage. Geological Society of America Abstracts with Programs 45. p. 192.

Catacosinos, P. A., and Daniels, P. A., Jr. 1991. Stratigraphy of Middle Proterozoic to Middle Ordovician formations in the Michigan Basin. In Catacosinos, P.A., and Daniels, P. A., Jr., eds. Early sedimentary evolution of the Michigan Basin. Geological Society of America Special Paper 256.

Catacosinos, P. A., Harrison, W. B., Reynolds, R. F., Westjohn, D. B., and Wollensak, M. S. 2000. Stratigraphic nomenclature for Michigan. Michigan Geological Survey. http://www.michigan.gov/documents/deq/2000CHRT_301468_7.PDF

Catanzaro, E. J. 1963. Zircon ages in southwestern Minnesota. Journal of Geophysical Research, v. 68, p. 2045–2047.

Cates, N. L., Ziegler, K., Schmitt, A. K., and Mojzsis, S. J. 2013. Reduced, reused, and recycled: detrital zircons define a maximum age for the Eoarchean (ca 3750–3780 Ma) Nuvvuagittuq Supracrustal Belt, Quebec (Canada). Earth and Planetary Science Letters, v. 362, p. 283–293.

Cawood, P. A., Strachan, R. A., Pisarevsky, S. A., Gladkochub, D. P., and Murphy, J. B. 2016. Linking collisional and accretionary orogens during Rodinia assembly and breakup: implications for models of supercontinent cycles. Earth and Planetary Science Letters, v. 449, p. 18–126.

Cercone, K. R., and Pollack, H. N. 1991. Thermal maturity of the Michigan Basin. Geological Society of America Special Paper 256. p. 1–11.

Chamberlain, K. R., Taylor, K., Evans, D. A. D., and Bleeker, W. 2015. Late Archean to Proterozoic reconstructions of Wyoming and Superior cratons. Geological Society of America Abstracts with Programs 47. p. 447.

Chandan, D., and Peltier, R. 2017. Regional and global climate for the mid-Pliocene using the University of Toronto version of CCSM4 and PlioMIP2 boundary conditions. Clim. Past, v. 13, p. 919–942.

Chandler, V. W. 1995. The west-central Minnesota earthquake of June 5, 1993: An opportunity to re-examine seismicity near the Morris fault. Seismological Research Letters, v. 66, p. 113–118.

Chandler, V. W., Boerboom, T. J., and Jirsa, M. A. 2007. Penokean tectonics along a promontory-embayment margin in east-central Minnesota. Precambrian Research, v. 157, p. 26–49.

Chapra, S. C., Dove, A., and Warren, G. J. 2012. Long-term trends of Great Lakes major ion chemistry. Journal of Great Lakes Research, v. 38, p. 550–560.

Cheney, E. S. 1996. Sequence stratigraphy and plate tectonic significance of the Transvaal succession of southern Africa and its equivalent in western Australia. Precambrian Research, v. 79, p. 3–24.

Chew, D. M., Cardona, A., and Miskovic, A. 2014. Tectonic evolution of western Amazonia from the assembly of Rodinia to its breakup. International Geology Review, v. 53, p. 1280–1296.

Chmielewski, L. M. 2017. Jacques Marquette and Louis Jolliet. Routledge Historical Americans Series. Routledge. 196 p.

Clifton, J. A., and Porter, F. W. 1987. Potawatomi. Indians of North America Series. Chelsea House. 98 p.

Coakley, B., and Gurnis, M. 1995. Far-field tilting of Laurentia during the Ordovician and constraints on the evolution of a slab under an ancient continent. Journal of Geophysical Research, v. 100, no. B4, p. 6313–6327.

Cockell, C., ed. 2007. An introduction to the Earth-life system. Cambridge University Press. 145 p.

Coe, A. L. 2002. The sedimentary record of sea-level change. Cambridge University Press. p. 57–98.

Coleman, J. L., Jr., and Cahan, S. M. 2012. Preliminary catalog of the sedimentary basins of the United States. US Geological Survey Open File Report 2012–1111. 27 p.

Committee on the Importance of Deep-Time Geologic Records. 2011. Understanding Earth's deep past. National Academies Press. 194 p.

Condie, K. C. 2016. Earth as an evolving planetary system. 3rd ed. Elsevier. 432 p.

Condie, K. C., and Kröner, A. 2008. When did plate tectonics begin? Evidence from the geologic record. In Condie, K. C., Pease, V., eds. When did plate tectonics begin on planet Earth? Geological Society of America Special Paper 440, p. 441–447.

Condron, A., and Winsor, P. 2012. Meltwater routing and the Younger Dryas: Proceedings of the National Academy of Sciences, v. 109, p. 19,929–19,933.

Cook, T. D., and Bally, A. W. 1975. Stratigraphic atlas of North and Central America. Shell Oil Company Exploration Department. 272 p.

Cooper, A., Turney, C., Hughen, K. A., Brook, B. W., McDonald, H. G., and Bradshaw, C. J. A. 2015. Abrupt warming events drove Late Pleistocene Holarctic megafaunal turnover. Science, v. 349, p. 602.

Corfu, F. 1993. The evolution of the Southern Abitibi Greenstone Belt in light of precise U–Pb geochronology. Economic Geology, v. 88, p. 1323–1340.

Corfu, F. 2013. A century of U-Pb geochronology: The long quest towards concordance. Geological Society of America Bulletin, v. 125, p 33–47.

Costa, J. E., and Schuster, R. L. 1987. The formation and failure of natural dams. US Geological Survey Open-File Report 87–392. 39 p.

Cowen, R. 2013. Common source for Earth and Moon water. Nature News. http://www.nature.com/news/common-source-for-earth-and-moon-water-1.12963

Craddock, J. P., Rainbird, R. H., Davis, W. J., Davidson, C., Vervoort, J. D., Konstantinou, A., Boerboom, T., Vorhies, S., Kerber, L., and Lundquist, B.. 2013. Detrital zircon geochronology and provenance of the Paleproterozoic Huron (~2.4–2.2 Ga) and Animikie (~2.2–18 Ga) basins, southern Superior Province. Journal of Geology, v. 121, p. 623–644.

Cui, P.-L., Sun, J.-G., Sha, D., Wang, X.-J., Zhang, P., Gu, A.-L., and Wang, Z.-Y. 2013. Oldest zircon xenocryst (4.17 Ga) from the North China craton. International Geology Review. doi:10.1080/00206814.2013.805925.

Ćuk, M., Hamilton, D., Lock S. J., and Stewart, S. T. 2016. Tidal evolution of the Moon from a high-obliquity, high-angular-momentum Earth. Nature, v. 539, p. 402–406.

Cullers, R. L., and Berendsen, P. 1998. The provenance and chemical variation of sandstones associated with the Mid-Continent Rift system, U.S.A. European Journal of Mineralogy, v. 10, p. 987–1002.

Cundari, R., Smyk, M., Campbell, D., and Puumala, M. 2018. Possible emplacement controls on diamond-bearing rocks north of Lake Superior. Institute on Lake Superior Geology Abstract Volume, p. 19–20.

Curry, B., Grimley, D. A., and McKay, E. D., III. 2011. Quaternary glaciations in Illinois. Developments in Quaternary Science, v. 15, p. 467–487.

Czeck, D. M., and Ormand, C. J. 2007. Geometry and folding history of the Baraboo Syncline: Implications for the Mazatzal orogeny in the north-central U.S. Precambrian Research, v. 157, p. 203–213.

Dalrymple, G. B. 2004. Ancient Earth, ancient skies: The age of the Earth and its cosmic surroundings. Stanford University Press. 264 p.

Dalziel, I. W. 1997. Neoproterozoic-Paleozoic geography and tectonics: Review, hypothesis, environmental speculation. Geological Society of America Bulletin, v. 109, p. 16–42.

Damyanov, N., Matthews, H. D., and Mysak, L. A. 2012. Observed decreases in the Canadian outdoor skating seaon due to recent winter warming. Environmental Research Letters, v. 7, 014028.

Daniels, P. A., Jr. 1982. Upper Precambrian sedimentary rocks: Oronto Group, Michigan-

Wisconsin. In Wold, R.J., Hinze, W.J., eds. Geology and tectonics of the Lake Superior Basin. Geological Society of America Memoir 156. p. 107–133.

Daniels, P.A., and Morton-Thompson, D. 2010. Michigan Basin Utica/Collingwood shales: A new resource play. American Association of Petroleum Geologists, Abstract Eastern Section Meeting. p. 38.

Davidson, A. 2008. Late Paleoproterozoic to Mid-Neoproterozoic history of northern Laurentia: an overview of central Rodinia. Precambrian Research, v. 160, p. 5–22.

Davies, G. F. 1992. On the emergence of plate tectonics. Geology, v. 20, p. 963–966.

Davis, D., and Green, J. 1997. Geochronology of the North American Midcontinent Rift in western Lake Superior and implications for its geodynamic evolution. Canadian Journal of Earth Sciences, v. 34, p. 476–488.

Davis, D. W. 2008. Sub-million-year age resolution of Precambrian igneous events by thermal extraction-thermal ionization mass spectrometer Pb dating of zircon: Application to crystallization of the Sudbury impact melt sheet. Geology, v. 36, p. 383–386.

Day, S. E. 2014. Assessing the style of advance and retreat of the Des Moines lobe using LiDAR topographic data. M.Sc. Thesis, Iowa State University, 79 p.

DeConto, R. M., Galeotti, S., Pagani, M., Tracy, D., Schaefer, K., Zhang, T., Pollard, D., and Beerling, D. J. 2012. Past extreme warming events linked to massive carbon release from thawing permafrost. Nature, v. 484, p. 87–91.

Derby, J. R., Raine, R. J., Runkel, A. C., and Smith M. P. 2014. Paleogeography of the Great American Carbonate Bank of Laurentia in the earliest Ordovician (Earth Termadocian): the Stonehenge transgression. In Derby, J., Fritz, R., Longacre, S., Morgan, W., and Sternbach, C. The Great American Carbonate Bank. American Association of Petroleum Geologists Memoir 98. p. 5–14.

Deschamps, P., Durand, N., Bard, E., Hamilin, B., Camoin, G., Thomas, A. L., Henderson, G. M., Okuno, J., and Yokoyama, Y. 2012. Ice-sheet collapse and sea-level rise at the Bøling warming 14,600 years ago. Nature, v. 483, p. 559–564.

De Vleeschouwer, D., Da Silva, A-C., Sinnesael, M., Chen, D., Day, J. E., Whalen, M. T., Guo, Z., and Claeys, P. 2017. Timing and pacing of the Late Devonian mass extinction event regulated by eccentricity and obliquity. Nature Communications, article 2268. https://www.nature.com/articles/s41467-017-02407-1

Dewane, T. J., and Van Schmus, W. R. 2007. U-Pb geochronology of the Wolf River batholith, north-central Wisconsin: Evidence for successive magmatism between 1484 and 1468 Ma. Precambrian Research, v. 157, p. 215–234.

Diaz, H. F., and Hughes, J. 1994. The medieval warm period. Kluwer Academic Publishers. 134 p.

Dickas, A. B., Mudrey, M. G., Jr., Ojakangas, R. W., and Shrake, D. L. 1992. A possible southeastern extension of the Midcontinent Rift system located in Ohio. Tectonics, v. 11, p. 1406–1414.

Dietz, R. S. 1964. Sudbury structure as an astrobleme. Journal of Geology, v. 73, p. 412–421.

Dilek, Y., and Newcomb, S. 2003. Ophiolite concept and the evolution of geological thought. Geological Society of America Special Paper 373, p. 1–16.

Dixon, E. J. 2013. Late Pleistocene colonization of North America from northeast Asia: New insights from large-scale paleogeographic reconstructions. Quaternary International, v. 285, p. 57–67.

Dodd, M. S., Papineau, D., Grenne, T., Slack, J. F., Rittner, M., Pirajno, F., O'Neil, J.O., and Little, C. R. S. 2017. Evidence for early life in Earth's oldest hydrothermal vent precipitates. Nature, v. 543, p. 60–64.

Domagal-Goldman, S. D., Poirier, B., and Wing, B. 2011. Mass-independent fractionation of sulfur isotopes: carriers and sources. https://nai.nasa.gov/media/medialibrary/2013/08/S-MIF_WorkshopSummary.pdf

Dominati, E. J., Patterson, M. G., and Mackay, A. D. 2010. A framework for classifying and quantifying the natural capital and ecosystem services of soils. Ecological Economics, v. 69, p. 1858–1868.

Dott, R. H., Jr., and Byers, C.W. 2016. Cambrian geology of the Baraboo Hills. Geological Society of America Field Guide 43. p. 43–51.

Doughty, M., Eyles, N., and Daurio, L. 2010. Ongoing neotectonic activity in the Timiskaming-Kipawa area of Ontario and Quebec. Geoscience Canada, v. 37. https://journals.lib.unb.ca/index.php/GC/article/view/18397.

Doughty, M., Eyles, N., and Eyles, C. 2012. High-resolution seismic reflection profiling of neotectonic faults in Lake Timiskaming, Timiskaming Graben, Ontario-Quebec, Canada. Sedimentology, v. 60, p. 983–1006.

Doughty, M., Eyles, N., Eyles, C., Wallace, K., and Boyce, J.I. 2014. Lake sediments as natural seismographs: Earthquake-related deformations (seismites) in central Canadian lakes. Sedimentary Geology, v. 313, p. 45–67.

Droser, M. L., and Gehling, J. G. 2015. The advent of animals: The view from the Ediacaran. Proceedings of the National Academy Science, v. 112, p. 4865–4870.

Drzyzga, S. A., Shortridge, A. M., and Schaetzl, R. J. 2012. Mappiong the phases of Glacial Lake Algonquin in the upper Great Lakes region, Canada and USA, using a geostatistical isostatic rebound model. Journal of Paleolimnology, v. 47, p. 357–371.

Dubé, B., and Gosselin, P. 2007. Greenstone-hosted quartz-carbonate vein deposits. In Goodfellow, W.D., ed. Mineral deposits of Canada: a synthesis of major deposit-types, district metallogeny, the evolution of geological provinces, and exploration methods. Geological Association of Canada, Mineral Deposits Division, Special Publication 5. p. 49–73.

Dyke, A. S. 2004. An outline of North American glaciation with emphasis on central and northern Canada: Developments in Quaternary Sciences, v. 2, part B, p. 373–424.

Dyson, F. 1999. Origins of life. 2nd ed. Cambridge University Press. 112 p.

Ebel, J. E., and Tuttle, M. 2002. Earthquakes in the eastern Great Lakes Basin from a regional perspective. Tectonophysics, v. 353, p. 17–30.

Edwards, C. T., Saltzman, M. R., Royer, D. L., and Fike, D. A. 2017. Oxygenation as a driver of the Great Ordovician Biodiversification Event. Nature Geoscience, v. 10, p. 925–929.

Egan, D. 2017. The death and life of the Great Lakes. Norton. 384 p.

Ege, J. R. 1984. Formation of solution subsidence sinkholes above salt beds. US Geological Survey Circular 897. 11 p.

Eldholm, O., and Coffin, M. 2000. Large igneous provinces and plate tectonics. American Geophysical Union Geophysical Monograph 121. p. 309–326.

Elle, A-W. 2009. Hudson's Bay Company adventures: The rollicking saga of Canada's fur traders. Heritage House. 349 p.

Elliott, T., and Stewart, S. 2013. Planetary science: Shadows cast on Moon's origin. Nature, v. 504, p. 90–91.

Ellis, C. J., Carr, D. H., and Loebel, T. J. 2011. The Younger Dryas and Late Pleistocene peoples of the Great Lakes region. Quaternary International, v. 242, p. 534–545.

Elmore, R. D. 1983. Precambrian non-marine stromatolites in alluvial fan deposits: The Copper Harbor Conglomerate, upper Michigan. Sedimentology, v. 30, p. 829–842.

Elmore, R. D. 1984. The Copper Harbor Conglomerate: A late Precambrian fining-upward alluvial fan sequence in northern Michigan. Geological Society of American Bulletin, v. 95, p. 610–617.

Elmore, R. D., Milavec, G. J., Imbus, S. W., and Engel, M. H. 1989. The Precambrian Nonesuch Formation of the North American Mid-Continent Rift: sedimentology and organic geochemical aspects of lacustrine deposition. Precambrian Research, v. 43, p. 191–213.

Emu, R., Smit, M. A., Schmitt, M., Kooijman, E., Schere, E. E., Sprung, P., Bleeker, W., and Mezger, K. 2018. Evidence for evolved Hadean crust from Sr isotopes in apatite within Eoarchean zircon from the Acasta Gneiss Complex. Geochmicial Cosmochimica Acta, v. 235, p. 450–462.

Environmental Protection Agency. 2017. Climate change indicators: Great Lakes water levels and temperatures. https://www.epa.gov/climate-indicators/great-lakes

Ernst, R., and Bleeker, W. 2010. Large igneous provinces (LIPs), giant dyke swarms, and mantle plumes: significance for breakup events within Canada and adjacent regions from 2.5 Ma to Present. Canadian Journal of Earth Sciences, v. 47, p. 695–739.

Erwin, D. 2014. Temporal acuity and the rate and dynamics of mass extinctions. Proceedings of the National Academy of Science, v. 111, p. 3203–3204.

Ettensohn, F. R., and Lierman, R. T. 2015. Using black shales to constrain possible tectonic and structural influence on foreland-basin evolution and cratonic yoking: Late Taconian orogeny, Late Ordovician Appalachian Basin, eastern USA. Geological Society of London Special Publication 413, SP413.5.

Evans, D. A. 2009. The paleomagnetically viable, long-lived, and all-inclusive Rodinia supercontinent reconstruction. Geological Society of London Special Publication 327. p. 371–404.

Evans, D. A., Li, Z. X., and Murphy, J. B. 2016. Supercontinent cycles through Earth history. Geological Society of London Special Publication 424. p. 1–14.

Eyles, N., 2010. Canadian Shield: The rocks that made Canada. Fitzhenry and Whiteside. 128 p.

Eyles, N., and Meriano, M. 2010. Road-impacted sediment and water in the Lake Ontario watershed and lagoon, city of Pickering, Ontario, Canada: An example of urban basin analysis. Sedimentary Geology, v. 224, p. 15–28.

Fairchild, L. M., Swanson-Hysell, N. L., Ramezani, J., Sprain, C. H. J., and Bowring, S. A.

2017. The end of Midcontinent Rift magmatism and the paleogeography of Laurentia. Lithosphere, v. 9, p. 117–133.

Farley, K. A., and McKeon, R. 2015. Radiometric dating and temperature history of banded iron formation-associated hematite, Gogebic iron range, Michigan, USA. Geology, v. 43, p. 1083–1086.

Farlow, J. O., Sunderman, J. A., Havens, J. J., Swinehart, A. L., Holman, J. A., Richards, R. L., Miller, N. G., Martin, R. A., Hunt, R. M., Jr., Storrs, G. W., Curry, B. B., Fluegeman, R. H., Dawson, M. R., and Flint, M. E. T. 2001. The Pipe Creek sinkhole biota: A diverse Late Tertiary continental fossil assemblage from Grant County, Indiana. American Midland Naturalist, v. 145, p. 367–378.

Farquhar, J., Bao, H., and Thiemens, M. 2000. Atmospheric influence of Earth's earliest sulfur cycle. Science, v. 289, p. 756–758.

Farrand, W. R. 1988. The glacial lakes around Michigan. Geological Survey Division, Bulletin 4, 10 p.

Faure, S., Godey, S., Fallara, F., and Trepanier, S. 2011. Seismic architecture of the Archean North American mantle and its relationship to diamondiferous kimberlite fields. Economic Geology, v. 106, p. 223–240.

Faust, T. H., Fujita, K., and Mackey, K.G. 1997. The September 2, 1994, central Michigan earthquake. Seismological Research Letters, v. 68, p. 460–465.

Feasby, D. G. 1997. Environmental restoration of uranium mines in Canada: Progress over 52 years. International Atomic Energy Agency, Vienna (Austria). p. 35–48; http://www.iaea.org/inis/collection/NCLCollectionStore/_Public/29/022/29022004.pdf

Fedorchuk, N. D., Dombos, S. Q., Corsetti, F., and Wilmeth, D. 2016. Early non-marine life: Evaluating the biogenicity of Mesoproterozoic fluvial-lacustrine stromatolites. Precambrian Research, v. 275, p. 105–118.

Finnegan, S., Bergmann, K., Eller, J. M., Jones, D. S., Fike, D. A., Eiseman, I., Hughes, N. C., Tripati, A. K., and Fischer, W. W. 2011. The magnitude and duration of Late Ordovician-Early Silurian glaciation. Science, v. 331, p. 903–906.

Fisher, D. C. 1984. Mastodon butchery by North American Paleo-Indians. Nature, v. 30, p. 271–272.

Fisher, D. C. 1995. Experiments on subaqueous meat caching. Current Research in the Pleistocene, v. 12, p. 77–80.

Fisher, D. C., 2009, Paleobiology and extinction of Proboscideans in the Great Lakes region of North America. In Haynes, G., ed. America megafaunal extinction at the end of the Pleistocene: Springer, p. 55–76.

Fisher, J. H., Barratt, M. W., Droste, J. B., and Shaver, R. H. 1988. Michigan Basin. In Sloss, L. L., ed. Sedimentary cover, North American craton. Geological Society of America, Geology of North America D-2.

Fisher, T. G., Weyer, K. A., Boudreau, A. M., Martin-Hayden, J. M., Krantz, D. E., and Breckenridge, A. 2012. Constraining Holocene lake levels and coastal dune activity in the Lake Michigan Basin. Journal of Paleolimnology, v. 47, p. 373–390.

Flowers, R. M., Ault, A. K., Kelley, S. A., Zhang, N., and Zhong, S. 2012. Epeirogeny or eustasy? Paleozoic-Mesozoic vertical motion of the North American continental interior from thermochronometry and implications for mantle dynamics. Earth and Planetary Science Letters, v. 317–318, p. 436–445.

Fox, C. 2001. A history of Michigan's Sand Dune Protection Act. Geological Society of America Abstracts with Programs 33. p. 2.

Fralick, P., Davis, D. W., Kissin, S. A. 2002. The age of the Gunflint Formation, Ontario, Canada: Single zircon U-Pb age determinations. Canadian Journal of Earth Sciences, v. 39, p. 1085–1091.

Fralick, P., and Riding, R. 2015. Steep Rock Lake: Sedimentology and geochemistry of an Archean carbonate platform. Earth-Science Reviews, v. 151, p. 132–175.

French, B. M. 1998. Traces of catastrophe: A handbook of shock-metamorphic effects in terrestrial meteorite impact structures. Lunar and Planetary Institute. 120 p.

French, B. M., Cordura, W. S., and Plescia, J. B. 2004. The Rock Elm meteorite impact structure, Wisconsin: geology and shock-metamorphic effects in quartz. Geological Society of America Bulletin, v. 116, p. 200–218.

Fretwell, J. D., Williams, J. S., and Redman, P. J. 1996. National water summary of wetland resources. US Geological Survey Water Supply Paper 2425. 431 p.

Friedman, G. M., and Kopaska-Merkel, D. C. 1991. Late Silurian pinnacle reefs of the Michigan Basin. In Catacosinos, P. A., and Daniels, P. A., Jr., eds. Early sedimentary evolution of the Michigan Basin. Geological Society of America Special Paper 256. p. 89–100.

Frieman, B. M., Kuiper, Y. D., Kelly, N. M., Monecke, T., and Kylander-Clark, A. 2017. Constraints on the geodynamic evolution of the southern Superior Province: U-Pb LA-ICP-MS analysis of detrital zircon in successor basins of the Archean Abitibi and Pontiac subprovinces of Ontario and Quebec, Canada. Precambrian Research, v. 292, p. 398–416.

Frost, C. D., and Frost, R. D. 2011. On ferroan (A-type) granitoids: their compositional variability and modes of origin: Journal of Petrology, v. 52, p. 39–53.

Froude, D. O., Ireland, T. R., Kinny, P. D., Williams, I. S., and Compston, W. 1983. Ion microprobe identification of 4100–4200 Myr-old zircons. Nature, v. 304, p. 617–618.

Fujita, K., and Sleep, N. 2012. Earthquakes. In Schaetzl, R., Darden, J., and Brandt, D., eds. Michigan geography and geology. Pearson Learning Solutions. p. 115–125.

Fullerton, D. S., Bush C. A., and Pennell, J. N. 2003. Map of surficial deposits and material in the eastern and central United States: U.S. Geological Survey Geologic Investigations Series I-2789.

Furnes, H., Banerjee, N. R., Muehlenbachs, K., Staudigel, H., and de Wit, M. 2004. Early life recorded in Archaean pillow lavas. Science, v. 304, p. 578–581.

Gaboury, D., and Pearson, V. 2008. Rhyolite geochemical signatures and association with volcanogenic massive sulfide deposits: Examples from the Abitibi Belt, Canada. Economic Geology, v. 103, p. 1531–1562.

Gaidos, E. J., Güdel, M., and Blake, G. A. 2000. The faint young sun paradox: An observational test of an alternative solar model. Geophysical Research Letters, v. 27, p. 501–504.

Gair, J. E. 1975, Bedrock geology and ore deposits of the Palmer quadrangle, Marquette County, Michigan: U.S. Geological Survey Professional Paper 769, 159 p.

Gair, J., and Thaden, R. E. 1968. Geology of the Marquette and Sands quadrangles, Marquette County, Michigan: U.S. Geological Survey Professional Paper 397, 87 p.

Gallagher, T. M., Sheldon, N. D., Mauk, J. L., Petersen, S. V., and Gueneli, N. 2017. Constraining the thermal history of the North American Midcontinent Rift system using carbonate clumped isotopes and organic thermal maturity indices. Precambrian Research, v. 294, p. 53–66.

Gao, C., McAndrews, J. H., Wang, Z., Menzies, J., Turton, C. L., Wood, B. D., Pei, J., and Kodors, C. 2012, Glacation of North America in the James Bay Lowland, Canada, 3.5 Ma: Geology, v. 40, p. 975–978.

Garzione, C. 2008. Surface uplift of Tibet and Cenozoic global cooling. Geology, v. 36, p. 1003–1004.

Geng, Y., Du, L., and Ren, L. 2012. Growth and reworking of the early Precambrian continental crust in the North China craton: Constraints from zircon Hf isotopes. Gondwana Research, v. 21, p. 517–529.

G-Farrow, C. E., and Mossman, D. J. 1988. Geology of Precambrian paleosols at the base of the Huronian Supergroup, Elliot Lake, Ontario, Canada. Precambrian Research, v. 42, p. 107–139.

Glukhovskii, M.Z., Kuz'min, M. I., Bayanova, T. B., Lyalina, L. M., Makrygina, V. A., and Shcherbakova, T. F. 2017. The first discovery of Hadean zircon in garnet granulite from the Sutam River (Aldan Shield). Doklady Earth Sciences, v. 476, p. 1026–1032.

Goldfarb, R. J., Baker, T., Dubé, B., Groves, D. I., Hart, C. J. R., and Gosselin, P. 2005. Distribution, character, and genesis of gold deposits in metamorphic terranes. Society of Economic Geologists 100th Anniversary Volume. p. 407–450.

Goldich, S. S. 1968. Geochronology in the Lake Superior region. Canadian Journal of Earth Sciences, v. 5, p. 715–724.

Goldich, S. S., Hedge, C. E. 1974. 3,800-Myr granitic gneisses in south-western Minnesota. Nature, v. 252, p. 467–467.

Goldich, S. S., Hedge, C. A., Stern, T. W. 1970. Age of the Morton and Motevideo gneisses and related rocks, southwestern Minnesota. Geological Society of America Bulletin, v. 81, p. 3671–3684.

Goldich, S. S., Nier, A. E., Baadsgaard, H., Hoffman, J. H., and Krueger, H. W. 1961. The Precambrian geology and geochronology of Minnesota. Minnesota Geological Survey Bulletin, v. 41, 193 p.

Gourcerol, B., Thurston, P.C., Kontak, D. J., Côtè-Mantha, O., and Biczok, J. 2016. Deposition setting Algoma-type banded iron formation. Precambrian Research, v. 281, p. 47–79.

Grammer, G. M., Harrison, W. B., III, and Barnes, D. A. 2018. Paleozoic stratigraphy and resources of the Michigan Basin. Geologic Society of America Special Paper 531, 361 p.

Gran, K. B. 2017. River incision, inverted long profiles, downstream coarsening, and terraces: what glaciation has wrought on incising Minnesota rivers and how it impacts management of rivers today. Geological Society of America Abstracts with Programs 49. p. 314.

Grandstaff, D. E. 1980. Origin of uraniferous conglomerates at Elliot Lake, Canada, and Witwatersrand, South Africa: implications for oxygen in the Precambrian atmosphere. Precambrian Research, v. 13, p. 1–26.

Great Lakes Environmental Research Laboratory. 2017. Historical ice cover. https://www.glerl.noaa.gov/data/ice/#historical

Green, J. C. 1982. Geology of the Keweenawan extrusive rocks. In Wold, R.J., Hinze, W.J., eds. Geology and tectonics of the Lake Superior Basin. Geological Society of America Memoir 156. p. 47–55.

Green, J. C., and Fitz, T. J., III. 1993. Extensive felsic lavas and rheoignimbrites in the Keweenawan Midcontinent Rift plateau volcanics, Minnesota: Petrographic and field recognition. J. Volcanology and Geothermal Research v. 54, 177–196.

Grey, K., and Sugitani, K. 2009. Palynology of Archean microfossils (c. 3.0 Ga) from the Mount Grant area, Pilbara craton, western Australia: Further evidence of biogenicity. Precambrian Research, v. 173, p. 60–69.

Griffin, W. L., O'Reilly, S. Y., Afonso, J. C., and Begg, G. C. 2009. The composition and evolution of lithospheric mantle: a re-evaluation and its tectonic implications. Journal of Petrology, v. 50, p. 1185–1204.

Grosch, E. G., Munoz, M., Mathon, O., and Mcloughlin, N. 2017. Earliest microbial trace fossils in Archaean pillow lavas under scrutiny: New micro-x-ray absorption near-edge spectroscopy, metamorphic and morphological constraints. Geological Society of London Special Publication 448. p. 57–70.

Grotzinger, J. P., and Knoll, A. H. 1999. Stromatolites in Precambrian carbonates: Evolutionary mileposts and environmental dipsticks. Annual Review of Earth and Planetary Sciences, v. 27, p. 313–358.

Gumsley, A. P. 2017. Validating the existence of the supercraton Vaalbara in the Mesoarcheaen to Palaeoproterozoic. PhD dissertation, Lund University, Faculty of Science, Department of Geology. Lithosphere and Biosphere Science. 130 p.

Gumsley, A. P., Chamberlain, K., Bleeker, W., Soderlund, U., Dekock, M. O., Larsson, E. R., and Bekker, A. 2017. Timing and tempo of the Great Oxidation Event. Proceedings of the National Academy of Sciences, v. 114, p. 1811–1816.

Gutierrez-Alonso, G., Johnston, S. T., Weil, A. B., Pastor-Galan, D., and Suarez, J. F. 2012. Buckling an orogen: The Cantabrian orocline. GSA Today, v. 22, p. 4–9.

Gutschick, R. C., and Sandberg, C. A. 1991. Late Devonian history of the Michigan Basin. In Catacosinos, P. A., and Daniels, P. A., Jr., eds. Early sedimentary evolution of the Michigan Basin. Geological Society of America Special Paper 256. p. 181–202.

Hallam, A. 1992. Phanerozoic sea-level changes. Columbia University Press. 266 p.

Halliday, A. N. 2008. A young Moon-forming giant impact at 70–110 million years accompanied by late-stage mixing, core formation, and degassing of the Earth. Philosophical Transactions the Royal Society, v. 366, p. 4163–4181.

Halls, H. 1972. Magnetic studies in northern Lake Superior. Canadian Journal of Earth Sciences, v. 9, p. 1349–1367.

Halls, H. 1974. A paleomagnetic reversal in the Osler Volcanic Group, northern Lake Superior. Canadian Journal of Earth Sciences, v. 11, p. 1200–1207.

Halls, H., and Pesonen, L. 1982. Paleomagnetism of Keweenawan rocks. Geological Society of America Memoir 156. p. 173–201.

Hamblin, W. K. 1958. Cambrian sandstones of northern Michigan. Michigan Geological Survey Publications. 146 p.

Hammer, P. T. C., Clowes, R. M., Cook, F. A., van der Velden, A. J., Vasudevan, K. 2007. The Lithprobe trans-continental lithospheric cross sections: Imaging the internal structure of the North American continent. Canadian Journal of Earth Sciences, v. 47, p. 821–857.

Han, T. M., and Runnegar, B. 1992. Megascopic eukaryotic algae from the 2.1-billion-year-old Negaunee iron-formation. Michigan. Science, v. 257, p. 232–235.

Hansen, M. C. 2009. Earthquakes in Ohio. Ohio Department of Natural Resources, Division of Geological Survey, Educational Leaflet 9.

Hapke, C. J., Malone, S., and Kratzman, M. 2009. National assessment of historical shoreline change: a pilot study of historical coastal bluff retreat in the Great Lakes, Erie, Pennsylvania. US Geological Survey Open-File Report 209–142. 34 p.

Haq, B. U., and Al-Qahtani, A. M. 2005. Phanerozoic cycles of sea-level change on the Arabian Platform. GeoArabia, v. 10/2, 127–160.

Haq, B. U., and Shutter, S. R. 2008. A chronology of Paleozoic sea-level changes. Science, v. 322, p. 64–68.

Harkins, S. 2009. Father Jacques Marquette. Profiles in American History. Mitchell Lane. 48 p.

Harrell, J. A., Hatfield, C. B., and Gunn, G. R. 1991. Mississippian system of the Michigan Basin: Stratigraphy, sedimentology, and economic geology. In Catacosinos, P. A. , Daniels, P. A., Jr. eds. Early sedimental evolution of the Michigan Basin. Geological Society of America Special Paper 256, p. 203–219.

Harrison, T. M. 2009. The Hadean crust: evidence from >4 Ga. Annual Review of Earth and Planetary Sciences, v. 37, p. 479–505.

Harrison, W. B., III., 2012. Hydrocarbon resources. In Schaetzl, R., Darden, J., and Brandt, D., eds. Michigan geography and geology. Pearson Learning Solutions, p. 126–138.

Harrison, W. B., III. 2014. Occurrence of potash-bearing strata (sylvite) in the Salia A-1 evaporite in the central Michigan Basin. AAPG Search and Discovery Article 10652. http://www.searchanddiscovery.com/pdfz/documents/2014/10652harrison/ndx_harrison.pdf.html

Hart, T. R., and MacDonald, C. A. 2007. Proterozoic and Archean geology of the Nipigon Embayment: Implications for emplacement of the Mesoproterozoic Nipigon diagase sills and mafic to ultramafic intrusions. Canadian Journal of Earth Sciences, v. 44, p. 1021–1040.

Hawkesworth, C. J., Cawood, P., and Dhuime, B. 2016. Tectonics and crustal evolutions. GSA Today, v. 26, p. 4–11.

Haynes, J. T. 1994. The Ordovician Deicke and Millbrig K-Bentonite beds of the Cincinnati Arch and the Southern Valley and Ridge Province. Geological Society of America Special Paper 290. p. 1–80.

Heaman, L. M., and Easton, R. M. 2005. Proterozoic history of the Lake Nipigon area, Ontario: constraints from U-Pb zircon and baddeleyite dating. In Easton, M., Hollings, P., eds. Institute on Lake Superior Geology Proceedings, 51st Annual Meeting, Nipigon, Ontario, Proceedings and Abstracts, v. 51, pt. 1, p. 24–25.

Heaman, L. M., Easton, R. M., Hart, T. R., Hollings, P., MacDonald, C. A., and Smyk, M. 2007. Further refinement to the timing of Mesoproterozoic magmatism, Lake Nipigon region, Ontario. Canadian Journal of Earth Sciences, v. 44, p. 1055–1086.

Heaman, L. M., and Kjarsgaard, B. A. 2000. Timing of eastern North American magmatism: Continental extension or the Great Meteor hotspot track? Earth and Planetary Science Letters, v. 178, p. 253–268.

Hegarty, K. A., Foland, S. S., Cook, A. C., Green, P. F., and Duddy, I. R. 2007. Direct measurement of timing: Underpinning a reliable petroleum system model for the Mid-Continent Rift system. American Association of Petroleum Geologists Bulletin, v. 91, p. 959–979.

Heinrich, H. 1988. Origin and consequences of cyclic ice rafting in the Northeast Atlantic Ocean during the past 130,000 years: Quaternary Research, v. 29, p. 142–152.

Hemming, S. R. 2004. Heinrich events: Massive late Pleistocene detritus layers of the North Atlantic and their global climate imprint. Reviews Geophysics, v. 42, p. 1005.

Hemming, S. R., Hanson, G. N., and McLennan, S. M. 1995. Precambrian crustal blocks in Minnesota: Neodymium isotope evidence from basement and metasedimentary rocks. In Sims, P. K., Carter, L. M. H., eds. U.S. Geological Survey Bulletin 1904. p. U1–U13.

Henry, L. G., McManus, J. F., Curry, W. B., Roberts, N. L., Piotrowski, A. M., and Keigwin, L. D. 2016. North Atlantic Ocean circulation and abrupt climate change during the last glaciation. Science, v. 353, p. 470–474.

Herrmann, A. D., Leslie, S. A., and MacLeod, K. G. 2011. Did intense volcanism trigger the first Late Ordovician icehouse? Palaios, v. 25, p. 831–836.

Heyl, A. V. 1968. The upper Mississippi Valley base-metal district. In Ridge, J. D., ed. Ore deposits of the United States, 1933–1967 (Graton-Sales vol.). American Institute of Mining, Metallurgical and Petroleum Engineers. p. 431–459.

Hiatt, E. E., Pufahl, P. K., and Edwards, C. T. 2015. Sedimentary phosphate and associated fossil bacteria in a Paleoptoterozoic tidal flat in the 1.85 Ga Michigamme Formation, Michigan, USA. Sedimentary Geology, v. 319, p. 24–39.

Hieshima, G. B., Zaback, D. A., Pratt, L. M. 1989. Petroleum potential of Precambrian Nonesuch Formation, Midcontinent Rift system. American Association of Petroleum Geologists Bulletin, v. 73, p. 363.

Hill, C., Corcoran, P. L., Aranha, R., and Longstaffe, F. J. 2016. Microbially induced sedimentary structures in the Paleproterozoic, upper Huronian Supergroup, Canada. Precambrian Research, v. 281, p. 155–165.

Hinze, W. J., Allen, D. J., Braile, L. W., and Mariano, J. 1997. The Midcontinent Rift system: A major Proterozoic continental rift. Geological Society of America Special Paper 312. p. 7–15.

Hinze, W. J., Wold, R. J., and O'Hara, N. W. 1982. Gravity and magnetic anomaly studies of Lake Superior. Geological Society of America Memoir 156. p. 203–214.

Hoffman, H. J., Pearson, D. A. B., and Wilson, B. H. 1980. Stromatolites and fenestral fabric in the Early Proterozoic Huronian Supergroup, Ontario. Canadian Journal of Earth Sciences, v. 17, p. 1351–1357.

Hoffman, P. F., Kaufman, A. J., Halverson, G. P., and Schrag, D. P. 1998. A Neoproterozoic snowball Earth. Science, v. 281 (5381), p. 1342–1346.

Holen, S. R., Deméré, T. A., Fisher, D. C., Fullagar, R., Paces, J. B., Jefferson, G. T., Beeton, J. M., Cerutti, R. A., Rountrey, A. N., Vescere, L., and Holen, K. A. 2017. A 130,000-year-old archaeological site in southern California, USA. Nature, v. 544, p. 479–483.

Holland, H. D. 1984. Chemical evolution of the atmosphere and oceans. Princeton University Press. 598 p.

Holland, H. D. 2002. Volcanic gases, black smokers, and the Great Oxidation Event. Geochimica Cosmochimica Acta, v. 66, p. 3811–3826.

Holling, Holling C. 1941. Paddle to the Sea: Houghton Mifflin, 32 p.

Hollings, P., Fralick, P., and Cousens, B. 2007. Early history of the Midcontinent Rift inferred from geochemistry and sedimentology of the Mesoproterozoic Osler Group, northwestern Ontario. Canadian Journal of Earth Sciences, v. 44, p. 389–412.

Hollings, P., Smyk, M., Heaman, L. M., and Halls, H. 2010. The geochemistry, geochronology, and paleomagnetism of the dikes and sills associated with the Mesoproterozoic Midcontinent Rift near Thunder Bay, Ontario, Canada. Precambrian Research, v. 183, p. 553–571.

Holm, D. K., Darrah, K. S., and Lux, D. R. 1998. Evidence for widespread ~1760 Ma metamorphism and rapid crustal stabilization of the early Proterozoic (1870–1820) Ma Penokean orogeny, Minnesota. American Journal of Science, v. 298, p. 60–81.

Holm, D. K., Schneider, D. A., and Coath, C.D. 1998. Age and deformation of Early Proterozoic quartzites in the southern Lake Superior region: Implications for extent of foreland deformation during final assembly of Laurentia. Geology, v. 26, p. 907–910.

Holm, D. K., Schneider, D. A., Rose, S., Mancuso, C., McKenzie, M., Foland, K. A., and Hodges, K. V. 2007. Proterozoic metamorphism and cooling in the southern Lake Superior region, North America, and its bearing on crustal evolution. Precambrian Research, v. 157, p. 106–126.

Holm, D. K., Van Schmus, R., MacNeill, L. C., Boerboom, T. J., Schweitzer, D., and Schneider, D. 2005. U-Pb zircon geochronology of Paleoproterozoic plutons from the northern midcontinent, USA: Evidence for subduction flip and continued convergence after geon 18 Penokean orogenesis: Geological Society of American Bulletin, v. 117, p. 259–272.

Houlihan, E., Runkel, A., Feinberg, J. M., Cowan, C.A., and Titus, S. 2015. Paleomagnetic age constraints of Midcontinent Rift strata, northern Minnesota. Geological Society of America Abstracts with Programs 47, p. 43.

Howell, P. D., and van der Pluijm, B. A. 1990. Early history of the Michigan Basin. Geology, v. 18, p. 1195–1198.

Huber, M., and Caballero, R. 2011. The Early Eocene equable climate problem revisited. Climates of the Past, v. 7, p. 603–633.

Huff, W. D., Kolatta, D. R., Bergstrom, S. M., and Zhang, Y. S. 1996. Large-magnitude

Middle Ordovician volcanic ash falls in North America and Europe: dimensions, emplacement, and post-emplacement characteristics. Journal of Volcanology and Geothermal Research, v. 73, p. 285–301.

Hull, P. M., Darroch, S., and Erwin, D. H. 2015. Rarity in mass extinctions and future of ecosystems. Nature, v. 528, p. 345–351.

Hunter, R. D., Panyushkina, I. P., Leavitt, S. W., Wiedenhoeft, A. C., and Zawiskie, J. 2006. A multiproxy environmental investigation of Holocene wood from a submerged conifer forest in Lake Huron, USA. Quaternary Research, v. 66, p. 667–77.

Hutchinson, D. R., White, R. S., Cannon, W. F., and Schulz, K. J. 1990. Keweenaw hot spot: Geophysical evidence for a 1.1 Ga mantle plume beneath the Midcontinent Rift system. Journal of Geophysical Research, v. 95, p. 10869–10884.

Iizuka, T., Horie, K., Komiya, T., Maruyama, S., Hirata, T., Hidaka, H., and Windley, B. F. 2006. 4.2 Ga zircon xenocryst in an Acasta gneiss from northwestern Canada: Evidence for early continental crust. Geology, v. 34, p. 245–249.

Imbus, S. W., Engel, M. H., and Elmore, R. D. 1990. Organic geochemistry and sedimentology of Middle Proterozoic Nonesuch Formation: Hydrocarbon assessment of a lacustrine rift deposit. In Katz, B.J., ed. Lacustrine basin exploration. American Association of Petroleum Geologists Memoir 50. p. 197–208.

Ivy-Ochs, S., and Kober, F. 2008. Surface exposure dating with cosmogenic nuclides. Quaternary Science Journal, v. 57, p. 179–209.

Izett, G. A., Cobban, W. A., Obradovich, J. D., and Dalrymple, G. B. 1998. 40Ar/39Ar age of the Manson impact structure, Iowa, and correlative impact ejecta in the Crow Creek member of the Pierre Shale (Upper Cretaceous), South Dakota and Nebraska. Geological Society of America Bulletin, v. 110, p. 361–376.

Jaebong, J., Urban, N. R., and Green, S. 1999. Release of copper from mine tailings on the Keweenaw Peninsula. Journal of Great Lakes Research, v. 25, p. 721–734.

Jambor, J. L. 1971. The silver-arsenide deposits of the Cobalt-Gowganda region, Ontario. Canadian Mineralogist, v. 11, p. 1–7.

James, H. L. 1955. Zones of regional metamorphism in the Precambrian of northern Michigan. Geological Society of America Bulletin, v. 66, p. 1455–1488.

James, H. L. 1958. Stratigraphy of pre-Keweenawan rocks in parts of northern Michigan. US Geological Survey Professional Paper 314-C. p. 27–44.

James, H. L., Dutton, C. E., Pettijohn, F. J., and Wier, K. L. 1968. Geology and Ore Deposits of the Iron River-Crystal Falls district, Iron Country, Michigan: U.S. Geological Survey Professional Paper 570, 134 p.

Jennings, C. E., and Johnson, M. D. 2011. The Quaternary of Minnesota. Developments in Quaternary Sciences, v. 15, p. 499–511.

Johnson, T. E., Brown, M., Goodenough, K. M., Clark, C., Kinny, P. D., and White, R. W. 2016. Subduction or sagduction: ambiguity in constraining the origin of ultramafic-mafic bodies in the Archean crust of NW Scotland. Precambrian Research, v. 283, p. 89–105.

Johnson, W. H., Hansel, A. K., Bettis, E.A., III, Karrow, P. F., Larson, G. J., and Lowell, T. V. 1997. Late Quaternary temporal and event classification, Great Lakes region, North America. Quaternary Research, v. 47, p. 1–12.

Jolly, W. T. 1978. Metamorphic history of the Archean Abitibi Belt: Metamorphism in the Canadian Shield. Geological Survey of Canada Paper 78–10. p. 63–77.

Kalliokoski, J. 1982. Jacobsville Sandstone. In Wold, R. J., Hinze, W. J., eds. Geology and tectonics of the Lake Superior Basin. Geological Society of America Memoir 156. p. 147–156.

Kamata, S., Sugita, S., Abe, Y., Ishihara, Y., Harada, Y., Morota, T., Namiki, N., Iwata, T., Hanada, H., Araki, K., Matsumoto, K., Tajika, E., Kuramoto, K., and Nimmo, F. 2015. Late Heavy Bombardment inferred from highly degraded impact basin structures. Icarus, v. 250, p. 492–503.

Kamber, B. S ., Whitehouse, M. J., Bolhar, R., and Moorbath, S. 2005. Volcanic resurfacing and the early terrestrial crust: Zircon U-Pb and REE constraints from the Isua Greenstone Belt, southern Greenland. Earth and Planetary Science Letters, v. 240, p. 276–290.

Karrow, P. F. 2000. A proposed diachronic revision of the Late Quaternary time-stratigraphic classification in the eastern and northern Great Lakes region. Quaternary Research, v. 54, p. 1–12.

Kasting, J. F., and Howard, M. T. 2006. Atmospheric composition and climate on the early Earth. Philosophical Transactions of the Royal Society B, v. 361, p. 1733–1742.

Kasting, J. F., and Ono, S. 2006. Palaeoclimates: The first two billion years. Philosophical Transactions of the Royal Society, ser. B, v. 361, p. 917–929.

Katz, M. R. 1955. The Black Swamp: A study in historical geography. Annals of the Association of American Geographers, v. 45, p. 1–35.

Kaufman, A. J., and Xiao, S. 2003. High CO_2 levels in the Proterozoic atmosphere estimated from analyses of individual microfossils. Nature, v. 425, p. 279–283.

Kean, W. F. 1981. Paleomagnetism of the late Ordovician Neda iron ore from Wisconsin, Iowa, and Illinois. Geophysical Research Letters, v. 8, p. 880–882.

Keays, R. R., and Lightfoot, P. C. 2015. Geochemical stratigraphy of the Keweenawan Midcontinent Rift volcanic rocks with regional implications for the genesis of associated Ni, Cu, Co and platinum group element sulfide mineralization. Economic Geology, v. 110, p. 1235–1267.

Keith, B. D., and Wickstrom, L. H. 1992. Lima-Indiana trend: Stratigraphic traps III. American Association of Petroleum Geologists Special Volume A025. p. 347–367.

Kelley, D. S. 2001. Black smokers: Incubators on the seafloor. American Museum of Natural History. http://www.amnh.org/learn/pd/earth/pdf/black_smokers_incubators.pdf

Kelly, W. C., and Nishioka, G. K. 1985. Precambrian oil inclusions in late veins and the role of hydrocarbons in copper mineralization at White Pine, Michigan. Geology, v. 13, p. 334–337.

Kemp, A. I. S., Wilde, S. A., Hawkesworth, C. J., Coath, C. D., Nemchin, A., Pidgeon, R. T., Vervoort, J. D., and DuFrane, S. A. 2010. Hadean crustal evolution revisited: New constraints from Pb–Hf isotope systematics of the Jack Hills zircons. Earth and Planetary Science Letters, v. 296, p. 45–56.

Kendall, A., Kesler, S. E., and Keoleian, G. A. 2008. Geologic vs. geographic constraints on cement resources. Resources Policy, v. 33, p. 160–167.

Kendall, A., Kesler, S. E., and Keoleian, G. A. 2010. Megaquarry versus decentralized mineral production: network analysis of cement production in the Great Lakes region, USA. Journal of Transport Geography, v. 18, p. 322–330.

Kennedy, G., and Mayer, T. 2002. Natural and constructed wetlands in Canada: An overview. Water Quality Resource Journal of Canada, v. 37, p. 295–325.

Kenny, G. G., Whitehouse, M. J., and Kamber, B. S. 2016. Differentiated impact melt sheets may be a potential source of Hadean detrital zircon. Geology, v. 44, p. 435–438.

Kerfoot, W. C., Urban, N. R., McDonald, C. P., Rossmann, R., and Zhang, H. 2016. Legacy mercury release during copper mining near Lake Superior. Journal of Great Lakes Research, v. 42, p. 50–61.

Kerr, M., and Eyles, N. 2001. Origin of drumlins on the floor or Lake Ontario and in upper New York State. Sedimentary Geology, v. 193, p. 7–20.

Kesler, S. E., and Simon, A. C. 2015. Mineral resources, economics, and the environment. Cambridge University Press. 434 p.

Ketchum, K. Y., Heaman, L. M., Bennett, G., and Hughes, D. J. 2013. Age, petrogenesis, and tectonic setting of the Thessalon volcanic rocks, Huronian Supergroup, Canada. Precambrian Research, v. 223, p. 144–172.

Kincare, K., and Larson, G. J. 2012. Evolution of the Great Lakes. In Schaetzl, R., Darden, J., and Brandt, D., eds. Michigan geography and geology. Pearson Learning Solutions, p. 174–190.

Kirschvink, J. L. 1992. Late Proterozoic low-latitude global glaciation: The snowball Earth. In Schopf, J. W., Klein, C., eds. The Proterozoic biosphere: A multidisciplinary study. Cambridge University Press. pp. 51–52.

Kjarsgaard, B. A., and Levinson, A. A. 2002. Diamonds in Canada. Gems and gemology, Fall 2002, p. 208–237.

Klewin, K. W., and Shirey, S. B. 1992. The igneous petrology and magmatic evolution of the Midcontinent Rift. Tectonophysics, v. 213, p. 33–40.

Knaeble, A. R. 2006. Landforms, stratigraphy, and lithologic characteristics of glacial deposits in central Minnesota. Minnesota Geological Survey Guidebook 22, 44 p.

Knoll, A. H., Barghoorn, E. S., and Awramik, S. M. 1978. New microorganisms from the Aphebian Gunflint iron formation, Ontario. Journal of Paleontology, v. 52, p. 976–992.

Koch, P. L., and Barnosky, A. D. 2006. Late Quaternary extinctions: State of the debate. Annual Review of Ecology, Evolution and Systematics, v. 37, p. 215–250.

Kolbert, E. 2014. The sixth extinction: An unnatural history. Bloomsbury Publishing. 231 p.

Konhauser, K. O., Hamade, T., Raiswell, R., Morris, R. C., Ferris, F. G., Southam, G., and Canfield, D. E. 2002. Could bacteria have formed the Precambrian banded iron formations? Geology, v. 12, p. 1079–1082.

Kopp, R. E., Kemp, A. C., Bitterman, K., Horton, B. P., Donnelly, J. P., Gehrels, W. R., Hay, C. C., Mitrovica, J. Z., Morrow, E. D., and Rahmstorf, S. 2016. Temperature-driven global sea-level variability in the Common Era. Proceedings of the National Academy of Science, http://www.pnas.org/content/pnas/113/11/E1434.full.pdf

Korochantseva, E. V., Trieloff, M., Lorenz, C. S., Buykin, A. I., Ivannova, J. A, Schwarz,

W. H., Hopp, J., and Jessberger, E. K. 2007. L-chondrite asteroid breakup tied to Ordovician meteorite shower by multiple isochron 40Ar-39Ar dating. Meteoritics and Planetary Science, v. 42, p. 113–130.

Krause, D. J. 1992. The making of a mining district: Keweenaw native copper, 1500–1870. Wayne State University Press. 198 p.

Krogh, T. E., Davis, D. W., and Corfu, F. 1984. Precise U-Pb zircon and baddeleyite ages for the Sudbury structure. In Pye, E. G., Naldrett, A. J., and Giblin, P. E., eds. Geology and ore deposits of the Sudbury structure. Ontario Geological Survey Special Volume 1. p. 431–446.

Kump, L. R. 2008. The rise of atmospheric oxygen. Nature, v. 451, p. 277–278.

LaBerge, G. L., ed. 1996. Volcanogenic massive sulfide deposits of northern Wisconsin: A commemorative volume. Institute on Lake Superior Geology Proceedings, 42nd Annual Meeting, Cable WI, v. 42, pt. 2, 179 p.

LaBerge, G. L., Cannon, W. F., Schulz, K. J., Klasner, J. S., and Ojakangas, R. W. 2003. Paleoproterozoic stratigraphy and tectonics along the Niagara suture zone, Michigan and Wisconsin. Institute on Lake Superior Geology Proceedings Volume 49, pt. 2, Field Trip Guidebook. 110 p.

LaBerge, G. L., Klasner, J. S., and Myers, P. E. 1998. New observations on the age and structure of Proterozoic quartzites in Wisconsin. U.S. Geological Survey Bulletin, v. 1904, p. B1–B16.

Lallensack, R. 2017. Global fingerprints of sea-level rise revealed by satellites. Nature News. https://www.nature.com/news/global-fingerprints-of-sea-level-rise-revealed-by-satellites-1.22588

Lang, A. H., Goodwin, A. H., Mulligan, R., Whitmore, D. R. E., Gross, G. A., Boyle, R. W., Johnstone, A. G., Chamberlain, J. A., and Rose, E. R. 1970. Economic minerals of the Canadian Shield. In geology and economic minerals of Canada. Geological Survey of Canada Economic Geology Report 1. p. 152–226.

Lang, N., and Wolff, E. W. 2011. Interglacial and glacial variability from the last 800 ka in marine, ice and terrestrial archives. Clim. Past, v. 7, p. 361–380.

Langford, F. F., and Morin, J. A. 1976. The development of Superior Province of northwestern Ontario by merging island arcs. American Journal of Science, v. 276, p. 1024–1034.

Larson, C. 1987. Geological history of the glacial Lake Algonquin and the Upper Great Lakes. U.S. Geological Survey Bulletin 1801, 36 p.

Larson, G. J. 2011. Ice-margin fluctuations at the end of the Wisconsin episode, Michigan, USA. Development in Quaternary Science, v. 15, p. 489–497.

Larson, G. J., and Schaetzl, R. 2001. Origin and evolution of the Great Lakes. J. Great Lakes Research, v. 27, p. 518–546.

Larson, P., and Mooers, H. 2003. Holocene drainage evolution of the Mississippi headwaters, Minnesota: implications for mid-Holocene eolian activity in the North American midcontinent. Geological Society of America Abstracts with Programs 35. p. 482.

Lavoie, D., Dietrich, J., Duchesne, M., Zhang, S., Pinet, N., Brake, V., Dewing, K., Asselin, E., Hu, K., Lajeunesse, P., and Roger, J. 2010. Geological setting and petroleum potential of the Paleozoic Hudson Platform, northern Canada. AAPG Search and Discovery Article 90172.

Lazorek, M., Carter, T. 2008. Oil and gas plays of Ontario. Ontario Petroleum Institute. http://www.ogsrlibrary.com/downloads/Ontario-Oil-Natural-Gas-Plays.pdf

Lenders, J. D. 2001. Long-term trends in the seasonal cycle of Great Lakes water levels. Journal of Great Lakes Research, v. 27, p. 342–353.

Leverington, D. W., Mann, J. D., and Teller, J. T. 2002. Changes in the bathymetry and volume of glacial Lake Agassiz between 9200 and 7700 ^{14}C yr B.P. Quaternary Research, v. 57, p. 244–252.

Lewis, C. F., Blasco, S. M., and Gareau, P. L. 2005. Glacial isostatic adjustment of the Laurentian Great Lakes Basin: Using the empirical record of strandline deformation for reconstruction of early Holocene Paleo-Lakes and discovery of a hydrologically closed phase. Geographie physique et Quaternaire, v. 59, p. 187–210.

Lewis, C. F., Cameron, G. D. M., Anderson, T. W. Heil, C. W., Jr., and Gareau, P. L. 2012. Lake levels in the Erie Basin of the Laurentian Great Lakes. J Paleolimnol, v. 47, p. 493–511.

Li, Z. X., Evans, D. A. D., and Murphy, J. B. 2016. Supercontinent cycle through earth history. Geological Society of London Special Publication 424. 289 p.

Lightfoot, P. A. 2017. Nickel sulfide ores and impact melts: Origin of the Sudbury Igneous Complex. Elsevier. 653 p.

Lisiecki, L. E., and Raymo, M. E. 2005. A Pliocene-Pleistocene stack of 57 globally distributed benthic $\delta^{18}O$ records (PDF). Paleoceanography, v. 20: PA1003. Bibcode:2005PalOc..20.1003L. doi:10.1029/2004PA001071.

Lister, A. M., and Stuart, A. J. 2008. The impact of climate change on large mammal distribution and extinction: Evidence from the last glacial/interglacial transition. Comptes Rendus Géosciences, v. 340, p. 615–620.

Liu, M., and Yang, Y. 2003. Extensional collapse of the Tibetan Plateau: Results of three-dimensional finite element modeling. Journal of Geophysical Research, v. 108, p. 2361.

Liu, X., Parker, G., Czuba, J. A., Oberg, K. J. Mier, J. M., Best, J. L., Parsons, D. R., Ashmore, P., Krishnappan B. G., and Garcia, M. H. 2012. Sediment mobility and bed armoring in the St. Clair River: Insights from hydrodynamic modeling. Earth Surface Processes and Landforms, v. 37, p. 957–970.

Lodge, R. W. D. 2016. Petrogenesis of intermediate volcanic assemblages from the Shebandowan Greenstone Belt, Superior Province: Evidence for subduction during the Neoarchean. Precambrian Research, v. 272, p. 150–167.

LoDuca, S. 2012. Paleozoic environments and life. In Schaetzl, R., Darden, J., and Brandt, D., eds. Michigan geography and geology. Pearson Learning Solutions, p. 40–59.

Long, D. G .F. 2004. The tectonostratigraphic evolution of the Huronian basement and the subsequent basin fill: Geological constraints on impact models of the Sudbury event. Precambrian Research, v. 129, p. 203–233.

Longo, A. A. 1984. A correlation for a middle Keweenawan flood basalt: The greenstone flow, Isle Royale and Keweenaw Peninsula, Michigan. M.S. thesis, Michigan Technological University. 198 p.

Loope, W. L., Loope, H. M., Goble, R. J., Fisher, T. G., Lytle, D. E., Legg, R. J., Wysocki, D. A., Hanson, P. R., and Young, A. R. 2012. Drought drove forest decline and dune building in eastern upper Michigan, USA, as the upper Great Lakes became closed basins. Geology, v. 40, p. 315–318.

Lovis, W. A., Arbogast, A. F., and Monaghan, G. W. 2012. The geoarchaeology of Lake Michigan coastal dunes. Environmental Research Series 2. Michigan State University Press. 223 p.

Lowell, T., Fisher, T., Hajdas, I., Glover, K., Loope, H., and Henry, T. 2009. Radiocarbon deglaciation chronology of the Thunder Bay, Ontario area and implications for ice sheet retreat patterns. Quaternary Science Reviews, v. 28, p. 1597–1607.

Luczaj, J. A., and Stieglitz, R. D. 2008. Geologic history of New Hope cave, Manitowoc County, Wisconsin. Wisconsin Speleologist, v. 6, p. 7–17.

Lyle, P. 2016. The abyss of time: A study in geological time and Earth history. Dunedin Academic Press. 216 p.

Lyons, T. W., Reinhard, C. T., and Planavsky, N. J. 2014. The rise of oxygen in Earth's early ocean and atmosphere. Nature, v. 506, p. 307–315.

Maas, R., and McCulloch, M. 1991. The provenance of Archean clastic metasediments in the Narryer Gneiss Complex, Western Australis: Trace element geochemistry, Nd isotopes, and U-Pb ages for detrital zircons. Geochimica Cosmochimica Acta, v. 55, p. 1915–1932.

MacGabhann, B. A. 2014. There is no such thing as the "Ediacaran biota." Geoscience Frontiers, v. 5, p. 53–62.

MacLeod, N. 2014. The geological extinction record: History, data, biases, and testing. Geological Society of America Special Paper 505. p. 1–29.

MacLeod, N. 2015. The great extinctions: What causes them and how they shape life. Firefly Books. 208 p.

MacPhee, R. D. E., and Marx, P. A. 1997. The 40,000 year plague: humans, hyperdisease, and first-contact extinctions. In Goodman, S. M., and Patterson, B. D. Natural Change and Human Impact in Madagascar. Smithsonian Institution Press. p. 169–217.

MacPherson, G. J. 2004. Calcium-aluminum-rich inclusions in chondritic meteorites. In Davis, A. M., ed. Treatise on Geochemistry. Vol. 1: Meteorites, comets, and planets. Elsevier. p. 201–246.

Maier, A. C., Cates, N. L., Trail, D., and Mojzsis, S. J. 2012. Geology, age, and field relations of Hadean zircon-bearing supracrustal rocks from Quad Creek, eastern Beartooth Mountains (Montana and Wyoming, USA). Chemical Geology, v. 312, p. 47–57.

Mainville, A., and Craymer, M. R. 2005. Present-day tilting of the Great Lakes region based on water level gauges. Geological Society of America Bulletin, v. 117, p. 1070–1080.

Malone, D. H., Stein, C. A., Craddock, J. P., Kley, J., Stein, S., and Malone, J. E. 2016. Maximum depositional age of the Neoproterozoic Jacobsville Sandstone, Michigan: Implications for the evolution of the Midcontinent Rift. Geosphere, v. 12, p. 1271–1282.

Marchi, S., Bottke, W. F., Elkins-Tanto, L. T., Bierhaus, M., Wuennemann, K., Morbidelli, A., and Kring, D. A. 2014. Widespread mixing and burial of Earth's Hadean crust by asteroid impacts. Nature, v. 511, p. 578–581.

Margold, M., Stokes, C. R., and Clark, C. D. 2015. Ice streams in the Laurentide Ice Sheet: Identification, characteristics and comparison to modern ice sheets. Earth-Science Reviews, v. 143, p. 117–146.

Marshall, C. R. 2006. Explaining the Cambrian "explosion" of animals. Annual Review of Earth and Planetary Sciences, v. 34, p. 355–384.

Martin, S. R. 1999. Wonderful power: The story of ancient copper working in the Lake Superior Basin. Wayne State University Press. p. 385.

Martini, A. M., Budai, J. M., Walter, L. M., and Schoell, M. 1996. Microbial generation of economic accumulations of methane within a shallow, organic-rich shale. Nature, v. 383, p. 155–158.

Matsch, C. L. 1983. River Warren, the southern outlet to Glacial Lake Agassiz. In Teller, J. T., and Clayton, L. eds. Glacial Lake Agassiz, Geological Association of Canada Special Paper 26, p. 231–244.

Mauk, J. L., Emsbo, P., and Theodorakos, P. 2015. Evaporated seawater formed sediment-hosted stratiform copper orebodies and second-stage copper mineralization in the Mesoproterozoic Nonesuch Formation of the Midcontinent Rift. Institute on Lake Superior Geology Abstracts, v. 61, p. 61–62.

Mauk, J. L., Kelly, W. C., Van der Pluijm, B. A., and Seasor, R. W. 1992. Relations between deformation and sediment-hosted copper mineralization: Evidence from the White-Pine of the Midcontinent Rift system. Geology, v. 20, p. 427–430.

Maurice, E. B. 2004. The last of the gentleman adventurers. Houghton Mifflin. 251 p.

May, G. S. 1967. Pictorial history of Michigan: The early years. Eerdmans. 239 p.

Maynard, J. B. 1986. Geochemistry of oolitic iron ores, an electron microprobe study. Economic Geology, v. 81, p. 1473–1483.

McBride, J. H., Leetaru, H. E., Bauer, R. A., Tingey, B. E., and Schmidt, S. E. 2007. Deep faulting and structural reactivation beneath the southern Illinois basin. Precambrian Research, v. 157, p. 289–313.

McElwain, J. C., and Punyasena, S. W. 2007. Mass extinction events and the plant fossil record. Trends in Ecology and Evolution, v. 22, p. 548–557.

McLelland, J. M., Selleck, B. W., and Bickford, M. E. 2010. Grenville Province, its Adirondack outlier, and the Mesoproterozoic inliers of the Appalachians. Geological Society of America Memoir 206, p. 21–49.

McLimans, R. K., Barnes, H. L., and Ohmoto, H. 1980. Sphalerite stratigraphy of the upper Mississippi Valley zinc-lead district, southwest Wisconsin. Economic Geology, v. 75, p. 351–361.

Medaris, L. G., Jr., Singer, B. S., Dott, R. H., Jr., Naymark, A., Johnson, C. M., and Schott, R.C. 2003. Late Paleoproterozoic climate, tectonics, and metamorphism in the southern Lake Superior region and Proto-North America: Evidence from Baraboo Interval Quartzites. Journal of Geology, v. 111, p. 243–257.

Medaris, L. G., Jr., Van Schmus, W. R., Loofboro, J., Stonier, P. J., Zhang, X., Holm, D. K.,

Singer, B. S., and Dott, R.H., Jr. 2007. Two Paleoproterozoic (Statherian) siliciclastic metasedimentary sequences in central Wisconsin: The end of the Penokean orogeny and cratonic stabilization of the southern Lake Superior region. Precambrian Research, v. 157, p. 188–202.

Merino, M., Keller, G. R., Stein, S., and Stein, C. 2013. Variations in Mid-Continent Rift magma volumes consistent with microplate evolution. Geophysical Research Letters, v. 40, p. 1513–1516.

Meyers, S. R., and Peters, S. E. 2011. A 56-million-year rhythm in North American sedimentation during the Phanerozoic. Earth Planetary Science Letters. doi:10.1016/j.epsl.2010.12.044.

Miall, A. D., ed. 2008. The sedimentary basins of the United States and Canada. Elsevier. p. 596.

Mickelson, D. M., and Colgan, P. M. 2003. The southern Laurentide Ice Sheet: Developments in Quaternarry Science, v. 1, p. 1–16.

Micklin, P. 2016. The future of the Aral Sea: Hope and despair. Environmental Earth Science, v. 75, p. 844.

Middleton, M. F. 2007. A model for the formations of intracratonic sag basins. Geophysical Journal International, v. 99, p. 665–676.

Milankovitch, M. 1941. Kanon der Erdbestrahlung und seine Andwendung auf das Eiszeitenproblem. R. Serbian Acad., Belgrade.

Miller, J. D. 2007. The Midcontinent Rift in the Lake Superior region: A 1.1 Ga large igneous province. http://www.largeigneousprovinces.org/07nov

Miller, J. D., Jr., and Chandler, V. W. 1997. Geology, petrology, and tectonic significance of the Beaver Bay complex, northeastern Minnesota. Geological Society of America Special Paper 312. p. 73–91.

Miller, J. D., and Nicholson, S. W. 2013. Geology and mineral deposits of the 1.1 Ma Midcontinent Rift in the Lake Superior region: An overview. In Miller, J., ed. Field guide to the copper-nickel-platinum group element deposits of the Lake Superior region. Precambrian Research Center Guidebook 13-01. p. 1–49.

Miller, K. G., Browning, J. V., and Wright, J. D. 2015. Sea-level change during hothouse, cool greenhouse, and icehouse worlds. American Geophysical Union Fall Meeting.

Miller, K. G., Korninz, M. A., Browning, J. V., Wright, J. D., Mountain, Gs. S., Katz, M. E., Sugarman, P. J., Cramer, B. S., Christie-Blick, N., and Pekar, S. F. 2005. The Phanerozoic record of global sea-level change. Science, v. 310, p. 1293–1298.

Miller, S. R., Mueller, P. A., Meert, J. G., and Kamenov, G. D. 2017. Detrital zircons reveal evidence of Hadean crust in the Singhbhum craton, India. Geological Society of America Abstracts with Programs 49. p. 134.

Mooers, H. D., Larson, P. C., and Marlow, L. R. 2005. Ice advances in the western Lake Superior region: A reevaluation of the St. Louis sublobe and the Marquette phase of the Superior lobe. Geological Society of America Abstracts with Programs, v. 37 (5), p. 92.

Mohajer, A., Eyles, N., and Rogojina, C. 1992. Neotectonic faulting in metropolitan Toronto: Implications for earthquake hazard assessment in the Lake Ontario region. Geology, v. 20, p. 1003–1006.

Mojzsis, S. J., Arrhenius, G., McKeegan, K. D., Harrison, T. M., Nutman, A.P., and Friend, C. R. L. 1996. Evidence of life on Earth before 3,800 million years ago. Nature, v. 384, p. 55–59.

Mojzsis, S. J., Harrison, T. M., and Pidgeon, R. T. 2001. Oxygen-isotope evidence from ancient zircons for liquid water at the Earth's surface 4,300 Myr ago. Nature, v. 409, p. 178–180.

Mojzsis, S. J., Morbidelli, A., Pahlevan, K., and Frank, E. A. 2013. Water on the primordial Earth. Mineralogical Magazine, v. 77, p. 1779.

Mole, D. R., Fiorentini, M. L., Thebaud, N., Cassidy, K. F., McCuaig, T. C., Kirkland, C. L., Romano, S. S., Doublier, M. P., Belousova, E. A., Barnes, S. J., and Miller, J. 2014. Archean komatiite volcanism controlled by the evolution of early continents. Proceedings of the National Academy of Sciences, v. 28, p. 10083–10088.

Morey, G. B. 1978. Lower and middle Precambrian stratigraphic nomenclature for east-central Minnesota. University of Minnesota Report of Investigations 21. 51 p.

Morey, G. B. 1983. Lower Proterozoic stratified rocks and Penokean orogeny in east-central Minnesota. In Medaris, L. G., Jr., ed. Early Proterozoic geology of the Lake Superior region. Geological Society of America Memoir 160. p. 97–122.

Morey, G. B. 1999. High-grade iron ore deposits of the Mesabi Range, Minnesota: Product of a continental-scale Proterozoic ground-water flow system. Economic Geology, v. 94, p. 133–142.

Morey, G. B., and Southwick, D. L. 1993. Stratigraphic and sedimentological factors controlling the distribution of epigenetic manganese deposits in iron-formation of the Emily district, Cuyuna iron range, east-central Minnesota. Economic Geology, v. 88, p. 104–122.

Mossler, J. H. 1992. Sedimentary rocks of Dresbachian age (Late Cambrian), Hollandale Embayment, southeastern Minnesota. Minnesota Geological Survey Report of Investigations 40. p. 70.

Mossler, J. H. 2008. Paleozoic stratigraphic nomenclature for Minnesota. Minnesota Geological Survey Report of Investigations 65. p. 76.

Mueller, P. A., and Wooden, J. L. 2012. Trace element and Lu-Hf systematics in Hadean-Archean detrital zircons: Implications for crustal evolution. Journal of Geology, v. 120, p. 15–29.

Murphy, J. B., and Nance, R. D. 2013. Speculations on the mechanisms for the formation and breakup of supercontinents. Geoscience Frontiers, v. 4, p. 185–194.

Murton, J. B., Bateman, M. D., Dallimore, S. R., Teller, J. T., and Yong, Z. 2010. Identification of Younger Dryas outburst flood path from Lake Agassiz to the Arctic Ocean. Nature, v. 464, p. 740–743.

Naafs, B. D. A., Hefter, J., and Stein, R. 2013. Millennial-scale ice rafting events and Hudson Strait Heinrich(-like) events during the late Pliocene and Pleistocene: A review. Quaternary Science Reviews, v. 80, p. 1–28.

Nance, R. D., and Murphy, J. B. 2013. Origins of the supercontinent cycle. Geoscience Frontiers, v. 4, p. 439–448.

Nance, R. D., Murphy, J. B., and Santosh, M. 2014. The supercontinent cycle: A retrospective essay. Gondwana Research, v. 25, p. 4–29.

Nance, R. D., Worsley, T. R., Moody, J. B. 1988. The supercontinent cycle. Scientific American, v. 259, p. 72–79.

Neff, B. P., and Nicholas, J. R. 2005. Uncertainty in the Great Lakes water balance. U.S. Geological Survey Scientific Investigations Report 2004–5100. 42 p.

Nemchin, A. A., Pidgeon, R. T., and Whitehouse, M. J. 2006. Re-evaluation of the origin and evolution of >4.2 Ga zircons from the Jack Hills metasedimentary rocks. Earth and Planetary Science Letters, v. 244, p. 218–233.

Nesbitt, E. A., Prothero, D. R., and Ivany, L. C. 2003. From greenhouse to icehouse. Columbia University Press. 376 p.

Newman, P. C. 2005. Company of adventurers: How the Hudson's Bay empire determined the destiny of a continent. Penguin Canada. 271 p.

NICE Working Group. 2007. Reinterpretation of Paleoproterozoic accretionary boundaries of the north-central United States based on new aeromagnetic-geologic compilation. Precambrian Research, v. 157, p. 71–79.

Nicholson, S., Cannon, W. F., and Schulz, K. J. 1992. Metallogeny of the Midcontinent Rift system of North America. Precambrian Research, v. 58, p. 353–386.

Nicholson, S., Shirey, S., Schultz, K., and Green, J. 1997. Rift-wide correlation of 1.1 Ma Midcontinent Rift system basalts: Implications for multiple mantle sources during rift development. Canadian Journal of Earth Sciences, v. 34, p. 504–520.

Nield, T. 2007. Supercontinent. Harvard University Press. 304 p.

Norman, M. D., Borg, L. E., Nyquist, L. E., Bogard, D. D. 2003. Chronology, geochemistry, and petrology of a ferroan noritic anorthosite clast from Descartes breccia 67215: Clues to the age, origin, structure, and impact history of the lunar crust. Meteoritics and Planetary Science, v. 38, p. 645–661.

Novotny, E. V., Murphy, D., and Stefan, H. G. 2008. Increase of urban salinity by road deicing salt. Science of the Total Environment, v. 406, p. 131–144.

O'Callaghan, J. W., Osinski, G. R., Lightfoot, P. C., Linnen, R. L., and Weirich, J. R. 2016. Reconstructing the geochemical signature or Sudbury breccia, Ontario, Canada: implications for its formation and trace metal content. Economic Geology, v. 111, p. 1705–1729.

O'Neil, J., Carlson, R.W., Francis, D., and Stevenson, R. 2008. Neodymium-142 evidence for Hadean mafic crust. Science, v. 321, p. 1828–1831.

Ochoa, D., Zavada, M. S., Liu, Y., and Farlow, J. O. 2016, Floristic implications of two contemporaneous inland upper Neogene sites in the eastern US: Pipe Creek sinkhole, Indiana, and the gray fossil site, Tennessee (USA). Palaeobiodiversity and Palaeoenvironments, v. 96, p. 239–254.

Ojakangas, R. W. 1988. Glaciation: An uncommon mega-event as a key to intracontinental and intercontinental correlation of Early Proterozoic basin fill, North American and Baltic cratons. In Kleinspehn, K. L, Paola, C., eds. New Perspectives in basin analysis. Springer Verlag, v. 431–444.

Ojakangas, R. W., and Dickas, A. B. 2002. The 1.1 Ma Midcontinent Rift system, central North America: Sedimentology of two deep boreholes, Lake Superior region. Sedimentary Geology, v. 147, p. 13–36.

Ojakangas, R. W., and Morey, G. B. 1982a. Keweenawan pre-volcanic quartz sandstones and related rocks of the Lake Superior region. Geological Society of America Memoir 156. p. 85–94.

Ojakangas, R. W., and Morey, G. B. 1982b. Keweenawan sedimentary rocks of the Lake Superior region: A summary. Geological Society of American Memoir 156. p. 157–164.

Ojakangas, R. W., Morey, G. B., and Green, J. C. 2001a. The Mesoproterozoic Midcontinent Rift system, Lake Superior region, U.S.A. In Eriksson, P., Catuneanu, O., and Martins-Neto, M., eds. The influence of magmatism, tectonics, sea level change, and palaeoclimate on Precambrian basin evolution: Change over time. Sedimentary Geology, v. 141, p. 421–442.

Ojakangas, R. W., Morey, G. B., and Southwick, D. L. 2001b. Paleoproterozoic basin development and sedimentation in the Lake Superior region, North America. Sedimentary Geology, v. 141–142, p. 319–341.

Ojakangas, R. W., Severson, M. J., and Jongewaard, P. K. 2011. Geology and sedimentology of the Paleoproterozoic Animikie Group: The Pokegama Formation, the Biwabik Iron Formation, and Virginia Formation of the eastern Mesabi iron range and Thomson Formation near Duluth, northeastern Minnesota. In Miller, J. D., Hudak, G. J., Wittkop, C., and McLaughlin, P. I., eds. Archean to Anthropocene: Field guides to the geology of the mid-continent of North America. Geological Society of America Field Guide 24. p. 101–120.

Ojakangas, R. W., Srinivasan, R., Hegde, V. S., Chandrakant, S. M., and Srikantia, S. V. 2014. The Talya Conglomerate and Archean (~2.7 Ga) Glaciomarine Formation, western Dharwar craton, southern India. Current Science, v. 106, p. 387–396.

Ojibwe People's Dictionary. 2018. http://ojibwe.lib.umn.edu/search?utf8=✓&q=goose&commit=Search&type=english

Olson, S. L., Kump, L. R, and, Kasting, J. F. 2013. Quantifying the areal extent and dissolved oxygen concentrations of Archean oxygen oases. Chemical Geology, v. 362, p. 35–43.

Olson, S. L., Reinhard, C. T., and Lyons, T. W. 2016. Limited role for methane in the mid-Proterozoic greenhouse. Proceedings of the National Academy of Science, v. 113, p. 11447–11452.

O'Shea, J. M., and Meadows, G. A. 2009. Evidence for early hunters beneath the Great Lakes: Proceedings of National Academy of Sciences, v. 106, p. 10120–10123.

Ostroff, A. G. 1964. The conversion of gypsum to anhydrite in aqueous salt solutions. Geochimica et Cosmochimica Acta, v. 28, p. 1363–1372.

Pagani, M., Caldeira, K., Archer, D., and Zachos, J. C. 2006. An ancient carbon mystery. Science, v. 314, p. 1556–1557.

Palacas, J. G. 1995. Superior Province: U.S. Geological Survey national oil and gas assessment. http://energy.cr.usgs.gov/oilgas/noga/index.htm.

Pantell, H. 1971. The story of the Ontonagon copper boulder. Mineralogical Record,

September–October. http://www.michigan.gov/documents/deq/GIMDL-GGOCB_302361_7.pdf

Pattison, L., and Bailey, D. G. 2016. Mineralogical and chemical composition of oolitic ironstones from the type locality, Clinton, New York. Geological Society of America Abstracts with Programs 48. Geological Society of America, Northeastern Section, 51st annual meeting.

Pavlov, A. A., Kasting, J. F., Brown, L. L., Rages, K. A., and Freedman, R. 2000. Greenhouse warming by CH_4 in the atmosphere of early Earth. Journal of Geophysical Research, v. 105, p. 11981–11990.

Pawlak, A., Eaton, D. W., Bastow, I. D., Kenday, J-M., Helffrich, G., Wookey, J., and Snyder, D. 2011. Crustal structure beneath Hudson Bay from ambient-noise tomography: implications for basin formation. Geophysical Journal International, v. 184, p. 65–84.

Pearson, D. G., and Wittig, N. 2008. Formation of Archaean continental lithosphere and its diamonds: The root of the problem. Geological Society of London Journal, v. 165, p. 895–914.

Pearson, V., and Daigneault, R. 2009. An Archean megacaldera complex: The Blake River Group, Abitibi Greenstone Belt. Precambrian Research, v. 168, p. 66–82.

Peck, W. H., Valley, J. W., Wilde, S. A., and Graham, C. M. 2001. Oxygen isotope ratios and rare earth elements in 3.3 to 4.4 Ga zircons: ion microprobe evidence for high $\delta^{18}O$ continental crust and oceans in the Early Archean. Geochimica et Cosmochimica Acta, v. 32, p. 4215–4229.

Percival, J. A., McNicoll, V., Brown, J. L., and Whalen, J. B. 2004 Convergent margin tectonics, central Wabigoon subprovince, Superior Province, Canada. Precambrian Research, v. 132, p. 213–244.

Percival, J. A., Sanborn-Barrie, M., Skulski, T., Stott, G. M., Helmstaedt, H., and White, D. J. 2006. Tectonic evolution of the western Superior Province from NATMAP and lithoprobe studies. Canadian Journal of Earth Sciences, v. 43, p. 1085–1117.

Percival, J. A., Skulski, T., Sanborn, Barrie, M., Stott, G., Leclair, A. D., Corkery, M. T., Boily, M. Geology and tectonic evolution of the Superior Province, Canada. Geological Association of Canada Special Paper 49- Chapter 6, p. 321–378.

Percival, J. A., and West, G. F. 1994. The Kapuskasing Uplift: A Geological and Geophysical Synthesis. Canadian Journal of Earth Sciences. v. 31, p. 1256–1286.

Pettijohn, F. J. 1952. Precambrian tillite, Menominee District, Michigan: Bulletin of the Geological Society of America. v. 63 p. 1289.

Philbrick, S. S. 1974. What future for Niagara Falls? Geological Society of America Bulletin, v. 85, p. 91–98.

Piercey, P., Schneidr, D. A., and Holm, D. K. 2007. Geochronology of Proterozoic metamorphism in the deformed Southern Province, northern Lake Huron region, Canada. Precambrian Research, v. 157, p. 127–143.

Pietrzak-Renaud, N. 2013. Sedimentary and metamorphic lithofacies of the Lower Negaunee iron formation, Marquette district, Michigan, USA. Canadian Journal of Earth Sciences, v. 50, p. 1165–1177.

Pietrzak-Renaud, N., and Davis, D. 2014. U-Pb geochronology of baddeleyite from the

Belleview metadiabase: Age and geotectonic implications for the Negaunee Iron Formation, Michigan. Precambrian Research, v. 250, p. 1–5.

Planavsky, N. J., Slack, J. F., Cannon, W. F., O'Connell, B., Isson, T. T., Asael, D., Jackson, J. C., Hardisty, D. S., Lyones, T. W., and Bekker, A. 2018. Evidence for episodic oxygenation in a weakly redox-buffered deep mid-Proterozoic ocean. Chemical Geology, v. 483, p. 581–594.

Pleger, T. C. 2002. A Brief introduction to the Old Copper Complex of the western Great Lakes, 4000–1000 BC. Proceedings of the Twenty-Seventh Annual Meeting of the Forest History Association of Wisconsin, October 5, 2002, p. 10–18.

Pohl, A., Donnadieu, Y., Le Hir, G., and Ferreira, D. 2017. The climatic significance of Late Ordovician–Early Silurian black shales. Paleoceanography and Paleoclimatology, v. 32, p. 397–423.

Polat, A. 2009. The geochemistry of Neoarchean (ca. 2700 Ma) tholeiitic basalts, transitional to alkaline basalts, and gabbros, Wawa Subprovince, Canada: Implications for petrogenetic and geodynamic processes. Precambrian Research, v. 168, p. 83–105.

Pollack, H. 2010. A world without ice. Penguin Random House. 304 p.

Pompeani, D. P., Abbott, M. B., Gain, D. J., DePasqual, S., and Finkenbinder, M. S. 2015. Copper mining on Isle Royale 6500–5400 years ago identified using sediment geochemistry from McCargoe Cove, Lake Superior. The Holocene, v. 25, p. 253–262.

Poulsen, H., Card, K., Mortensen, J., and Robert, F. 1992. Plate tectonics and the mineral wealth of the Canadian Shield. Geotimes, v. 37, p. 19–21.

Poulton, S., Fralick, P. W., and Canfield, D. E. 2004. The transition to a sulphidic ocean ~1.84 billion years ago. Nature, v. 431, p. 173–177.

Pregitzer, K. S., Reed, D. R., Bornhorst, T. J., Foster, D. R., Mroz, G. D., Mclachlan, J. S., Lakes. P. E., Stokke, D. D., Martin, P. E., and Brown, S. E. 2000. A buried spruce forest provides evidence at the stand and landscape scale for the effects of environment on vegetation at the Pleistocene/Holocene boundary. Journal of Ecology, v. 88, p. 45–53.

Prest, V. K., Donaldson, J. A., and Mooers, H. D. 2000. The omar story: The role of omars in assessing glacial history of west-central North America: Geographie physique et Quaternaire, v. 54, p. 257–270.

Prest, V. K., Grant, D. R., and Rampton, V. N. 1968. Glacial Map of Canada: Geological Survey of Canada, Map 1253A, Scale 1:5,000,000.

Prevec, S. A., Lightfoot, P. C., and Keays, R. R. 2000. Evolution of the sublayer of the Sudbury Igneous Complex: Geochemical, Sm-Nd isotopic, and petrologic evidence. Lithos, v. 51, p. 271–292.

Quinn F. 1985. Temporal effects of St. Clair River dredging on Lakes St. Clair and Erie water levels and connecting channel flow. Journal of Great Lakes Research, v. 11, p. 400–403.

Quirke, T. T., and Collins, W. H. 1930. The disappearance of the Huronian. Geological Survey of Canada Memoir 160. 127 p.

Raatz, W. D., and Ludvigson, G. A. 1996. Depositional environments and sequence stratigraphy of upper Ordovician epicontinental deep water deposits, eastern Iowa and southern Minnesota. Geological Society of American Special Paper 306. p. 143–159.

Randall, L. 2016. Dark matter and the dinosaurs: The astounding interconnectedness of the universe. Ecco. 432 p.

Randall, L., and Reece, M. 2014. Dark matter as a trigger for periodic comet impacts. Physical Review Letters v. 112, 161301 (1–5).

Rasmussen, B., Bekker, A., and Fletcher, I. R. 2013. Correlation of Paleoproterozoic glaciations based on U-Pb zircon ages for tuff beds in the Transvaal and Huronian supergroups. Earth and Planetary Research Letters, v. 382, p. 173–180.

Rasmussen, B., Zi, J. W., Sheppard, S., Krapez, B., and Muhling, J. R. 2016. Multiple episodes of hematite mineralization indicated by U-Pb dating of iron-ore deposits, Marquette Range, Michigan, USA. Geology, v. 44, p. 547–550.

Raup, D., and Sepkoski, J., Jr. 1982. Mass extinctions in the marine fossil record. Science, v. 215, p. 1501–1503.

Raymo, M. R., and Huybers, P. 2008. Unlocking the mysteries of the ice ages. Nature, v. 451, p. 284–285.

Retallack, G. J., Krinsley, D. H., Fischer, R., Razink, J. J., and Langworthy, K. A. 2016. Archean coastal-plain paleosols and life on land. Gondwana Research, v. 40, p. 1–20.

Richards, B. H., Wolfe, P. J., and Potter, P. E. 1997. Pre–Mount Simon basins of western Ohio. Geological Society of American Special Paper 312. p. 243–259.

Richards, M. A., Alvarez, W., Self, S., Karlsstrom, L., Renne, P. R., Manga, M., Sprain, C. J., Smit, J., Vanderkluysen, L., and Gibson, S. A. 2015. Triggering of the largest Deccan eruptions by the Chicxulub impact. GSA Bulletin, v. 127, p. 1507–1520.

Riding, R. 2000. Microbial carbonates: The geological record of calcified bacterial-algal mats and biofilms. Sedimentology, v. 47, p. 179–214.

Riller, U. 2005. Structural characteristics of the Sudbury impact structure, Canada: Impact-induced versus orogenic deformation, a review. Meteoritics and Planetary Science, v. 40, nr. 11, p. 1723–1740.

Rimando, R. E., and Benn, K. 2005. Evolution of faulting and paleo-stress field within the Ottawa Graben, Canada. Journal of Geodynamics, v. 39, p. 337–360.

Ripley, E. M. 1986. Origin and concentration mechanisms of copper and nickel in Duluth Complex sulfide zones: A dilemma. Economic Geology, v. 81, p. 974–978.

Ripley, E. M. 2014. Ni-Cu-PGE mineralization in the Partridge River, South Kawishiwi, and Eagle intrusions: A review of contrasting styles of sulfide-rich occurrences in the Midcontinent Rift system. Economic Geology, v. 109, p. 309–324.

Rivers, T. 2015. Tectonic setting and evolution of the Grenville orogeny: An assessment of progress over the last 40 years. Geoscience Canada, v. 42, p. 77–124.

Rivers, T., Culshaw, N., Hynes, A., Indares, A., Jamieson, R., and Martignole, J. 2012. The Grenville orogen: a post-LITHOPROBE perspective. Chapter 3 in Percival, J. A., Cook, F. A., Clowes, R. M., eds. Tectonic Styles in Canada: The LITHOPROBE perspective. Geological Association of Canada Special Paper 49. p. 97–238.

Robert, F., Poulsen, K. H., Cassidy, K. F., and Hodgson, C. J. 2005. Gold metallogeny of the Superior and Yilgarn cratons. Society of Economic Geologists 100th Anniversary Volume. p. 1001–1033.

Roberts, N. M. W., Van Kranendonk, M., Parman, S., and Clift, P.D. 2015. Continent formation through time. Geological Society of London Special Publications 389. p. 1–16.

Robertson, J. A., and Card, K. D. 1972. Geology and scenery, north shore of Lake Huron region. Ontario Geological Survey Geological Guidebook 4. p. 224.

Rogala, B., Fralick, P. W., Heaman, L. M., and Metsaranta, R. 2007. Lithostratigraphy and chemostratigraphy of the Mesoproterozoic Sibley Group, northwest Ontario, Canada. Canadian Journal of Earth Science, v. 44, p. 1131–1149.

Rogers, J. J. W., and Santosh, M. 2002. Configuration of Columbia, a Mesoproterozoic supercontinent. Gondwana Research, v. 5, p. 5–22.

Rogowski, A. J., and Farlow, J. O. 2012. Relative abundance of vertebrate microfossils of the Pipe Creek Sinkhole (late Neogene, Grant County, IN). Geological Society of America Abstracts with Program 44. p. 62.

Rollinson, H. 2010. Coupled evolution of Archean continental crust and subcontinental lithospheric mantle. Geology, v. 38, p. 1083–1086.

Romano, D., Holm, D. K., and Foland, K. A. 2000. Determining the extent and nature of Mazatzal-related overprinting of the Penokean orogenic belt in the southern Lake Superior region, north-central USA. Precambrian Research, v. 104, p. 25–46.

Roscoe, S. M. 1969. Huronian rocks and uraniferous conglomerates. Geological Survey of Canada Paper 68–40. p. 205.

Roscoe, S. M. 1973. The Huronian Supergroup, a Paleoaphebian succession showing evidence of atmospheric evolution. In Young, G. M., ed. Huronian stratigraphy and sedimentation. Geological Association of Canada Special Paper 12. p. 31–47.

Roscoe, S. M., and Card, K. D. 1992. Early Proterozoic tectonics and metallogeny of the Lake Huron region of the Canadian Shield. Precambrian Research, v. 58, p. 99–119.

Roscoe, S. M., and Card, K. D. 1993. The reappearance of the Huronian in Wyoming: rifting and drifting of ancient continents. Canadian Journal of Earth Sciences 30. p. 2475–2480.

Rosemeyer, T. 1999. The history, geology, and mineralogy of the White Pine mine, Ontonagon County, Michigan. Rocks and Minerals, v. 74, p. 160–176.

Rosing, M. T., Bird, D. K., Sleep, N. H., Bjerrum, C. J. 2010. No climate paradox under the faint early sun. Nature, v. 464, p. 744–747.

Roussell, D. H., and Brown, H. 2009. A field guide to the geology of Sudbury, Ontario. Ontario Geological Survey Open File Report 6243. p. 77.

Rovere, A., and Vacchi, P. S. M. 2016. Eustatic and relative sea level changes. Current Climate Change Reports, v. 2, p. 221–231.

Rowan, E. L., and Goldhaber, M. B. 1995. Duration of mineralization and fluid-flow history of the upper Mississippi Valley zinc-lead district. Geology, v. 23, p. 609–612.

Rowley, D. B. 2017. Earth's constant mean elevation: Implication for long-term sea level controlled by oceanic lithosphere dynamics and a Pitman world. Journal of Geology, v. 125, p. 141–153.

Rubie, D. C., Nimmo, F., and Melosh, H. J. 2015. Formation of the Earth's core. Treatise on Geophysics, v. 9, p. 43–79.

Rullkötter, J., Meyers, P. A., Schaefer, R. G., and Dunham, K.W. 1986. Oil generation in the Michigan Basin: A biological marker and carbon isotope approach. Organic Geochemistry, v. 10, p. 359–375.

Runkel, A. C., McKay, R. M., and Palmer, A. R. 1998. Origin of a classic cratonic sheet sandstone: Stratigraphy across the Sauk II–Sauk III boundary in the upper Mississippi Valley. Geological Society of America Bulletin, v. 110, p. 188–210.

Russell, M. J. 2017. Life is a verb, not a noun. Geology, v. 45, p. 1143–1144.

Rutte, D., Ratschbacher, L., Schneider, S., Stübner, K., Stearns, M.A., Gulzar, M. A., and Hacker, B. R.. 2017. Building the Pamir-Tibetan Plateau: Crustal stacking, extensional collapse, and lateral extrusion in the central Pamir, 1: geometry and kinematics, Tectonics, v. 36. doi:10.1002/2016TC004293.

Rybczynski, N., Gosse, J. C., Harington, C. R., Wogelius, R. A., Hidy, A. J., Buckley, M. 2013. Mid-Pliocene warm period deposits in the high arctic yield insight into camel evolution. Nature Communications, v. 4, Article 1550.

Sado, E. V., and Carswell, B. F. 1987. Surficial geology of northern Ontario: Ontario Geological Survey, Map 2518, scale 1:1,200,000.

Sagan, C., and Mullen, G. 1972. Earth and Mars: Evolution of atmosphere and surface temperatures. Science, v. 177, p. 52–56.

Sage, R. P., and Sage, V. L. 2006. Glacial lake Algonquin and Nipissing shoreline bedrock features: Mackinac Island, Michigan: Field Trip Guidebook, v. 52, part 2. Institute of Lake Superior Geology, 28 p.

Sandom, C., Faurby, S., Sandel, B., and Svernning, J-C. 2014. Global late Quaternary megafauna extinctions linked to humans, not climate change. Proceedings of the Royal Society B, v. 281, 20133254.

Sanford, B. V., and Grant A. C. 1990. New findings relating to the stratigraphy and structure of the Hudson Platform. Geological Survey of Canada Paper 90–1D. p. 17–30.

Schaetzl, R. J., Drzyzga, S. A., Weisenborn, B. N., Kincare, K. A., Lepczyk, X. C., Shein, K., Dowd, C. M., and Linker, J. 2002. Measurement, correlation, and mapping of glacial Lake Algonquin shoreline in northern Michigan. Annals Association American Geographers, v. 92, p. 399–415.

Schaetzl, R. J., Lepper, K., Thomsa, S. E., Grove, L., Treiber, E., Farmer, A., Fillmore, A., Jordan, L., Dickerson, B., and Alme, K. 2017. Kame deltas provide evidence for a new glacial lake and suggest early glacial retreat from central Lower Michigan, USA. Geomorphology, v. 280, p. 167–178.

Schmidt, R. G. 1963. Geology and ore deposits of the Cuyuna North Range, Minnesota. USGeological Survey Professional Paper 407. p. 91.

Schmitt, G. 2016. Menominee Indians. Arcadia Publishing. 128 p.

Schmitz, B., Harper, D. T. A., Peucker-Ehrenbrink, B., Stouge, S., Alwamark, C., Cronholm, A., Begstrom, S., Tassinari, M., and Wang, X. 2008. Asteroid breakup linked to the Great Ordovician Biodiversification Event. Nature Geoscience, v. 1, p. 49–53.

Schmitz, M. D., Bowring, S. A., Southwick, D. L., Boerboom, T. J., and Wirth, K. R. 2006. High-precision U-Pb geochronology in the Minnesota River Valley subprovince and its bearing on Neoarchean to Paleoproterozoic evolution of the southern Superior Province. Geological Society of America Bulletin, v. 118, p. 82–93.

Schneider, D. A., Bickford, M. E., Cannon, W. F., Schulz, K. J., and Hamilton, M. A. 2002. Age of volcanic rocks and syndepositional iron formations, Marquette Range Supergroup: Implications for the tectonic setting of Paleoproterozoic iron formations of the Lake Superior region. Canadian Journal of Earth Sciences, v. 39, p. 999–1012.

Schneider, D. A., Holm, D. K., O'Boyle, C., Hamilton, M., and, Jercinovic, M. 2004. Paleoproterozoic development of a gneiss dome corridor in the southern Lake Superior region, USA. In Whitney, D. L., Teyssier, C., and Siddoway, C. S., eds. Gneiss domes in orogeny. Geological Society of America Special Paper 380. p. 339–357.

Schopf, J. W., Kudryavtsev, A. B., Czaja, A. D., and Tripathi, A. B. 2007. Evidence of Archean life: Stromatolites and microfossils. Precambrian Research, v. 158, p. 141–155.

Schulte, P., ed., 2011. The Great Lakes water agreements. In Gleick, P. H., ed. The world's water. Springer International. p. 165–170.

Schulz, K. J. 1984. Volcanic rocks of northeastern Wisconsin. Field Trip Guidebook, Thirtieth Annual Institute on Lake Superior Geology. p. 51–80.

Schultz, K. J., and Cannon, W. F. 2007. The Penokean orogeny in the Lake Superior region. Precambrian Research, v. 157, p. 4–25.

Schwartz, J. J., Stewart, E. K., and Medaris, G. L., Jr. 2018. Detrital zircons in the Waterloo quartzite, Wisconsin: Implications for the ages of deposition and folding of supermature quartzites in the southern Lake Superior region. Proceedings of the 64th Institute on Lake Superior Geology, p. 95–96.

Schwerdtner, W. M., Rivers, T., Tsolas, J., Waddington, D. H., Page, S., and Yang, J. 2016. Transtensional origin of multi-order cross-folds in a high-grade gneiss complex, southwestern Grenville Province: Formation during post-peak gravitational collapse. Canadian Journal of Earth Science, v. 53, p. 1511–1538.

Scotese, C. R. 2009. Late Proterozoic plate tectonics and palaeogeography: A tale of two supercontinents, Rodinia and Pannotia. Geological Society of London Special Publication 326. p. 67–83.

Sekine, Y., Tajika, E., Ohkouchi, N., Ogawa, N. O., Goto, K., Tada, R., Yamamoto, S., and Kirschvink, J. L. 2010. Anomalous negative excursion of carbon isotope in organic carbon after the last Paleoproterozoic glaciation in North America. Geochemistry, Geophysics, Geosystems, v. 11, Q08019. doi:10.1029/2010GC003210.

Sepkoski, J. J., Jr. 1996. Patterns of Phanerozoic extinction: A perspective from global databases. In Walliser, O. H., ed. Global events and event stratigraphy in the Phanerozoic. Springer-Verlag. p. 35–51.

Servais, T., Harper, D. A. T., Li, J., Munnecke, A., Owen, A. W., and Sheehand, P. M. 2009. Understanding the Great Ordovician biodiversification event (GOBE): Influences of paleogeography, paleoclimate, or paleoecology? GSA Today, v. 19, p. 4–10.

Sheldon, N. D. 2013. Causes and consequences of low atmospheric pCO_2 in the Late Mesoproterozoic. Chemical Geology, v. 362, p. 224–231.

Shen, B., Dong, L., Xiao, S., and Kowalewski, M. 2008. The Avalon explosion: Evolution of ediacara morphospace. Science, v. 319, p. 81–84.

Shen, S. Z., Crowley, J. L., Wang, Y., Bowring, S. A., Erwin, D. H., Sadler, P. M., Cao, C. Q., Rothman, D. H., Henderson, C. M., Ramezani, J., Zhang, H., Shen, Y., Wang, X. D., Wang, W., Mu, L., Li, W. Z., Tang, Y. G., Liu, X. L., Liu, L. J., Zeng, Y., Jiang, Y. F., Jin, Y. G. 2011. Calibrating the end-Permian mass extinction. Science, v. 334, p. 1367–1372.

Sherwood, S. C., and Huber, M. 2010. An adaptability limit to climate change due to heat stress. Proceedings of the National Academy of Science, v. 107, p. 9552–9555.

Shibaike, Y., Takamori, S., and Ida, S. 2016. Excavation and melting of the Hadean continental crust by Late Heavy Bombardment. Icarus, v. 266, p. 189–203.

Shulman, M. 1966. The billion-dollar windfall. McGraw-Hill. 239 p.

Simo, J. A., Emerson, N. R., Byers, C. W., and Lugvigson, G. A. 2003. Anatomy of an embayment in an Ordovician epeiric sea, upper Mississippi Valley, USA. Geology, v. 31, p. 545–548.

Sims, P. K., Card, K. D., Morey, G. B., and Peterman, Z. E. 1980. The Great Lakes tectonic zone: A major crustal structure in central North America. Geological Society of America Bulletin, v. 91, pt. 1, p. 690–698.

Sims, P. K., Carter, L. M. H., eds. 1996. Archean and Proterozoic geology of the Lake Superior region, U.S.A. US Geological Survey Professional Paper 1556. 115 pp.

Sims, P. K., Van Schmus, W. R., Schulz, K. J., Peterman, Z. E. 1989. Tectono-stratigraphic evolution of the Early Proterozoic Wisconsin magmatic terranes of the Penokean orogen. Canadian Journal of Earth Sciences, v. 26, p. 2145–2158.

Skinner, C. L. 2008. The upper country: French enterprise in the colonial Great Lakes. Johns Hopkins University Press. 224 p.

Slack, J. F., and Cannon, W. F. 2009. Extraterrestrial demise of banded iron formation 1.85 billion years ago. Geology, v. 37, p. 1011–1015.

Sleep, N. 2018. Cratonic basins with reference to the Michigan Basin. Geological Society of London. Special Publication 472, p. 231–256.

Sloan, R. E. 1987. Tectonic, biostratigraphy and lithostratigraphy of the Middle and Lower Ordovician of the upper Mississippi Valley. University of Minnesota Report of Investigations 35. p. 7–20.

Sloss, L. L. 1963. Sequences in the cratonic interior of North America. Geological Society of America Bulletin, v. 74, p. 93–113.

Smith, E. I., 1983, Geochemistry and evolution of the early Proterozoic, post-Penokean rhyolites, granites and related rocks of south-central Wisconsin, USA. Geological Society of America, Memoir 160, p. 113–128.

Soller, D. R. 2001. Map showing the thickness and character of Quaternary sediments in the glaciated United State east of the Rocky Mountains. U.S. Geological Survey Miscellaneous Investigations Series Map I-1970-E.

Sonnenfeld, P., and Al-Aasm, I. 1991. The Salina evaporites in the Michigan Basin. In Catacosinos, P. A., and Daniels, P. A., Jr., eds. Early sedimentary evolution of the Michigan Basin. Geological Society of America Special Paper 256. p. 139–153.

Southwick, D. L., Morey, G. B. 1991. Tectonic imbrication and foredeep development in the Penokean orogen, east-central Minnesota: An interpretation based on regional geophysics and results of test drilling. US Geological Survey Bulletin 1904-C. p. C1–C17.

Southwick, D. L., Morey, G. B., and McSwiggen, P. L. 1988. Geologic map (Scale 1:250,000) of the Penokean orogen, central and eastern Minnesota, and accompanying text. Minnesota Geological Survey, Report of Investigations 37.

Southwick, D. L., Morey, G. B.,; and Mossler, J. H. 1986. Fluvial origin of the lower Proterozoic Sioux Quartzite, southwestern Minnesota. Geological Society of America Bulletin, v. 97, p. 1432–1441.

Sprague, D. D., Michel, F. A., and Vermaier, J. C. 2016. The effects of migration of ca. 100-year-old arsenic-rich mine tailing in Cobalt, Ontario, Canada. Environmental Earth Sciences, v. 75, p. 405. doi.org/10.1007/s12665-015-4898-1.

Stanley, S. M. 2016. Estimates of the magnitudes of major marine mass extinctions in Earth history. Proceedings of the National Academy of Science. http://www.pnas.org/content/113/42/E6325.full.pdf

Stark, T. J. 1997. The East Continent Rift Complex: Evidence and conclusions. Geological Society of America Special Paper 312. p. 253–266.

Stauffer, C. R., and Thiel, G. A. 1944. The iron ores of southeastern Minnesota. Economic Geology, v. 39, p. 327–339.

Steffen, W., Grinevald, J., Crutzen, P., and McNeill, J. 2011. The Anthropocene: Conceptual and historical perspectives. Philosophical Transactions of the Royal Society, v. A369, p. 842–867.

Stein, C. A. Kley, J., Stein, S., Craddock, J.P., and Malone, D. H. 2015. Age of the Jacobsville Sandstone and implication for the evolution of the Midcontinent Rift. Geological Society of America Annual Meeting Abstract/Paper 261842.

Stein, C. A., Stein, S., Merino, M., Keller, G. R., Flesch, L. M., and Jurdy, D. M. 2014. Was the Midcontinent Rift part of a successful seafloor-spreading episode? Geophysical Research Letters, v. 41, p. 1–6.

Stein, C. A., Stein, S., Elling, R., Keller, R. G., and Kley, J. 2018. Is the "Grenville Front" in the central United States really the Midcontinent Rift? GSA Today, v. 28, p. 4–10.

Stern, R. J. 2008. Modern-style plate tectonics began in Neoproterozoic time: An alternative interpretation of Earth's tectonic history. In Condie, K., and Pease, V., eds. When did plate tectonics begin? Geological Society of America Special Paper 440. 265–280.

Stern, R. J. 2013. When did plate tectonics begin on Earth and what came before? GSA. Blog. http://geosociety.wordpress.com/2013/04/28/when-did-plate-tectonics-begin-onearth-and-what-came-before/

Stevenson, R. K., Henry, P., and Gariepy, C. 2009. Isotopic and geochemical evidence for differentiation and crustal contamination from granitoids of the Berens River subprovince, Superior Province, Canada. Precambrian Research, v. 168, p. 123–133.

Stewart, E. K., and Mauk, J. L. 2017. Sedimentology, sequence-stratigraphy, and geochemical variations in the Mesoproterozoic Nonesuch Formation, northern Wisconsin, USA. Precambrian Research, v. 294, p. 111–132.

Stiff, B. J., and Hansel. A. K. 2004. Quaternary glaciations in Illinois. In Ehlers, J., and Gibbard, P. L., eds. Quaternary Glaciations: Extent and Chronology, Part II, Elsevier, p. 71–82.

Stockwell, C. H. 1962. A tectonic map of the Canadian Shield. In Tectonics of the Canadian Shield, Stevenson, J. S., ed. Royal Society of Canada, Special Publications 4. p. 6–15.

Stockwell, C. H. 1964. Fourth report on structural provinces, orogenies, and time-classification of rocks of the Canadian Precambrian Shield. In Age determinations and geological studies. Pt. 2: Geological studies. Geological Survey of Canada Paper 64–17. p. 1–21.

Stott, G. M., Corkery, M. T., Percival, J. A., Simard, M., and Goutier, J. 2010. A revised terrane subdivision of the Superior Province. In Summary of field work and other activities, 2010. Ontario Geological Survey Open File Report 6260. p. 20-1–20-10.

Strik, G., Blake, T. S., Zegers, T. E., White, S. H., and Langereis, C. G. 2003. Palaeomagnetism of flood basalts in the Pilbara craton, western Australia: Late Archaean continental drift and the oldest known reversal of the geomagnetic field. Journal of Geophysical Research: Solid Earth, v. 108 (B12), p. 2551.

Sugg, Z. 2007. Assessing U.S. farm drainage: Can GIS lead to better estimates of subsurface drainage extent? World Resources Institute. http://pdf.wri.org/assessing_farm_drainage.pdf

Suszek, T. 1997. Petrography and sedimentation of the Middle Proterozoic (Keweenawan) Nonesuch Formation, western Lake Superior region, Midcontinent Rift system. Geological Society of America Special Paper 312. pp. 195–210.

Swanson-Hysell, N. L., Burgess, S. D., Maloof, A. C., and Bowring, S. A. 2014a. Magmatic activity and plate motion during the latent stage of Midcontinent Rift development. Geology, v. 42, p. 475–478.

Swanson-Hysell, N. L., Vaughan, A. A., Mustain, M. R., and Asp, K. E. 2014b. Confirmation of progressive plate motion during the Midcontinent Rift's early magmatic stage from the Osler Volcanic Group, Ontario, Canada. Geochemistry, Geophysics, Geosystems, v. 15, p. 2039–2047.

Swenson, J. B., Person, M., Raffensperger, J. P., Cannon W. F., Woodruff, L. G., and Berndt, M. E. 2004. A hydrogeologic model of stratiform copper mineralization in the Midcontinent Rift system, northern Michigan, USA. Geofluids, v. 4, p. 1–22.

Swezey, C. S. 2008. Regional stratigraphy and petroleum systems of the Michigan Basin, North America. U.S. Geological Survey Scientific Investigations Map 2978.

Swezey, C. S., Hatch, J. R., East, J. A., Hayba, D. O., and Repetski, J. E. 2015. Total petroleum systems of the Michigan Basin: Petroleum geology and geochemistry and assessment of undiscovered resources. Chapter 2 of US Geological Survey Michigan Basin Province Assessment Team. Geologic assessment of undiscovered oil and gas resources of the U.S. portion of the Michigan Basin. US Geological Survey Digital Data Series DDS–69–T. 162 p. doi. org/10.3133/ds69T.

Syverson, K. M., and Colgan, P. M. 2004. The Quaternary of Wisconsin: A review of stratigraphy and glaciation history: Quaternary Glaciations—Extent and Chronology, Part II, p. 295–311.

Tabor, C. R. 2016. From Greenhouse to Icehouse: Understanding Earth's Climate Extremes Through Models and Proxies: Unpublished PhD Dissertation, University of Michigan, 215 p.

Telford, P. G. 1988. Devonian stratigraphy of the Moose River Basin, James Bay lowland, Ontario, Canada. AAPG Memoir 14. p. 123–132.

Teller, J. T. 2013. Lake Agassiz during the Younger Dryas. Quaternary Research, v. 80, p. 361–369.

Teller, J. T., and Clayton, L., eds. 1983. Glacial Lake Agassiz. Geological Association of Canada Special Paper 26, 451 pp.

Teller, J. T., Leverington, D. W., and Mann, J. D. 2002. Freshwater outbursts to the oceans from glacial Lake Agassiz and their role in climate change during the last deglaciation. Quaternary Science Reviews, v. 21, p. 879-887.

Tera, F., Papanastassiou, D. A., and Wasserburg, G. J. 1974. Isotopic evidence for a terminal lunar cataclysm. Earth and Planetary Science Letters, v. 22, p. 1–21.

Theis, N. J. 1979. Uranium-bearing and associated minerals and their geochemical and sedimentological context, Elliot Lake, Ontario. Geological Survey of Canada Bulletin 304. p. 50.

Therriault, A. M., Fowler, A. D., and Grieve, R. A. F. 2002. The Sudbury Igneous Complex: A differentiated impact melt sheet. Economic Geology, v. 97, p. 1521–1540.

Thomas, M. D., and Teskey, D. J. 1994. An interpretation of gravity anomalies over the Midcontinent Rift, Lake Superior, constrained by GLIMPCE seismic and aeromagnetic data. Canadian Journal of Earth Sciences, v. 31, p. 682–697.

Thurston, P. C. 2015. Igneous rock associations 19.Greenstone belts and granite-greenstone terranes: Constraints on the nature of the Archean world. Geoscience Canada, v. 42, p. 437–484.

Thurston, P. C., Ayer, J. A., Goutier, J., Hamilton, M. A. 2008. Depositional gaps in the Abitibi Greenstone Belt stratigraphy: A key to exploration for syngenetic mineralization. Economic Geology, v. 103, p. 1097–1134.

Thurston, P. C., and Breaks, F. W. 1978. Metamorphic and tectonic evolution of the Uchi-English River subprovince in metamorphism in the Canadian Shield. In Fraser, J. A., and Heywood W. W., eds. Geological Survey of Canada Paper 78–10. p. 49–62.

Tinkham, D. K., and Marshak, S. 2004. Precambrian dome-and-keel structure in the Penokean orogenic belt of northern Michigan, USA. Geological Society of America Special Paper 380. p. 321–338.

Titze, D. J., and Austin, J. A. 2014. Winter thermal structure of Lake Superior. Limnology and Oceanography, v. 59, p. 1336–1348.

Tohver, E., Holm, D. K., van der Pluijm, B., Essene, E. J., and Cambray, F.W. 2007. Late Paleoproterozoic (geon 18 and 17) reactivation of the Neoarchean Great Lakes Tectonic Zone, northern Michigan, USA: evidence from kinematic analysis, thermobarometry, and 40Ar/39Ar geochronology. Precambrian Research, v. 157, p. 144–168.

Tohver, E., Teixeira, W., van der Pluijm, B., Geraldes, M. C., Bettencourt, J. S., and Rizzotto, F. 2004. Restored transect across the exhumed Grenville orogeny of Laurentia and Amazonia with implications for crustal architecture. Geology, v. 34, p. 669–672.

Torsvik, T. H., Smethurst, M. A. S., Meert, J. G., van der Voo, R., McKerrow, W. S., Brasier, M. D., Sturt, B.A., and Walderhaug, H. J. 1996. Continental break-up and collision in the Neoproterozoic and Paleozoic: A tale of Baltica and Laurentia. Earth-Science Reviews, v. 40, p. 29–258.

Trehu, A., Morel-a-Huissier, P., Meyer, R., Hajnal, Z., Karl, J., Mereu, R., Sexton, J., Shay, J., Chan, W.-K., Epili, D., Jefferson, T., Shih, X.-R., Wendling, S., Milkereit,

B., Green, A., and Hutchinson, D. 1991. Imaging the Midcontinent Rift beneath Lake Superior using large aperture seismic data. Geophysical Research Letters, v. 18, p. 625–628.

Tremblay, A., Roden-Tice, M. K., Brandt, J. A., and Megan, T. W. 2012. Mesozoic fault reactivation along the St. Lawrence Rift system, eastern Canada: Thermochronologic evidence from apatite fission-track dating. GSA Bulletin, v. 125, p. 794–810.

Tzedakis, P. C., Channell, J. E. T., Hodell, D. A., Kleiven, H. F., and Skinner, L. C. 2012. Determining the natural length of the current interglacial. Nature Geoscience, v. 5, p. 138–141.

US Geological Survey Michigan Basin Province Assessment Team. 2015. Geologic assessment of undiscovered oil and gas resources of the U.S. portion of the Michigan Basin. U.S. Geological Survey Digital Data Series DDS-69-T. 4 chaps., variously paged. https://dx.doi.org/10.3133/ds69T

Valley, J. M., Peck, W. H., King, E. M., and Wilde, S.A. 2002. A cool early Earth. Geology, v. 30, p. 351–354.

Valley, J. W., Cavosie, A. J., Ushikubo, T., Reinhard, D. A., Lawrence, D. F., Larson D. J., Clifton, P. H., Kelly T. F., Wilde, S. A. Moser, D. E., and Spicuzza, M. J. 2014. Hadean age for a post-magma-ocean zircon confirmed by atom-probe tomography. Nature Geoscience, v. 7, p. 219–223.

Vallini, D. A., Cannon, W. F., and Schulz, K. J. 2006. Age constraints for Palaeoproterozoic glaciation in the Lake Superior region: Detrital zircon and hydrothermal xenotime ages for the Chocolay Group, Marquette Range Supergroup. Canadian Journal of Earth Sciences, v. 43, p. 571–591.

Van der Voo, R. 2004. Presidential Address: Paleomagnetism, oroclines, and growth of the continental crust. GSA Today, v. 14, p. 4–9.

Van Schmus, W. R. 1980. Chronology of igneous rocks associated with the Penokean orogeny in Wisconsin. In Morey, G., Hanson, G., eds. Selected studies of Archean gneisses and lower Proterozoic rocks, southern Canadian Shield. Geological Society of America Special Paper 182. p. 159–168.

Van Schmus, W. R., and Anderson, J. L. 1977. Gneiss and migmatite of Archean age in the Precambrian basement of central Wisconsin. Geology, v. 5, p. 45–48.

Van Schmus, W. R., Green, J. C., Halls, H. C. 1982. Geochronology of Keweenawan rocks of the Lake Superior region: A summary. Geological Society of America Memoir 156. p. 165–178.

Van Schmus, W. R., Maass, R. S. 1983. Hatfield gneiss, Lake Arbutus dam. In Brown, B. A., ed. Three billion years of geology: A field trip through the Archean, Proterozoic, Paleozoic, and Pleistocene geology of the Black River Falls area of Wisconsin. Geological Survey of Wisconsin Field Trip Guide Book 9. p. 38–41.

Van Schmus, W. R., Schneider, D. A., Holm, D. K., Dodson, S., and Nelson, B. K. 2007. New insights into the southern margin of the Archean-Proterozoic boundary in the north-central United States based on U-Pb, Sm-Nd, and Ar-Ar geochronology. Precambrian Research, v. 157, p. 80–105.

Velbel, M. A. 2012. The "lost interval": Geology from the Permian to the Pliocene. In

Schaetzl, R., Darden, J., and Brandt, D., eds. Michigan geography and geology. Pearson Learning Solutions, p. 60–68.

Vervoort, J., Wirth, K., Kennedy, B., Sandland, T., and Harpp, K. 2007. The magmatic evolution of the Midcontinent Rift: New geochronologic and geochemical evidence from felsic magmatism. Precambrian Research, v. 157, p. 235–268.

Voice, P. J., and Harrison, W. B., III. 2016. Clinton-type deposits in the Michigan Basin: A preliminary discussion of the significance of Lower Silurian hematitic, phosphatic grainstones in western Michigan. Geological Society of America Abstracts with Programs 48, p. 35.

Wacey, D., Menon, S., Green, L., Gerstmann, D., Kong, C., Mcloughlin, N., Saunders, M., and Brasier, M. 2012. Taphonomy of very ancient microfossils from the ~3400 Ma Strelley Pool Formation and ~1900 Ma Gunflint Formation: New insights using a focused ion beam. Precambrian Research, v. 220–221, p. 234–250.

Waight, T. 2014. Sedimentary rocks (Rb-Sr geochronology). Encyclopedia of Scientific Dating Methods. p. 1–7. https://link.springer.com/referenceworkentry/10.1007/978-94-007-6326-5_117-1

Walker, E. C., Sutcliffe, R. H., Shaw, C. S. J., Shore, G. T., and Penczak, R. S. 1993. Precambrian geology of the Coldwell Alkali Complex. Ontario Geological Survey Open File Report 5868. p. 30.

Walling, D. E. 2006. Human impact on land-ocean sediment transfer by the world's rivers. Geomorphology, v. 79, p. 192–216.

Wallman, K., and Aloisi, G. 2012. The global carbon cycle: Geological processes. In Knoll, A. H., Canfield, D. E., and Konhauser, K. O., eds. Fundamentals of geobiology. Blackwell. p. 20–35.

Walter, M. R., and Heys, G. R. 1985. Links between the rise of the metazoa and the decline of the stromatolites. Precambrian Research, v. 29, p. 149–174.

Wang, Y., Lesher, M. C., Lightfoot, P. C., Pattison, E. F., and Golightly, J. P. 2018. Shock metamorphic features in mafic and ultramafic inclusion in the Sudbury Igneous Complex: Implications for their origin and impact excavation: Geology, v. 46, p. 443–446.

Ward, Peter D. 2000. Rivers in time: The search for clues to Earth's mass extinctions. Columbia University Press, 320 p.

Warren, W. W. 2009. History of the Ojibway people. 2nd ed. Minnesota Historical Society. 316 p. Originally published in 1885.

Waters, C. N., Zalasiewicz, J., Summerhayes, C., Barnosky, A. D., Poirier, C., Galuszka, A., Cearreta, A., Edgeworth, M., Ellis, E. C., Ellis, M., Jeandel, C., Leinfelder, R., McNeill, J. R., Richter, D. D., Steffen, W., Syvitski, J., Vidas, D., Wagreich. M., Williams, M., Zhisheng, A., Grinevald J., Odada, E., Oreskes, N., and Wolfe, A. P. 2016. The Anthropocene is functionally and stratigraphically distinct from the Holocene. Science, v. 351, no, 6269, p. 137–141.

Weary, D. J., and Doctor, D. H. 2014. Karst in the United States: A digital map compilation and database. U.S. Geological Survey Open-File Report 2014–1156. 23 p. doi.org/10.3133/ofr20141156.

Webb, K., Smith, B. S., Paul, J., and Hetman, C. 2004. Geology of the Victor kimberlite, Attawapiskat, northern Ontario, Canada: Cross-cutting and nested craters. Lithos, v. 76, p. 29–50.

Weiblen, P. W. 1982. Keweenawan intrusive igneous rocks. Geological Society of America Memoir 156. p. 57–81.

Weidendorfer, D., Schmidt, M. W., and Mattsson, H. B. 2017. A common origin for carbonatite magmas. Geology, v. 45, p. 507–510.

Weil, A. B., Van der Voo, R., Niocaill, C. M., and Meert, J. G. 1998. The Proterozoic supercontinent Rodinia: Paleomagnetically derived reconstructions for 1100 to 800 Ma. Earth and Planetary Science Letters, v. 154, p. 13–24.

Weisman, A. 2008. The world without us. Picador-Thomas Dunne. 438 p.

White, D. J., Musacchio, G., Helmstaedt, H. H., Harrap, R. M., Thurston, P. C., van der Velden, A., and Hall, K. 2003. Images of a lower-crustal oceanic slab: Direct evidence for tectonic accretion in the Archean western Superior Province. Geology, v. 31, p. 997–1000.

White, W. S. 1968. The native-copper deposits of northern Michigan. In Ridge, J. D., ed. Ore deposits of the United States, 1933–1967. American Institute of Mining, Metallurgical, and Petroleum Engineers. p. 303–325.

White, W.S., and Wright, J. C. 1954. The White Pine copper deposit, Ontonagon County, Michigan. Economic Geology, v. 49, p. 675–716.

Whitmeyer, S. J., and Karlstrom, K. E. 2007. Tectonic model for the Proterozoic growth of North America: Geosphere, v. 3, p. 220–259.

Wiechert, U., Halliday, A. N., Lee, D.-C., Snyder, G. A., Taylor, L. A., and Rumble, D. 2001. Oxygen isotopes and the Moon-forming giant impact. Science, v. 294, p. 345–349.

Wilde, S. A., Valley, J. W., Peck, W. H., and Graham, C. M. 2001. Evidence from detrital zircons for the existence of continental crust and ocean on the Earth 4.4 Gyr ago. Nature, v. 409, p. 175–179.

Wilkinson, B. H. 2005. Humans as geologic agents: A deep time perspective. Geology, v. 33, p. 161–164.

Wilson, T. P., and Long, D. T. 1993. Geochemistry and isotope chemistry of Michigan Basin brines: Devonian formations. Applied Geochemistry, v. 8, p. 81–100.

Wingate, M. T. D. 1998. A palaeomagnetic test of the Kaapvaal-Pilbara (Vaalbara) connection at 2.78 Ga. South African Journal of Geology, v. 101, p. 257–274.

Wirth, H. S. 2016. Detrital zircon U-Pb geochronology of the Neoproterozoic Jacobsville Sandstone in the footwall of the Keweenawan thrust, Michigan, USA. Geological Society of America Abstracts with Programs 48, p. 128.

Wisconsin Geological Survey. 2011. Bedrock stratigraphic units in Wisconsin. Open file report 2006-06, https://wgnhs.uwex.edu/pubs/000872/

Witzke, B. J., and Kolata, D. R. 1989. Changing structural and depositional patterns, Ordovician Champlainian and Cincinnatian series of Iowa-Illinois. In Ludvigson, A., Bunker, B. J., eds. New perspectives on the Paleozoic history of the upper Mississippi Valley: An examination of the Plum River fault zone. Iowa Department of Nature Resources. p. 55–77.

Witzke, B. J., Ludvigson, G. A., and Day, J. 1996. Paleozoic sequence stratigraphy: Views from the North American craton. Geological Society of America Special Paper 306, 446 p.

Wright, H. E., Jr. 1990. Geologic history of Minnesota rivers. Minnesota Geological Survey Educational Series 7. p. 20.

Wyche, S., Nelson, D. R., Riganti, A. 2004. 4350–3130 Ma detrital zircons in the Southern Cross Granite-Greenstone terrane, western Australia: implications for the early evolution of the Yilgarn craton. Australian Journal of Earth Sciences, v. 51, p. 31–45.

Wyman, D., and Kerrich, R. 2010. Mantle plume–volcanic arc interaction: consequences for magmatism, metallogeny, and cratonization in the Abitibi and Wawa subprovinces, Canada. Canadian Journal of Earth Sciences, v. 47, p. 565–589.

Yansa, C. H., and Adams, K. M. 2012. Mastodons and Mammoths in the Great Lakes Region, USA and Canada: New Insights into their Diets as they Neared Extinction. Geography Compass, v. 10, p. 1–14.

Yeomans, D. K. 2013. Near-Earth objects: Finding them before they find us. Princeton University Press, 172 p.

Young, G. M. 2013. Climatic catastrophes in Earth history: Two great Proterozoic glacial episodes. Geological Journal, v. 48, p. 1–21.

Young, G. M. 2014. Contradictory correlations of Paleoproterozoic glacial deposits: Local, regional, or global controls? Precambrian Research, v. 247, p. 33–44.

Young, G. M. 2015. Did prolonged two-stage fragmentation of the supercontinent Kerorland lead to arrested orogenesis of the southern margin of the Superior Province? Geoscience Frontiers, v. 6, p. 419–435.

Young, G. M., Long, D. G. F., Fedo, C. M., and Nesbitt, H. W. 2001. The Paleoproterozoic Huronian Basin: Product of a Wilson cycle accompanied by glaciation and meteorite impact. Sedimentary Geology, v. 141–142, p. 233–254.

Yu, S-Y., Colman, S. M., Lowell, T. V., Milne, G. A., Fisher T. G., Breckenridge, A., Boyd, M., and Teller, J. T. 2010. Freshwater outburst from Lake Superior as a trigger for the cold event 9300 years ago. Nature, v. 328, p. 1262–1264.

Zhang, H-F., Wang, J-L., Zhou, D-W., Yang, Y-H., Zhang, G-W., Santosh, M., Yu, H., and Zhang, J. 2014. Hadean to Neoarchean episodic crustal growth: Detrital zircon records in Paleproterozoic quartzites from the southern North China craton. Precambrian Research, v. 254, p. 245–257.

Zhao, G., Sanzhong, L., Sun, M., and Wilde, S. A. 2011. Assembly, accretion, and breakup of the Palaeo-Mesoproterozoic Columbia Supercontinent: record in the North China craton revisited. International Geology Review, v. 53, p. 1331–1356.

Zhao, M., Reinhard, C. T., and Planavsky, N. 2017. Terrestrial methane fluxes and Proterozoic climate. Geology, v. 46, p. 139–142.

Zieg, M. J., and Marsh, B. D. 2005. The Sudbury Igneous Complex: Viscous emulsion differentiation of a superheated impact melt sheet. GSA Bulletin, v. 117, p. 1427–1450.

Zolnai, A. I., Price, R. A., and Helmstaedt, H. 1984. Regional cross-section of the Southern Province adjacent to Lake Huron, Ontario: Implications for the tectonic significance of the Murray fault zone. Canadian Journal of Earth Sciences, v. 21, p. 447–456.

Zoltai, S. C. 1965. Glacial features of the Quetico-Nipigon area, Ontario. Canadian Journal of Earth Sciences, v. 2, p. 247–258.

INDEX

Page numbers in italics refer to the illustrations.

Printed and bound by CPI Group (UK) Ltd, Croydon, CR0 4YY

09/06/2025

14685632-0001